Springer Monographs in Mathematics

Springer

London
Berlin
Heidelberg
New York
Barcelona
Budapest
Hong Kong
Milan
Paris
Santa Clara
Singapore
Tokyo

Michael Crabb   Ioan James

Fibrewise Homotopy Theory

 Springer

Michael Charles Crabb
Department of Mathematics
University of Aberdeen
Aberdeen
AB24 3UE
UK

Ioan Mackenzie James
Mathematical Institute
Oxford University
24-29 St Giles
Oxford
OX1 3LB
UK

British Library Cataloguing in Publication Data
Crabb, M.C.
  Fibrewise homotopy theory. - (Springer monographs in mathematics
  1. Homotopy theory 2. Fibre bundles (Mathematics)
  I. Title II. James, Ioan Mackenzie
  514.2'4

Library of Congress Cataloging-in-Publication Data
Crabb, M.C. (Michael Charles)
  Fibrewise homotopy theory / Michael Crabb and Ioan James
    p. cm. -- (Springer monographs in mathematics)
  Includes bibliographical references (p. -   ) and index.
  ISBN-13: 978-1-4471-1267-9   e-ISBN-13: 978-1-4471-1265-5
  DOI: 10.1007/978-1-4471-1265-5

  1. Homotopy theory. 2. Fibre bundles (Mathematics) I. James, I.M.
(Ioan Mackenzie), 1928-   . II. Title. III. Series.
QA612.7.C685 1998   98-7022
514'.24--dc21   CIP

Mathematics Subject Classification (1991): 55-02, 55M20, 55M30, 55P25, 55P42, 55P99, 55Q15, 55R65, 55R99, 57R22, 19L20

Typesetting: Camera-ready by the authors and Michael Mackey

12/3830-543210 Printed on acid-free paper

Preface

Topology occupies a central position in the mathematics of today. One of the most useful ideas to be introduced in the past sixty years is the concept of fibre bundle, which provides an appropriate framework for studying differential geometry and much else. Fibre bundles are examples of the kind of structures studied in fibrewise topology.

Just as homotopy theory arises from topology, so fibrewise homotopy theory arises from fibrewise topology. In this monograph we provide an overview of fibrewise homotopy theory as it stands at present. It is hoped that this may stimulate further research. The literature on the subject is already quite extensive but clearly there is a great deal more to be done.

Efforts have been made to develop general theories of which ordinary homotopy theory, equivariant homotopy theory, fibrewise homotopy theory and so forth will be special cases. For example, Baues [7] and, more recently, Dwyer and Spalinski [53], have presented such general theories, derived from an earlier theory of Quillen, but none of these seem to provide quite the right framework for our purposes. We have preferred, in this monograph, to develop fibrewise homotopy theory more or less *ab initio*, assuming only a basic knowledge of ordinary homotopy theory, at least in the early sections, but our aim has been to keep the exposition reasonably self-contained.

Fibrewise homotopy theory has attracted a good deal of research interest in recent years, and it seemed to us that the time was ripe for an expository survey. The subject is at a less mature stage than equivariant homotopy theory, to which it is closely related, but even so the wealth of material available makes it impossible to cover everything. For example, we do not deal with the recent work [51] of Dror Farjoun on the localization of fibrations.

This monograph is divided into two parts. The first provides a survey of fibrewise homotopy theory, beginning with an outline of the basic theory and proceeding to a selection of applications and more specialized topics. The second part is concerned with the stable theory; the emphasis is on theory appropriate for geometric applications, and it is hoped that the account will be accessible to readers who may not already be experts in the classical stable theory. Part II does assume a certain familiarity with the basic ideas from Part I, but is written in such a way that the reader interested mainly in the stable theory should be able to begin with Part II and refer back to

Part I as necessary. More details on the contents of specific sections can be found in the Introductions to the two parts. Cross-referencing within each part is by section number. We have not attempted a complete bibliography of publications related to fibrewise homotopy theory; those which are cited in either Part I or Part II are listed at the end of Part II. Similarly, the index at the end of the book covers both parts.

Certain sections are based on previously published work, and where appropriate this is mentioned in the text. We are grateful to the publishers in question for permission to include this material.

Our thinking on fibrewise homotopy theory has been influenced by the work of many colleagues, but we owe a special debt to those with whom we have collaborated on joint papers (both published and unpublished). We are grateful to our co-authors for sharing their insight with us. MCC would like to record, in particular, his thanks to Andrew Cook, Karlheinz Knapp and Wilson Sutherland.

Contents

Part I. A Survey of Fibrewise Homotopy Theory

Introduction

The basic ideas of fibrewise homotopy theory seem to have occurred in the late 1960s to several people independently. Thus J.C. Becker [8], J.F. McClendon [106], L. Smith [125] and I.M. James [78] all made use of the theory in work published in 1969 or 1970, while several others, such as L. Hodgkin and J.-P. Meyer, were also well aware of its possibilities. I.M. James first published a systematic account of the basic theory in 1985 [85] but this was largely based on much earlier work [78] put aside when he became aware that so many others were thinking on the same lines. Five years later, after further research, he returned to the subject in [86]. Although the present exposition is to some extent based on these earlier accounts it mainly consists of new material.

There is some truth in the observation that once the correct definitions have been formulated any well-organized and methodical account of the relevant homotopy theory, such as [44], can be converted to fibrewise homotopy theory by writing in the word 'fibrewise' wherever it makes sense and adjusting formulae accordingly. Yet even at the most elementary level it is necessary to exercise care and not jump to conclusions, just as it is in the case of equivariant homotopy theory. One might hope that some completely routine way may be found of producing fibrewise versions of results in ordinary homotopy theory but it would be an exaggeration to say that this is possible at present, although Heller [74] suggests that a way may be found.

On the question of terminology, we find it best on the whole to try and use the term *fibrewise* throughout. For example we now prefer the term *fibrewise pointed space* to the alternatives such as *sectioned space, ex-space*, etc. One reason is that fibrewise corresponds closely to the French *fibré* and the German *faserweise*. However, excessive repetition of the term fibrewise may seem monotonous and so we make the convention that it governs the words which come after it so that the expression *fibrewise compact Hausdorff space*, for example, means *fibrewise compact, fibrewise Hausdorff, fibrewise space*. Of course, the time may come when it will be possible to leave out the term fibrewise, just as one does in vector bundle theory, and to simplify the notations accordingly. However, experience suggests that to do so at the present time is liable to cause confusion.

In the exposition which follows we assume that the reader is familiar with the basic notions of ordinary homotopy theory, as set out in [44], for example. Routine fibrewise versions of proofs of well-known results in the ordinary theory are generally omitted. Otherwise, and except for certain examples, the exposition is fairly self-contained.

The text is divided into three chapters, each consisting of a number of sections. Chapter 1 is concerned with the category of fibrewise spaces and fibrewise maps, classified by fibrewise homotopy. As we shall see, it is not always obvious what is the most appropriate fibrewise version of a concept in ordinary homotopy theory. Chapter 2 is concerned with the category of fibrewise pointed spaces and fibrewise pointed maps, classified by fibrewise pointed homotopy. In the ordinary theory not a great deal of attention is usually paid to the difference between the pointed theory and the non-pointed theory but in the fibrewise version the difference is vital. More specialized topics are considered in Chapter 3. Several of the sections are closely modelled on material which has appeared elsewhere: Sections 17 and 20 are edited versions of [92] and [63], respectively; Sections 19 and 21 are based on [90] and [89], respectively, and Sections 22 and 23 have been extracted from [31].

Chapter 1. An Introduction to Fibrewise Homotopy Theory

1 Fibrewise spaces

Basic notions

Let us work over a (topological) base space B. A *fibrewise space* over B consists of a space X together with a map $p : X \to B$, called the *projection*. Usually X alone is sufficient notation. We regard any subspace of X as a fibrewise space over B by restricting the projection. When p is a fibration we describe X as *fibrant*.

We regard B as a fibrewise space over itself using the identity as the projection. We regard the topological product $B \times T$, for any space T, as a fibrewise space over B using the second projection.

Let X be a fibrewise space over B. For each point b of B the *fibre* over b is the subset $X_b = p^{-1}b$ of X; fibres may be empty since we do not require p to be surjective. Also for each subspace B' of B we regard $X_{B'} = p^{-1}B'$ as a fibrewise space over B' with projection p' determined by p.

Fibrewise spaces over B constitute a category with the following definition of morphism. Let X and Y be fibrewise spaces over B with projections p and q, respectively. A *fibrewise map* $\phi : X \to Y$ is a map in the ordinary sense such that $q \circ \phi = p$, in other words such that $\phi X_b \subseteq Y_b$ for each point b of B. If $\phi : X \to Y$ is a fibrewise map over B then the restriction $\phi_{B'} : X_{B'} \to Y_{B'}$ is a fibrewise map over B' for each subspace B' of B. Thus a functor is defined from the category of fibrewise spaces over B to the category of fibrewise spaces over B'.

Equivalences in the category of fibrewise spaces over B are called *fibrewise topological equivalences* or *fibrewise homeomorphisms*. If ϕ, as above, is a fibrewise topological equivalence over B then $\phi_{B'}$ is a fibrewise topological equivalence over B' for each subspace B' of B. In particular ϕ_b is a topological equivalence for each point b of B. However, this necessary condition for a fibrewise topological equivalence is obviously not sufficient. To see this take $Y = B$ to be a non-discrete space and take X to be the same set with the discrete topology and the identity as projection; the identity function has no continuous inverse.

A fibrewise map $\phi : X \to Y$ is said to be *fibrewise constant* if $\phi = t \circ p$ for some section $t : B \to Y$. The same example as in the previous paragraph shows that a fibrewise map may be constant on each fibre but not fibrewise constant.

Fibrewise product and coproduct

Given an indexed family $\{X_j\}$ of fibrewise spaces over B the *fibrewise product* $\prod_B X_j$ is defined as a fibrewise space over B, and comes equipped with a family of fibrewise projections

$$\pi_j : \prod_B X_j \to X_j.$$

The fibres of the fibrewise product are just the products of the corresponding fibres of the factors. The fibrewise product is characterized by the following Cartesian property: for each fibrewise space X over B the fibrewise maps

$$\phi : X \to \prod_B X_j$$

correspond precisely to the families of fibrewise maps $\{\phi_j\}$, where

$$\phi_j = \pi_j \circ \phi : X \to X_j.$$

For example if $X_j = X$ for each index j the *diagonal*

$$\Delta : X \to \prod_B X$$

is defined so that $\pi_j \circ \Delta = 1_X$ for each j.

If $\{X_j\}$ is as before the *fibrewise coproduct* $\coprod_B X_j$ is also defined, as a fibrewise space over B, and comes equipped with a family of fibrewise insertions

$$\sigma_j : X_j \to \coprod_B X_j.$$

The fibres of the fibrewise coproduct are just the coproducts of the corresponding fibres of the summands. The fibrewise coproduct is characterized by the following cocartesian property: for each fibrewise space X over B the fibrewise maps

$$\psi : \coprod_B X_j \to X$$

correspond precisely to the families of fibrewise maps $\{\psi_j\}$, where

$$\psi_j = \psi \circ \sigma_j : X_j \to X.$$

For example if $X_j = X$ for each index j the *codiagonal*

$$\nabla : \coprod_B X \to X$$

is defined so that $\nabla \circ \sigma_j = 1_X$ for each j.

The notations $X \times_B Y$ and $X \sqcup_B Y$ are used for the fibrewise product and fibrewise coproduct in the case of a family $\{X, Y\}$ of two fibrewise spaces, and similarly for finite families generally. When $X = Y$ the switching maps

$$X \times_B X \to X \times_B X, \qquad X \sqcup_B X \to X \sqcup_B X$$

are defined with components (π_2, π_1) and (σ_2, σ_1), respectively.

Given a map $\alpha : B' \to B$, for any space B', we can regard B' as a fibrewise space over B. For each fibrewise space X over B we denote by $\alpha^* X$ the fibrewise product $X \times_B B'$, regarded as a fibrewise space over B' using the second projection, and similarly for fibrewise maps. Thus α^* constitutes a functor from the category of fibrewise spaces over B to the category of fibrewise spaces over B'. When B' is a subspace of B and α the inclusion this is equivalent to the restriction functor described earlier.

By a *fibrewise topology*, on a fibrewise set X over B, we mean any topology on X such that the projection p is continuous. By a *fibrewise basis*, for a fibrewise topology, we mean a collection $\mathcal{U}$ of subsets of X which forms a basis for a topology after augmentation by the topology induced by p. In other words, the open sets of X are the unions of intersections of members of $\mathcal{U}$ and sets of the form X_W, where W is open in B. For example, consider the product $B \times T$, where T is a space. A fibrewise basis for the fibrewise topology is given by the collection of products $B \times U$, where U runs through the open sets of T.

The term *fibrewise sub-basis* is used in a similar sense or we may, on occasion, say that the fibrewise topology is generated by a family of subsets, meaning that finite intersections of members of the family form a fibrewise basis.

Note that in checking the continuity of fibrewise functions, where the fibrewise topology of the codomain is generated in this way, it is sufficient to verify that the preimages of fibrewise subbasic open sets are open.

Fibre bundles

A fibrewise space X over B is said to be *trivial* if X is fibrewise homeomorphic to $B \times T$ for some space T, and then a fibrewise homeomorphism $\phi : X \to B \times T$ is called a *trivialization* of X. A fibrewise space X over B is said to be *locally trivial* if there exists an open covering of B such that X_V is trivial over V for each member V of the covering. A locally trivial fibrewise space is the simplest form of fibre bundle or bundle of spaces. As Dold [45] has shown, the theory of fibre bundles is improved if it is confined to the class of numerable bundles, i.e. bundles which are trivial over every member of some numerable covering of the base. Derwent [40] and tom Dieck [42] have pointed out that such a covering may be taken to be countable, thus facilitating inductive arguments.

A more sophisticated form of the notion of fibre bundle involves a topological group G, the structural group. A principal G-bundle over the base space B is a locally trivial fibrewise space P over B on which G acts freely. Moreover, the action is fibre-preserving, so that each of the fibres is homeomorphic to G. Such a principal G-bundle P over B determines a functor

$P_\#$ from the category of G-spaces to the category of fibre bundles over B. Specifically $P_\#$ transforms each G-space A into the associated bundle $P \times_G A$ with fibre A, and similarly with G-maps. We refer to $P_\#$ as the *associated bundle functor*.

The theory of fibre bundles is dealt with in the standard textbooks such as Steenrod [128] or Bredon [19], where a large variety of examples are discussed. Some of these will be appearing in the course of our work.

From our point of view it is only natural to proceed a stage further and develop a fibrewise version of the theory of fibre bundles, as in [95]. Thus let X and T be fibrewise spaces over B. By a *fibrewise fibre bundle* over X, with fibrewise fibre T, we mean a fibrewise space E together with a fibrewise map $p : E \to X$ which is locally fibrewise trivial, in the sense that there exists a covering of X such that E_V is fibrewise homeomorphic to $V \times_B T$, over B, for each member V of the covering. This is the simplest form of the definition, but of course there is a more sophisticated form, involving a fibrewise structural group. Details are given in Section 8 below.

Classes of fibrewise spaces

There are various classes of fibrewise spaces which will appear in the work we shall be doing later, for example, the class of fibrewise open spaces, where the projection is open. To be of any interest to us such a class must be invariant, so that a fibrewise space which is fibrewise homeomorphic to a member of the class is also a member of the class. It must also be natural, in the sense that pull-backs of a member are also members. Furthermore, fibrewise products of members are also members, at least finite fibrewise products. Fibre bundles are such a class.

In fibrewise topology the existence of local sections is a condition of some importance, but more usually it is the existence of local slices which is required.

Definition 1.1 The fibrewise space X over B is *locally sliceable* if for each point b of B and each point x of X_b there exists a neighbourhood W of b and a section $s : W \to X_W$ such that $s(b) = x$.

The condition implies that p is open since if U is a neighbourhood of x in X then $s^{-1}(X_W \cap U) \subseteq pU$ is a neighbourhood of b in W. In other words, locally sliceable fibrewise spaces are fibrewise open.

There are fibrewise versions of all the usual separation conditions of topology, in fact the number of different fibrewise separation conditions which can reasonably be defined is quite large. For our purposes, however, only two or three are of real significance.

Definition 1.2 The fibrewise space X over B is *fibrewise Hausdorff* if the diagonal embedding

$$\Delta : X \to X \times_B X$$

is closed.

Equivalently, for each point b of B and each pair x, x' of distinct points of X_b there exist disjoint neighbourhoods of x, x' in X.

Subspaces of fibrewise Hausdorff spaces are fibrewise Hausdorff. The following two properties of fibrewise Hausdorff spaces are worth mentioning.

Proposition 1.3 *Let $\phi : X \to Y$ be a fibrewise map, where X and Y are fibrewise spaces over B. If Y is fibrewise Hausdorff the fibrewise graph of ϕ is closed in $X \times_B Y$.*

Proposition 1.4 *Let $\phi, \psi : X \to Y$ be fibrewise maps, where X and Y are fibrewise spaces over B. If Y is fibrewise Hausdorff the coincidence set $K(\phi, \psi)$ of ϕ and ψ is closed in X.*

These results follow easily from the definition.

From the viewpoint of fibrewise topology it seems natural to revise some of the terminology of ordinary topology. For example

Definition 1.5 The fibrewise space X over B is *fibrewise discrete* if the projection p is a local homeomorphism.

Clearly, fibrewise discrete spaces are locally sliceable and hence fibrewise open. An attractive characterization of this class of fibrewise spaces is given by

Proposition 1.6 *Let X be a fibrewise space over B. Then X is fibrewise discrete if and only if* (i) *X is fibrewise open and* (ii) *the diagonal embedding*

$$\Delta : X \to X \times_B X$$

is open.

Corollary 1.7 *Let $\phi : X \to Y$ be a fibrewise map, where X is fibrewise open and Y is fibrewise discrete over B. Then the fibrewise graph*

$$\Gamma : X \to X \times_B Y$$

of ϕ is an open embedding.

Corollary 1.8 *Let $\phi, \psi : X \to Y$ be fibrewise maps, where X and Y are fibrewise spaces over B. If Y is fibrewise discrete the coincidence set $K(\phi, \psi)$ of ϕ and ψ is open in X.*

Another fibrewise separation condition we shall need is as follows.

Definition 1.9 The fibrewise space X over B is *fibrewise regular* if for each point b of B, each point x of X_b and each neighbourhood V of x in X there exists a neighbourhood W of b in B and a neighbourhood U of x in X_W such that the closure $X_W \cap \bar{U}$ of U in X_W is contained in V.

When the fibrewise topology of X is given in terms of a fibrewise sub-basis it is sufficient if the condition for fibrewise regularity is satisfied for fibrewise subbasic neighbourhoods V. Subspaces of fibrewise regular spaces are also fibrewise regular, as can easily be shown.

Fibrewise open means that the projection is open, fibrewise closed that the projection is closed. Because fibrewise products of fibrewise closed spaces are not, in general, fibrewise closed, the class of fibrewise closed spaces is only of minor importance. A stronger condition is needed, as in

Definition 1.10 The fibrewise space X over B is *fibrewise compact* if the projection is proper.

In other words X is fibrewise compact if X is fibrewise closed and every fibre of X is compact. One can also characterize the condition in terms of coverings, as follows.

Proposition 1.11 *The fibrewise space X over B is fibrewise compact if and only if for each point b of B and each covering $\mathcal{U}$ of X_b by open sets of X there exists a neighbourhood W of b in B such that a finite subfamily of $\mathcal{U}$ covers X_W.*

Proposition 1.12 *Let X be fibrewise compact over B. Suppose that X is fibrewise discrete. Then $X \to B$ is a finite covering space.*

For consider a point $b \in B$. Choose for each $x \in X_b$ an open neighbourhood U_x in X such that $p(U_x)$ is open in B and the restriction of p is a homeomorphism $U_x \to p(U_x)$. Since the intersection of U_x with the fibre X_b is precisely $\{x\}$, it follows from Proposition 1.11 that X_b is finite and that there is an open neighbourhood W of b in B such that $X_W \subseteq \bigcup_x U_x$. Let V be the open subset

$$V = W \cap \bigcap_x p(U_x) \subseteq B.$$

Then $X_V \to V$ is trivial.

The images of fibrewise compact spaces under fibrewise maps are also fibrewise compact. This follows at once from the definition; with a little more effort we obtain the useful

Proposition 1.13 *Let $\phi : X \to Y$ be a fibrewise map, where X is fibrewise compact and Y is fibrewise Hausdorff. Then ϕ is proper.*

Proposition 1.14 *Let X be fibrewise regular over B and let K be a fibrewise compact subset of X. Let b be a point of B and let V be a neighbourhood of K_b in X. Then there exists a neighbourhood W of b in B and a neighbourhood U of K_W in X_W such that the closure $X_W \cap \bar{U}$ of U in X_W is contained in V.*

There is just one more class of fibrewise spaces we need to consider here.

Definition 1.15 The fibrewise space X over B is *fibrewise locally compact* if for each point b of B and each point x of X_b there exists a neighbourhood W of b in B and a neighbourhood U of x in X_W such that the closure $X_W \cap \bar{U}$ of U in X_W is fibrewise compact over W.

It is easy to see that fibrewise compact spaces are fibrewise locally compact, also that closed subspaces of fibrewise locally compact spaces are fibrewise locally compact. We conclude with two results which are not quite so obvious; proofs may be found in Section 3 of [86]. Recall that we make the convention that the term 'fibrewise' governs everything that follows it. For example 'fibrewise locally compact Hausdorff space' means a fibrewise space which is both fibrewise locally compact and fibrewise Hausdorff.

Proposition 1.16 *Let X be fibrewise locally compact Hausdorff over B. Then X is fibrewise regular.*

Proposition 1.17 *Let X be fibrewise locally compact regular over B. Then for each point b of B, each compact subset C of X_b, and each neighbourhood V of C in X, there exists a neighbourhood W of b in B and a neighbourhood U of C in X_W such that the closure $X_W \cap \bar{U}$ of U in X_W is fibrewise compact over W and contained in V.*

Fibrewise quotients

By a *fibrewise quotient map* we mean a fibrewise map which is a quotient map in the ordinary sense. Fibrewise products of fibrewise quotient maps are not necessarily fibrewise quotient maps. We prove

Proposition 1.18 *Let $\phi : X \to Y$ be a fibrewise quotient map, where X and Y are fibrewise spaces over B. Then the fibrewise product*

$$\phi \times 1 : X \times_B T \to Y \times_B T$$

is a fibrewise quotient map, for all fibrewise locally compact regular T.

For let $U \subseteq X \times_B T$ be open and saturated with respect to $\psi = \phi \times 1$. We have to show that ψU is open in $Y \times_B T$. So let $(y, t) \in \psi U$, where $y \in Y_b$,

$t \in T_b$, $b \in B$, and pick $x \in \phi^{-1}(y) \subseteq X_b$. We have $(x, t) \in U$, since U is saturated. Consider the subset N of T_b given by

$$\{x\} \times N = (\{x\} \times T_b) \cap U.$$

Now N is open in T_b, since U is open in $X \times_B T$, and so $N = M \cap T_b$, where M is open in T. Since T is fibrewise locally compact there exists, by Proposition 1.17, a neighbourhood $K \subseteq M$ of t in T_W such that K is fibrewise compact over W. Consider the subset

$$V = \{\xi \in X_W \mid \{\xi\} \times_W K \subseteq U\}$$

of X_W. We have $(y, t) \in \phi V \times_W K \subseteq \psi U$. So to prove that ψU is a neighbourhood of (y, t) in $Y \times_B T$ it is sufficient to prove that ϕV is a neighbourhood of y in Y.

In fact V is open in X. For let $\xi \in V$ so that $\{\xi\} \times_W K_\beta \subseteq U$, where $\beta = p(\xi)$ and $p : X \to B$. Since K is fibrewise compact over W the projection

$$X_W \times_W K \to X_W \times_W W \to X_W$$

is closed. Since U is a neighbourhood of the inverse image $\{\xi\} \times K_\beta$ of ξ under the projection there exists a neighbourhood $W' \subseteq W$ of β and a neighbourhood V' of ξ in $X_{W'}$ such that $V' \times_{W'} K_{W'} \subseteq U$. This implies that $V' \subseteq V$, by the definition of V, and so V is open.

Moreover, V is saturated. For $V \subseteq \phi^{-1} \phi V$, as always. Also

$$\phi^{-1} \phi V \times_W K = \psi^{-1} \psi (V \times_W K) \subseteq \psi^{-1} \psi U = U.$$

Therefore $\phi^{-1} \phi V \subseteq V$, by the definition of V, and so $\phi^{-1} \phi V = V$. Thus V is saturated, as well as open, and so ϕV is open. Since $y \in \phi V$ this completes the proof.

Given a fibrewise space X over B a *fibrewise equivalence relation* on X is given by a subset R of the fibrewise product $X \times_B X$. We refer to the fibrewise set X/R of equivalence classes, with the quotient topology, as the *fibrewise quotient space*. Of course, fibrewise maps $X/R \to Z$, for any fibrewise space Z, correspond precisely to invariant fibrewise maps $X \to Z$.

We describe a fibrewise map $\phi : (X, A) \to (X', A')$ as a *fibrewise relative homeomorphism* if (i) A is closed in X, (ii) ϕ maps $X - A$ bijectively onto $X' - A'$, and (iii) X' is a fibrewise quotient space of X under ϕ.

In general there is no simple condition at the level of X which implies that X/R is fibrewise Hausdorff. Suppose, however, that $R = (\phi \times \phi)^{-1} \Delta Z$ for some fibrewise map $\phi : X \to Z$, where Z is fibrewise Hausdorff. Then the induced fibrewise map $X/R \to Z$ is injective and so X/R is fibrewise Hausdorff.

Consider a space D and a closed subspace E of D. For any fibrewise space X over B let $\Phi_B(X)$ denote the push-out of the cotriad

$$X \times D \xleftarrow{2} X \times E \xrightarrow{\pi_2} E,$$

and similarly for fibrewise maps. Thus an endofunctor Φ_B of our category is defined. It can be shown, as in [86], that $\Phi_B(X)$ is fibrewise Hausdorff whenever X is fibrewise Hausdorff. For $(D, E) = (I, \{0\})$, where $I = [0, 1] \subseteq \mathbb{R}$, the endofunctor is known as the *fibrewise cone* and denoted by C_B. When $(D, E) = (I, \{0, 1\})$ the endofunctor is known as the *fibrewise suspension* and denoted by Σ_B. For example $\Sigma_B(X) = B \times I$ when $X = B$, and $\Sigma_B(X) = B \times \dot{I}$ when $X = \emptyset$. Note that the associated bundle functor $P_\#$ discussed earlier, from the category of G-spaces to the category of fibrewise spaces, transforms the equivariant cone into the fibrewise cone and the equivariant suspension into the fibrewise suspension. For example, taking G to be the orthogonal group $O(n)$, the fibrewise cone of an $(n-1)$-sphere bundle is the associated n-ball bundle, and the fibrewise suspension is the associated n-sphere-bundle.

More generally let X_i $(i = 0, 1)$ be a fibrewise space. Consider the fibrewise equivalence relation on the coproduct

$$X_0 \sqcup (X_0 \times I \times X_1) \sqcup X_1$$

which identifies (x_0, t, x_1), with x_t whenever $t = 0$ or 1. The fibrewise set $X_0 *_B X_1$ of equivalence classes, with the quotient topology, is called the *fibrewise join* of X_0 and X_1. For example, if X_t is the sphere-bundle associated with E_t, where E_t is a euclidean bundle over B, then $X_0 *_B X_1$ is the sphere-bundle associated with the Whitney sum $E_0 \oplus E_1$. When $X_0 = S^{n-1} \times B$ and $X_1 = X$ we may identify $X_0 *_B X_1$ with the n-fold fibrewise suspension $\Sigma_B^n(X)$ of the fibrewise space X.

It should be noted that the fibrewise join is not in general associative, with the quotient topology. However if, following Milnor [113], we replace this by the coarsest topology which makes the coordinate functions

$$t : X_0 *_B X_1 \to B \times I,$$

$$x_0 : t^{-1}(B \times [0, 1)) \to X_0,$$

$$x_1 : t^{-1}(B \times (0, 1]) \to X_1$$

continuous then associativity holds without restriction. Furthermore, the topologies coincide when X_0 and X_1 are fibrewise compact Hausdorff.

Fibrewise mapping-spaces

Finally, let us turn to the problem of constructing a right adjoint to the fibrewise product. One has to impose a topology with the necessary properties on the fibrewise set

$$\mathrm{map}_B(X, Z) = \coprod_{b \in B} \mathrm{map}(X_b, Z_b),$$

where X and Z are fibrewise spaces over B. Although this can be done in general, as we shall see later, the case when $X = B \times T$, for some space T, admits of simpler treatment. In fact maps of $\{b\} \times T$ into Z_b can be regarded as maps of T into Z, in the obvious way, and so $\mathrm{map}_B(B \times T, Z)$ can be topologized as a subspace of $\mathrm{map}(T, Z)$, with the compact-open topology. It is easy to check that for any fibrewise space Y over B a fibrewise map

$$Y \times T = (B \times T) \times_B Y \to Z$$

determines a fibrewise map

$$Y \to \mathrm{map}_B(B \times T, Z),$$

through the standard formula, and that the converse holds when T is compact Hausdorff. In fact this special case is sufficient for the great majority of situations where fibrewise mapping-spaces are used in what follows.

Some examples

The reader may wish to treat the following examples, related to the text of this section, as exercises.

Example 1.19. Let $\phi : X \to Y$ be an open and closed fibrewise surjection where X and Y are fibrewise spaces over B. Let $\lambda : X \to \mathbb{R}$ be a continuous real-valued function which is fibrewise bounded above, in the sense that λ is bounded above on each fibre of X. Then $\mu : Y \to \mathbb{R}$ is continuous, where

$$\mu(\eta) = \sup_{\xi \in \phi^{-1}(\eta)} \lambda(\xi).$$

Example 1.20. Let $\phi : X \to Y$ be a fibrewise function, where X and Y are fibrewise spaces over B. Suppose that X is fibrewise open and that the product

$$id \times \phi : X \times_B X \to X \times_B Y$$

is open. Then ϕ itself is open.

Example 1.21. Let X be a closed subspace of $B \times \mathbb{R}^n$, $(n \geq 0)$, regarded as a fibrewise space over B under the first projection. Then X is fibrewise compact if X is fibrewise bounded, in the sense that there exists a continuous real-valued function $\lambda : B \to \mathbb{R}$ such that X_b is bounded by $\lambda(b)$ for each point b of B.

Example 1.22. Let $\phi : X \to Y$ be a fibrewise function, where X and Y are fibrewise spaces over B. Then, if X is fibrewise compact and the product

$$id \times \phi : X \times_B X \to X \times_B Y$$

is proper, ϕ is proper.

Example 1.23. Let $\phi : X \to Y$ be an open fibrewise surjection, where X and Y are fibrewise spaces over B. Suppose that the preimage of the diagonal of Y with respect to $\phi \times \phi$ is closed in $X \times_B X$. Then Y is fibrewise Hausdorff.

Example 1.24. Let $\phi : X \to Y$ be a fibrewise function, where X and Y are fibrewise spaces over B. Suppose that X is fibrewise Hausdorff and that the fibrewise graph of ϕ in $X \times_B Y$ is fibrewise compact. Then ϕ is continuous.

Example 1.25. Let $\phi : X \to Y$ be a fibrewise map, where X and Y are fibrewise discrete over B. Then ϕ is a local homeomorphism.

Example 1.26. Let X be a fibrewise space over B and let A be an open subspace of X. If A is fibrewise open, then the projection $\pi : X \to X/_B A$ is open.

Example 1.27. Let $\phi : X \to Y$ be a proper fibrewise function, where X and Y are fibrewise spaces over B. If X is fibrewise Hausdorff and fibrewise regular then so is ϕX.

Example 1.28. If $B = CX$ is the cone on the non-empty space X, there is no fibrewise map from X, regarded as a fibrewise space with the constant map to the apex of the cone as projection, into X, regarded as a fibrewise space with the inclusion as projection.

2 Fibrewise transformation groups

Fibrewise topological groups

Let us continue to work over a base space B. We describe a fibrewise space G as a *fibrewise topological group* or *fibrewise group-space* over B if G is equipped with fibrewise maps

$$m : G \times_B G \to G, \qquad e : B \to G, \qquad u : G \to G$$

such that the following three conditions are satisfied.

$$m \circ (m \times 1) = m \circ (1 \times m) : G \times_B G \times_B G \to G,$$
$$m \circ (c \times 1) \circ \Delta = 1 = m \circ (1 \times c) \circ \Delta : G \to G,$$
$$m \circ (u \times 1) \circ \Delta = c = m \circ (1 \times u) \circ \Delta : G \to G.$$

Here, as usual, c denotes the fibrewise constant determined by e. We refer to m as the *fibrewise multiplication*, to e as the *neutral section* and to u as the *fibrewise inversion*. The three conditions imply, and are implied by, the statement that the fibre G_b is a topological group, with multiplication m_b, neutral element $e(b)$, and inversion $u(b)$, for all points b of B.

For example, B itself is a fibrewise topological group, with the identity as fibrewise multiplication, neutral section and fibrewise inversion. More generally, $B \times T$ is a fibrewise topological group for each topological group T.

For a more interesting example consider the cylinder $I \times T$, where $I = [0,1] \subseteq \mathbb{R}$ and T is a (discrete, additive) Abelian group. The fibrewise topological group structure on $I \times T$, as a fibrewise space over I, induces a fibrewise topological group structure on the quotient fibrewise space $(I \times T)/R$ over the circle $I/\dot{I}$, where R identifies $(t,0)$ with $(-t,1)$ for all $t \in T$.

For another example, let G_0 be a compact Hausdorff topological group. Then the unreduced suspension $\Sigma(G_0)$ of G_0 forms a fibrewise topological group over $\Sigma(pt) = I$. More generally, the join $G_0 * X$ is a fibrewise topological group over the cone $(pt) * X$ for all spaces X.

For another type of example, consider a vector bundle E over B, and take $G = \mathrm{Aut}_B(E)$, the bundle formed by automorphisms of the fibres. More generally, let P be a principal Γ-bundle over B, where Γ is a topological group. Then the associated adjoint bundle $P \times_\Gamma \Gamma$ with fibre Γ is a fibrewise topological group, where Γ acts on itself by conjugation. See [4] for some applications.

Returning to the general case, let G be a fibrewise topological group over B. We describe a subspace H of G as a *subgroup* if $m(H \times_B H) \subseteq H$ and $uH \subseteq H$, in other words if H_b is a subgroup of G_b for each point b of B. The term *normal subgroup* is defined similarly. For example, the image of the neutral section is always a normal subgroup, the fibrewise trivial subgroup.

Of course a subgroup of a fibrewise topological group is itself a fibrewise topological group, with the induced fibrewise multiplication, neutral section and fibrewise inversion.

A subgroup H of the fibrewise topological group G determines a pair of fibrewise equivalence relations on G. In one case the relation is the preimage of the division function

$$d : G \times_B G \to G,$$

where $d(g, g') = g(g')^{-1}$. Then we refer to the fibrewise quotient space G/H as the *fibrewise right factor space*. In the other case we use the division function $u \circ d$ instead of d and obtain the *fibrewise left factor space* $H\backslash G$. Of course u induces a fibrewise homeomorphism between G/H and $H\backslash G$. When H is normal we have $G/H = H\backslash G$. Note that if H is a subgroup of G the neutral section of G induces a section of G/H which may also be referred to as the neutral section. Also the neutral section of G/H is closed (respectively open) if and only if H is closed (respectively open) in G. Similarly in the case of $H\backslash G$.

Let G and G' be fibrewise topological groups with fibrewise multiplications

$$m : G \times_B G \to G, \qquad m' : G' \times_B G' \to G'.$$

By a *fibrewise homomorphism* of G into G' we mean a fibrewise map $\phi : G \to G'$ such that

$$m' \circ (\phi \times \phi) = \phi \circ m.$$

The condition implies that $\phi \circ e = e'$ and $\phi \circ u = u' \circ \phi$, where e' denotes the neutral section and u' the fibrewise inversion in the structure of G'. The terms *fibrewise isomorphism* and *fibrewise automorphism* are defined similarly.

By the *fibrewise kernel* of a fibrewise homomorphism $\phi : G \to G'$ we mean the preimage of the neutral section of G'. Note that the fibrewise kernel is always a normal subgroup of G. More generally, the preimage of a normal subgroup is always a normal subgroup.

Given a section $s : B \to G$ of the fibrewise topological group G, a fibrewise automorphism of G is defined which is given on G_b by conjugation with respect to $s(b)$ for each point b of B. We may refer to this operation as *fibrewise conjugation* with respect to s. Subgroups K and L of G are said to be *fibrewise conjugate* if L is the image of K under fibrewise conjugation with respect to some section s of G. Fibrewise conjugacy constitutes an equivalence relation between the subgroups of G. The fibrewise conjugacy class of a subgroup H of G is denoted by $[H]$.

Fibrewise transformation groups

We continue to work over the base space B. Let G be a fibrewise topological group and let E be a fibrewise space. By a *fibrewise action* of G on the right of E we mean a fibrewise map $r : E \times_B G \to E$ such that the following two conditions are satisfied.

$$r \circ (r \times 1) = r \circ (1 \times m) : E \times_B G \times_B G \to E,$$
$$r \circ (1 \times c) \circ \Delta = 1 : E \to E;$$

here $c : E \to G$ denotes the fibrewise constant given by the neutral section. These conditions imply, and are implied by, the statement that E_b is a right G_b-space for each point b of B. We describe E, with this structure, as a *fibrewise right G-space. Fibrewise left G-spaces* are defined similarly.

For example, we can regard G itself as a fibrewise right G-space by taking r to be the fibrewise multiplication. More generally, we can regard any fibrewise topological group containing G as a subgroup as a fibrewise right G-space, or equally well as a fibrewise left G-space.

Given a fibrewise action r of G on the right of E we usually denote $r(\xi, g)$ by $\xi.g$, where $\xi \in E_b$, $g \in G_b$ and $b \in B$. Also when $E' \subseteq E$ and $G' \subseteq G$ we denote $r(E' \times_B G')$ by $E'.G'$.

If E is a fibrewise right G-space the fibrewise action determines a fibrewise equivalence relation on E. The resulting fibrewise quotient space E/G is called the *fibrewise right orbit space*. If E is a fibrewise left G-space the *fibrewise left orbit space* $G \backslash E$ is defined similarly.

Fibrewise (right or left) G-spaces over a base form a category in which the morphisms are *fibrewise G-maps*, i.e. fibrewise maps which are equivariant with respect to the fibrewise action. The equivalences of the category are called *fibrewise G-equivalences*. Note that if a fibrewise G-map is a homeomorphism then it is necessarily a fibrewise G-equivalence.

Let E be a fibrewise right G-space and let F be a fibrewise left G-space. A fibrewise action of G on the right of $E \times_B F$ is defined by

$$(\xi, \eta).g = (\xi.g, g^{-1}.\eta).$$

We denote the fibrewise orbit space $(E \times_B F)/G$ by $E \times_G F$, for simplicity, and refer to it as the *fibrewise mixed product* of E and F.

Let E be a fibrewise right G-space. Consider the fibrewise map

$$\theta : E \times_B G \to E \times_B E$$

given by $\theta(\xi, g) = (\xi, \xi.g)$. We describe the fibrewise action as *proper* if θ is proper. If the fibrewise action of G on E is proper then so is the fibrewise action of G on each invariant subspace of E and so is the fibrewise action of each subgroup of G on E.

Proposition 2.1 *Let E be a proper fibrewise right G-space, where G is fibrewise Hausdorff. Then E is fibrewise Hausdorff.*

For since G is fibrewise Hausdorff the fibrewise graph $E \to E \times_B G$ of the fibrewise constant $c : E \to G$ is closed, by Proposition 1.3. Postcomposing with the proper fibrewise map θ yields the diagonal $\Delta : E \to E \times_B E$, which is therefore closed. Hence E is fibrewise Hausdorff, as asserted.

Note that for any fibrewise G-space E the fibrewise graph

$$\Gamma : E \times_B G \to E \times_{E/G} E \times_B G$$

of the action is an embedding. If E is fibrewise Hausdorff over E/G the fibrewise graph is closed and so Γ is proper. If in addition G is fibrewise compact then the projection

$$\pi : E \times_{E/G} E \times_B E \to E \times_{E/G} E$$

is proper, hence $\theta = \pi\Gamma$ is proper. Thus we have

Proposition 2.2 *Let E be a fibrewise G-space, where G is fibrewise compact. If E is fibrewise Hausdorff then E is a proper fibrewise G-space.*

Let $s : B \to E$ be a section of the fibrewise right G-space E. We may compare the fibrewise constant map $c = s \circ p : G \to E$ with the composition $a = r \circ (c \times 1) \circ \Delta$, where

$$G \xrightarrow{\Delta} G \times_B G \xrightarrow{c \times 1} E \times_B G \xrightarrow{r} E.$$

The coincidence set $K(a,c) = K$ is a subgroup of G, called the *fibrewise stabilizer* of s. Note that a induces an injective fibrewise map of G/K into E, with image $(sB).G$.

If the fibrewise action is proper the fibrewise map $a : G \to E$ is proper, since it can be identified with the restriction

$$sB \times_B G \to B \times_B E$$

of θ to sB, and so a induces a fibrewise homeomorphism between G/K and $(sB.G)$.

Let us say that the fibrewise action of G on the right of E is *free* if G_b acts freely on E_b for each point b of B. In that case a fibrewise function $d : R \to G$ is defined, where $R \subseteq E \times_B E$ is the image of θ and where $d(\xi, \eta)$, for $(\xi, \eta) \in R$, is the unique element $g \in G$ such that $\eta = \xi.g$. We may refer to d as the *division function*. Of course d is continuous if $E = \Gamma$, a fibrewise topological group containing G as a subgroup, and the fibrewise action is given by fibrewise multiplication.

Proposition 2.3 *Let E be a fibrewise G-space, where the fibrewise action is free. The fibrewise action is proper if and only if* (i) *R is closed in $E \times_B E$ and* (ii) *the division function $d : R \to G$ is continuous.*

For if d is continuous a fibrewise map $R \to E \times_B G$ is given by π_1 in the first factor, by d in the second. Postcomposing with θ gives the inclusion $R \subseteq E \times_B E$. Hence θ is homeomorphic when R is closed. For the converse just reverse the argument.

Note that R is closed in $E \times_B E$ if E/G is fibrewise Hausdorff since R is just the preimage of the diagonal of E/G with respect to

$$\pi \times \pi : E \times_B E \to E/G \times_B E/G.$$

Fibrewise open groups

If we are to develop the theory of fibrewise topological transformation groups beyond this very elementary level some restrictions need to be imposed, so from now on we concentrate our attention on the family of fibrewise open groups. The family includes, for example, fibrewise topological groups of the form $B \times T$, where T is a topological group. It also includes fibrewise discrete groups, where the projection is open. Obviously, an open subgroup of a fibrewise open group is also fibrewise open.

Proposition 2.4 *Let G be a fibrewise topological group and let H be a fibrewise open subgroup of G. Then the projection $\pi : G \to G/H$ is open.*

For consider the fibrewise homeomorphism

$$\psi : G \times_B H \to G \times_B H$$

given by $\psi(g, h) = (g.h, h)$. The first projection $\pi_1 : G \times_B H \to G$ is open, since H is fibrewise open, and so the restriction $m' = \psi \circ \pi_1$ of the fibrewise multiplication m to $G \times_B H$ is open. In particular, if $V \subseteq G$ is open then so is $\pi^{-1}\pi(V) = m'(V \times_B H)$. Hence π is open, as asserted.

It is an important consequence of Proposition 2.4 that G/H is a fibrewise left G-space, with the fibrewise action induced by fibrewise multiplication, whenever H is fibrewise open. Furthermore, when H is normal, as well as fibrewise open, a fibrewise topological group structure on G/H is defined so that the natural projection $\pi : G \to G/H$ is an open fibrewise homomorphism of G onto G/H with fibrewise kernel H. Conversely, suppose that G' is the image of G under an open fibrewise homomorphism $\phi : G \to G'$. Then ϕ induces a fibrewise isomorphism between G/H and G', where H is the fibrewise kernel of ϕ.

Proposition 2.5 *Let G be a fibrewise topological group and let H be a fibrewise open subgroup of G. Then G/H is fibrewise Hausdorff if and only if H is closed in G.*

For the division map d induces a fibrewise map

$$d' = d(\pi \times \pi)^{-1} : G/H \times_B G/H \to G,$$

by Proposition 2.4. Since the preimage of $G - H$ under d' is the complement

$$(G/H) \times_B (G/H) - \Delta(G/H)$$

of the diagonal the assertion follows at once.

Also recall from Proposition 1.6 that a fibrewise open space is fibrewise discrete if and only if the diagonal is open in the fibrewise square. So by a similar argument to that used to prove Proposition 2.5 we obtain

Proposition 2.6 *Let G be a fibrewise open topological group and let H be a fibrewise open subgroup of G. Then G/H is fibrewise discrete if and only if H is open in G.*

Proposition 2.7 *Let E be a fibrewise right G-space, where G is fibrewise open. Then the natural projection $\pi : E \to E/G$ is open.*

For the fibrewise action r can be expressed as the composition

$$E \times_B G \xrightarrow{\alpha} E \times_B G \xrightarrow{\pi_1} E,$$

where $\alpha(\xi, g) = (\xi.g, g)$. Now π_1 is open, since G is fibrewise open, and α is a fibrewise homeomorphism. Thus r is open, in particular if $U \subseteq E$ is open then so is $\pi^{-1}\pi(U) = U.G$. Hence π is open, as asserted.

Let H be a fibrewise open normal subgroup of the fibrewise topological group G, and let E be a fibrewise G-space. Then E/H is a fibrewise G-space, with the induced fibrewise action, and hence is a fibrewise (G/H)-space. By transitivity of fibrewise quotient topologies the canonical bijection

$$(E/H)/(G/H) \to E/G$$

is a fibrewise homeomorphism.

Another straightforward consequence of Proposition 2.7 is the associative law for fibrewise mixed products, as follows.

Proposition 2.8 *Let D, E, F be fibrewise spaces and let G, H be fibrewise open groups. Suppose that G acts fibrewise on the right of D and on the left of E, while H acts fibrewise on the right of E and on the left of F. Also suppose that the action of G on the left of E commutes with the action of H on the right of E. Then G acts fibrewise on the left of $E \times_H F$, and H acts fibrewise on the right of $D \times_G E$, so that the identity on $D \times_B E \times_B F$ induces a fibrewise homeomorphism between $(D \times_G E) \times_H F$ and $D \times_G (E \times_H F)$.*

Proposition 2.9 *Let E be a fibrewise space. Let G and H be fibrewise open groups. Suppose that G acts fibrewise on the left of E while H acts fibrewise on the right of E, and that the actions commute. Then there are induced fibrewise actions of G on the left of E/H and of H on the right of $G\backslash E$ such that the identity on E induces a fibrewise homeomorphism between $G\backslash(E/H)$ and $(G\backslash E)/H$.*

Again the proof is straightforward and will be omitted. For the same reason we omit the proof of

Proposition 2.10 *Let E, F be fibrewise spaces, and let G, H be fibrewise open topological groups. Suppose that G acts fibrewise on the right of E and on the left of F. Also suppose that H acts fibrewise on the right of F and that the actions of G and H on F commute with each other. Then there is an induced action of H on the right of $E \times_G F$ such that the identity on $E \times_G F$ induces a fibrewise homeomorphism between $(E \times_G F)/H$ and $E \times_G (F/H)$.*

It follows, in particular, that if E is any fibrewise right G-space then $E \times_G (G/H)$ is fibrewise homeomorphic to $(E \times_G G)/H$ and hence to E/H.

After this point in the exposition we make the convention that fibrewise actions are always on the right, unless otherwise stated, and similarly with fibrewise orbit spaces.

Proposition 2.11 *Let E be a fibrewise G-space, where G is fibrewise discrete. Then the natural projection $\pi : E \to E/G$ is a local homeomorphism.*

The proof is very similar to that of Proposition 2.4. Among the fibrewise G-spaces where G is fibrewise discrete a special rôle is played by those where the fibrewise action is both free and proper. Such fibrewise actions will be called *fibrewise properly discontinuous.*

Fibrewise open groups which are also fibrewise compact enjoy further properties which are of considerable importance.

Proposition 2.12 *Let E be a fibrewise G-space, where G is fibrewise compact. Then the natural projection $\pi : E \to E/G$ is proper.*

To show that π is closed we use a similar argument to that used to prove Proposition 2.4. To complete the proof, let T be a fibrewise space, regarded as a fibrewise G-space with trivial action. Then the natural projection from $E \times_B T$ to $(E \times_B T)/G$ is closed. However, the latter is equivalent to $E/G \times_B T$, as a fibrewise space, and the natural projection is equivalent to $\pi \times 1$, which is therefore closed. Since this is true for all T we conclude that π is proper, as asserted.

Corollary 2.13 *Let G be a fibrewise topological group and let H be a fibrewise open compact subgroup of G. Then the projection $\pi : G \to G/H$ is proper.*

Proposition 2.14 *Let E be a fibrewise G-space, where E is fibrewise Hausdorff and G is fibrewise compact. Then the fibrewise action is proper.*

To see this we express θ as the composition

$$E \times_B G \xrightarrow{\ \Gamma\ } E \times_B E \times_B G \xrightarrow{\ \pi_{12}\ } E \times_B E,$$

where Γ is the fibrewise graph of the fibrewise action and π_{12} is the canonical projection. Now π_{12} is proper, since G is fibrewise compact, and Γ is closed, since E is fibrewise Hausdorff, and so proper also. Therefore θ is proper, as asserted.

Lemma 2.15 *Let E be a fibrewise G-space, where G is fibrewise compact. Let A be a subspace of E and let s be a section of G such that $A \subseteq (sB).A$. Then $A = (sB).A$.*

To see this observe that $A_b \subseteq g.A_b$, for each point b of B, where $g = s(b)$. Since G_b is compact it follows from a standard result (see (4.25) of [85], for example) that $A_b = g.A_b$. Therefore $A = (sB).A$, as asserted. This implies

Proposition 2.16 *Let H be a subgroup of the fibrewise compact group G. Suppose that $H.(sB) \subseteq (sB).H$ for some section s of G. Then $H.(sB) = (sB).H$.*

Given subgroups K, L of the fibrewise compact group G, let us say that $[K] \leq [L]$ if K is fibrewise conjugate in G to a subgroup of L. Then Proposition 2.16 implies that the relation $\leq$ is a partial ordering of fibrewise conjugacy classes of subgroups.

Still assuming that G is fibrewise compact, let E be a fibrewise Hausdorff fibrewise G-space. Given a section s of E consider the corresponding fibrewise map of G/K onto $(sB).G$, where K denotes the fibrewise stabilizer of s. Now G/K is fibrewise compact and $(sB).G$ is fibrewise Hausdorff, hence the fibrewise map is a homeomorphism. In this case, therefore, the partial ordering of classes of fibrewise stabilizers described above leads to a corresponding ordering of classes of 'fibrewise orbits' $(sB).G$, where s runs through the sections of E, and hence to a notion of 'fibrewise orbit type'.

Some illustrations of the way fibrewise transformation groups can be investigated will be found in Part II, Section 13.

3 Fibrewise homotopy

Basic notions

Fibrewise homotopy is an equivalence relation between fibrewise maps. Specifically, consider fibrewise maps $\theta, \phi : X \to Y$, where X and Y are fibrewise spaces over B. A *fibrewise homotopy* of θ into ϕ is a homotopy $f_t : X \to Y$ in the ordinary sense which is a fibrewise map at each stage. If there exists a fibrewise homotopy of θ into ϕ we say that θ is *fibrewise homotopic* to ϕ and write $\theta \simeq_B \phi$. In this way an equivalence relation is defined on the set of fibrewise maps from X to Y, and the set of equivalence classes is denoted by $\pi_B[X; Y]$. Formally, π_B constitutes a binary functor from the category of fibrewise spaces to the category of sets, contravariant in the first entry and covariant in the second.

We say that two sections s_0 and s_1 of a fibrewise space $X \to B$ are *vertically homotopic* if they are homotopic through sections, that is, if there is a homotopy s_t, $0 \leq t \leq 1$, where each map s_t is a section. The fibrewise space X is said to be *vertically connected* if it has just one vertical homotopy class of sections.

Recall from Section 1 that for each principal G-bundle P over B the associated bundle functor $P_\#$ is defined, from the category of G-spaces to the category of fibrewise spaces over B. This not only transforms G-maps into fibrewise maps but also transforms G-homotopies into fibrewise homotopies.

The operation of composition for fibrewise maps induces a function

$$\pi_B[Y; Z] \times \pi_B[X; Y] \to \pi_B[X; Z],$$

for any fibrewise spaces X, Y, Z over B. Moreover, there are natural equivalences between $\pi_B[X \sqcup_B Y; Z]$ and $\pi_B[X; Z] \times \pi_B[Y; Z]$, and between $\pi_B[X; Y \times_B Z]$ and $\pi_B[X; Y] \times \pi_B[X; Y]$.

Postcomposition with a fibrewise map $\psi : Y \to Z$ induces a function

$$\psi_* : \pi_B[X; Y] \to \pi_B[X; Z],$$

while precomposition with a fibrewise map $\phi : X \to Y$ induces a function

$$\phi^* : \pi_B[Y; Z] \to \pi_B[X; Z].$$

Similar notation is used in the case of fibrewise homotopy classes rather than fibrewise maps.

Note that fibrewise maps $\phi : X \to Y$, where X and Y are fibrewise spaces over B, correspond precisely to sections of the fibrewise product $X \times_B Y$, regarded as a fibrewise space over X. Similarly, fibrewise homotopy classes of fibrewise maps correspond precisely to vertical homotopy classes of sections. Now if Y is a fibre space over B then $X \times_B Y$ is the induced fibre space over X and so the classification of sections by vertical homotopy coincides with their classification as maps by ordinary homotopy. This is a useful little trick.

The fibrewise map $\phi : X \to Y$ is called a *fibrewise homotopy equivalence* if there exists a fibrewise map $\psi : Y \to X$ such that

$$\psi \circ \phi \simeq_B 1_X, \qquad \phi \circ \psi \simeq_B 1_Y.$$

Thus an equivalence relation between fibrewise spaces is defined; the equivalence classes are called *fibrewise homotopy types*.

A fibrewise homotopy into a fibrewise constant is called a *fibrewise null-homotopy*. A fibrewise space is said to be *fibrewise contractible* if it has the same fibrewise homotopy type as the base space, in other words, if the identity is fibrewise null-homotopic. A subset U of a fibrewise space X is said to be *fibrewise categorical* if the inclusion $U \to X$ is fibrewise null-homotopic.

There is a point which should be made here about fibrewise quotient spaces. As we saw in Section 1, if X is a fibrewise space with fibrewise equivalence relation R the fibrewise maps of X/R into Z correspond precisely to the invariant fibrewise maps of X to Z. Since $I = [0, 1]$ is compact we can regard $(X/R) \times I$ as a fibrewise quotient space of $X \times I$, in the obvious way, and so the correspondence extends to fibrewise homotopies.

For example, consider the fibrewise cone $C_B(X)$ on the fibrewise space X. A fibrewise null-homotopy of the identity on $C_B(X)$ is induced by the fibrewise homotopy

$$f_t : X \times I \to X \times I \qquad (t \in I),$$

where $f_t(x, s) = (x, st)$ $(x \in X, s \in I)$. Therefore $C_B(X)$ is fibrewise contractible.

Let $\phi : X \to Y$ and $\psi : Y \to X$ be fibrewise maps such $\psi \circ \phi \simeq_B 1_X$. Then ψ is said to be a *left inverse* of ϕ, up to fibrewise homotopy, and ϕ to be a *right inverse* of ψ, up to fibrewise homotopy. Note that if ϕ admits both a left inverse ψ and a right inverse ψ', up to fibrewise homotopy, then $\psi \simeq_B \psi'$ and so ϕ is a fibrewise homotopy equivalence.

Examples can easily be given to show that fibrewise maps may be homotopic, as ordinary maps, but not fibrewise homotopic. Thus take $B = I$ and $X = (I \times \{0,1\}) \cup (\{0\} \times I)$, with the first projection. Although X is contractible, as an ordinary space, it is not fibrewise contractible since the fibres over points of $(0,1]$ are not contractible.

It can be shown, however, that for a large class of fibrewise spaces X there exists an integer m such that for each fibrewise map $\phi : X \to X$ which is null-homotopic on each fibre the m-fold composition $\phi \circ \ldots \circ \phi$ is fibrewise null-homotopic. Details are given in [82] and [108]. Another result which might be mentioned here concerns the group $G(X)$ of fibrewise homotopy classes of fibrewise homotopy equivalences of X with itself. Under the same conditions it is shown, in [83] and [108], that $G_1(X)$ is nilpotent of class less than m, where $G_1(X)$ denotes the normal subgroup of $G(X)$ consisting of classes of fibrewise homotopy equivalences which are homotopic to the identity on each fibre.

Later we shall be discussing a variety of different problems which have attracted research interest. At this stage, however, we mention just two more. Take X to be an orthogonal $(n-1)$-sphere bundle over a finite connected complex B, for some $n \geq 1$. Since the centre of the orthogonal group $O(n)$ contains the scalar -1 we have a fibrewise map $c : X \to X$ given by the antipodal transformation in each fibre. The antipodal transformation has degree $(-1)^n$ and so is homotopic to the identity when n is even. Under what conditions is it true that the fibrewise map c is fibrewise homotopic to the identity? Questions of this type are investigated in [82] and [84]. Another problem in the same area is considered by Noakes [118], as follows. Each fibrewise self-map ϕ of X has a certain degree $d(\phi)$ on the fibres. When n is even, Noakes shows that the degrees which can be so realized are precisely the integers which are congruent to one modulo some number which depends on X.

Some examples

Example 3.1. Take $B = [0,1] \subseteq \mathbb{R}$ and

$$E = [0,1] \times \{0,1\} \cup \{0\} \times [0,1] \subseteq \mathbb{R} \times \mathbb{R},$$

regarded as a fibrewise space using the first projection. The section $s : B \to E$ given by $s(t) = (t,0)$ is a homotopy equivalence but not a fibrewise homotopy equivalence.

Example 3.2. Suppose that the subspace X of $B \times \mathbb{R}^n$ is *fibrewise star-like* in the sense that for some section $s : B \to X$ the line segment

$$(b, (1-t)s(b) + tx) \qquad (0 \leq t \leq 1)$$

is entirely contained within the fibre X_b for each point $x \in X_b$, $b \in B$. Then X is fibrewise contractible.

4 Fibrewise cofibrations

Basic notions

Let A be a fibrewise space over the base space B. By a *fibrewise cofibre space* under A we mean a fibrewise space X together with a fibrewise map $u : A \to X$ having the following fibrewise homotopy extension property. Let E be a fibrewise space, let $f : X \to E$ be a fibrewise map, and let $g_t : A \to E$ be a fibrewise homotopy of $f \circ u$. Then there exists a fibrewise homotopy $h_t : X \to E$ of f such that $g_t = h_t \circ u$. For example, the push-out $T \vee_B A$ is a fibrewise cofibre space under A for each fibrewise space T.

Instead of describing X as a fibrewise cofibre space under A we may describe u as a *fibrewise cofibration*. An important special case is when A is a subspace of X and u is the inclusion. In that case we describe (X, A) as a *fibrewise cofibred pair* when the condition is satisfied. Note that $(X, \emptyset)$ and (X, X) are fibrewise cofibred pairs.

By taking E in the condition to be the product $B \times T$, for any space T, we see that a fibrewise cofibration $u : A \to X$ is a cofibration in the ordinary sense, in particular u is necessarily injective.

It is easy to check that the fibrewise coproduct of fibrewise cofibred spaces under A is also a fibrewise cofibred space under A.

Proposition 4.1 *Let (X, A) be a closed fibrewise pair. Then (X, A) is fibrewise cofibred if and only if $(X \times \{0\}) \cup (A \times I)$ is a fibrewise retract of $X \times I$.*

For suppose that u is a fibrewise cofibration. In the fibrewise homotopy extension condition take E to be $(X \times \{0\}) \cup (A \times I)$, take $f : X \to E$ to be the obvious fibrewise map and take $g : A \times I \to E$ to be the obvious fibrewise homotopy. Then the extension $h : X \times I \to E$ is a fibrewise retraction.

To prove the converse, suppose that there exists a fibrewise retraction $r : X \times I \to (X \times \{0\}) \cup (A \times I)$. Given a fibrewise space E, fibrewise map $f : X \to E$ and fibrewise homotopy $g : A \times I \to E$, as in the fibrewise homotopy extension property, we combine f and g to obtain a fibrewise map of $(X \times \{0\}) \cup (A \times I)$ into E. Then precomposition with r gives the desired extension.

In fact a similar result to Proposition 4.1 holds even when A is not closed in X, with the mapping cylinder of the inclusion replacing the subspace of $X \times I$.

For example, consider the associated bundle functor $P_\#$, as in Section 1, from the category of G-spaces to the category of fibrewise spaces. From

Proposition 4.1 we see that $P_\#$ transforms closed cofibred pairs in the equivariant sense into closed cofibred pairs in the fibrewise sense. In particular, taking G to be the orthogonal group $O(n)$, we see that the pair consisting of an n-ball bundle over B and the associated $(n-1)$-sphere bundle is fibrewise cofibred.

Corollary 4.2 *Let (X, A) be a closed fibrewise cofibred pair over B. Then the pair*

$$(X, A) \times_B T = (X \times_B T, A \times_B T)$$

is also fibrewise cofibred, for all fibrewise spaces T.

Fibrewise Strøm structures

We now come to an important characterization of closed fibrewise cofibred pairs. It is a fibrewise version of the corresponding characterization in the ordinary theory. There are several variants of the condition, of which we prefer the one due to Strøm, as follows.

Let (X, A) be a closed fibrewise pair. A *fibrewise Strøm structure* on (X, A) is a pair (α, h) consisting of a map $\alpha : X \to I$ which is zero throughout A together with a fibrewise homotopy $h : X \times I \to X$ rel A of 1_X such that $h(x, t) \in A$ whenever $t > \alpha(x)$.

Proposition 4.3 *Let (X, A) be a closed fibrewise pair. Then (X, A) is fibrewise cofibred if and only if (X, A) admits a fibrewise Strøm structure.*

For suppose that (X, A) is fibrewise cofibred, so that there exists a fibrewise retraction

$$r : X \times I \to (X \times \{0\}) \cup (A \times I),$$

as in Proposition 4.1. Since I is compact a map $\alpha : X \to I$ is given by

$$\alpha(x) = \sup_{t \in I} |\pi_2 r(x, t) - t| \qquad (x \in X).$$

Then $(\alpha, \pi_1 r)$ constitutes a fibrewise Strøm structure on (X, A).

Conversely, let (α, h) be a fibrewise Strøm structure on (X, A). Then a fibrewise retraction

$$r : X \times I \to (X \times \{0\}) \cup (A \times I)$$

is given by

$$r(x, t) = \begin{cases} (h(x, t), 0) & t \le \alpha(x), \\ (h(x, t), t - \alpha(x)) & t \ge \alpha(x). \end{cases}$$

Hence (X, A) is fibrewise cofibred, by Proposition 4.1.

One of the main applications of the above characterization is to prove the fibrewise product theorem, as follows.

Theorem 4.4 *Let (X, X') and (Y, Y') be closed fibrewise cofibred pairs over* B. *Then the closed fibrewise pair*

$$(X, X') \times_B (Y, Y') = (X \times_B Y, X' \times_B Y \cup X \times_B Y')$$

is also fibrewise cofibred.

To see this choose fibrewise Strøm structures (α, h) on (X, X') and (β, k) on (Y, Y'). Define $\gamma : X \times_B Y \to I$ by $\gamma(x, y) = \min(\alpha(x), \beta(y))$, and define $\ell : (X \times_B Y) \times I \to X \times_B Y$ by

$$\ell(x, y, t) = (h(x, \min(t, \beta(y))), k(y, \min(t, \alpha(x)))).$$

Then (γ, ℓ) constitutes a fibrewise Strøm structure for the closed fibrewise pair $(X, X') \times_B (Y, Y')$, as required.

For example, consider the endofunctor Φ_B of our category determined by a closed cofibred pair (D, E). We see from Theorem 4.4 that if the closed fibrewise pair (X, A) is fibrewise cofibred then so is the closed fibrewise pair $(\Phi_B(X), \Phi_B(A))$. In particular, this is true for the fibrewise cone C_B and the fibrewise suspension Σ_B.

There is a weaker form of the concept of fibrewise cofibration which is also important, as follows. Let A be a fibrewise space over the base space B. By a *weak fibrewise cofibre space* under A we mean a fibrewise space X together with a fibrewise map $u : A \to X$ such that X has the same fibrewise homotopy type under A as a fibrewise cofibration $u' : A \to X'$. This implies, and is implied by, a weak form of the fibrewise homotopy extension property: for each fibrewise space E, fibrewise map $f : X \to E$, and fibrewise homotopy $g_t : A \to E$ of $f \circ u$, there exists a fibrewise homotopy $h_t : X \to E$ such that $g_t = h_t \circ u$ and such that h_0 is fibrewise homotopic to f over B. The properties of weak fibrewise cofibrations are similar to those of fibrewise cofibrations.

5 Fibrewise fibrations

Basic notions

At this stage we change our point of view somewhat. Although we continue to work over the base space B we not only consider fibrewise spaces over B but also fibrewise spaces over those fibrewise spaces. Thus let X be a fibrewise space over B. By a *fibrewise fibre space* over X we mean a fibrewise space E together with a fibrewise map $p : E \to X$ with the following fibrewise homotopy lifting property. Let A be a fibrewise space, let $f : A \to E$ be a fibrewise map, and let $g_t : A \to X$ be a fibrewise homotopy such that $g_0 = p \circ f$. Then there exists a fibrewise homotopy $h_t : A \to E$ of f such that $g_t = p \circ h_t$. We emphasize that fibrewise here means over B, not over X. For example, the fibrewise product $T \times_B X$ is a fibrewise fibre space over X for each fibrewise space T.

Instead of describing E as a fibrewise fibre space over X we may describe the projection $p : E \to X$ as a *fibrewise fibration*. We also describe the pull-back $\tau^* E$ as a *fibrewise fibre*, for each section τ of X over B.

Note that if p is a fibration, in the ordinary sense of the term, then p is a fibrewise fibration. This is in contrast to the situation in the case of cofibrations. However, fibrewise maps exist which are fibrewise fibrations but not fibrations in the ordinary sense. For example take $X = B$; then every fibrewise space over B is a fibrewise fibre space.

Consider the *fibrewise free path-space*

$$\mathcal{P}_B(X) = \operatorname{map}_B(B \times I, X),$$

which comes equipped with a family of projections

$$\rho_t : \mathcal{P}_B(X) \to X \qquad (0 \le t \le 1),$$

given by evaluation at t. It is a formal exercise in the use of adjoints to show that ρ_t is a fibrewise fibration for $t = 0, 1$. Let us regard $\mathcal{P}_B(X)$ as a fibrewise space over X using ρ_0 as projection. Then for any fibrewise space E and fibrewise map $p : E \to X$ the *fibrewise mapping path-space* $W_B(p)$ is defined as the pull-back $p^* \mathcal{P}_B(X)$. Since $\mathcal{P}_B(X)$ is a fibrewise fibre space over X it follows quite formally that $W_B(p)$ is a fibrewise fibre space over E. By the cartesian property we have a fibrewise map

$$k : \mathcal{P}_B(E) \to W_B(p),$$

with components $\mathcal{P}_B(p)$ and ρ_0. The following characterization of fibrewise fibrations is fundamental.

Proposition 5.1 *The fibrewise map $p : E \to X$ is a fibrewise fibration if and only if the fibrewise map $k : \mathcal{P}_B(E) \to W_B(p)$ admits a right inverse.*

For suppose that p is a fibrewise fibration. Take $W = W_B(p)$ as the domain, in the fibrewise homotopy lifting condition, take ρ_0 to be f and take $g_t = \rho_t \circ \mathcal{P}_B(p)$. The condition implies the existence of a fibrewise homotopy $h_t : W \to E$ of ρ_0 such that $g_t = p \circ h_t$. The right adjoint $W \to \mathcal{P}_B(E)$ of the fibrewise homotopy is a right inverse of k as required.

Conversely, suppose that k admits a right inverse. By taking the left adjoint we obtain a fibrewise homotopy $h_t : W \to E$, as above. Let A be a fibrewise space, let $f : A \to E$ be a fibrewise map, and let $g_t : A \to X$ be a fibrewise homotopy such that $g_0 = p \circ f$. Then g_t determines a fibrewise map $g : A \to \mathcal{P}_B(X)$ which combines with f to give a fibrewise map $\ell : A \to W$. Now $h_t \circ \ell : A \to E$ is a fibrewise homotopy of f over g_t, as required.

We have already seen, in Section 4, that the associated bundle functor $P_\#$ transforms cofibrations in the equivariant sense into cofibrations in the fibrewise sense. Now we can see, using Proposition 5.1, that the functor also transforms fibrations in the equivariant sense into fibrations in the fibrewise sense. The proof is a straightforward exercise in the use of adjoints.

Proposition 5.2 *Let $p : E \to X$ be a fibrewise map, where E is a fibrewise space. Then the evaluation map*

$$\rho_1 : W_B(p) \to X$$

is a fibrewise fibration.

To see this let us express fibrewise maps into $W = W_B(p)$ in the usual way by giving their components in $\mathcal{P}_B(X)$ and E; the former we regard as fibrewise homotopies. Let A be a fibrewise space, let $f : A \to W$ be a fibrewise map, and let $g_t : A \to E$ be a fibrewise homotopy of $\rho_1 \circ f$. Consider the fibrewise map

$$H : A \times I \times I \to X$$

given by

$$H(a, s, t) = \begin{cases} f'(a, 2s(2 - t)^{-1}) & (s \leq 1 - \tfrac{1}{2}t) \\ g(a, 2s - 2 + t) & (s \geq 1 - \tfrac{1}{2}t) \end{cases}$$

where f' is the fibrewise homotopy given by the first component of f. Take H as the first component of a fibrewise deformation of f in which the first component remains stationary. Then we obtain a fibrewise homotopy of f over g_t, as required.

The fibrewise homotopy lifting property has a number of useful consequences. For example, let E and F be fibrewise spaces over X with projections p and q, respectively. Let $\phi : E \to F$ be a fibrewise map such that $q \circ \phi$ is fibrewise homotopic to p. If q is a fibrewise fibration then it follows from the fibrewise homotopy lifting property that ϕ is fibrewise homotopic to a fibrewise map ψ such that $q \circ \psi = p$.

The fibrewise Dold theorem

One of the most important results of the classical theory is due to Dold. This provides a bridge between ordinary homotopy theory and fibrewise homotopy theory. The fibrewise version of Dold's theorem provides a bridge between fibrewise homotopy theory over B and fibrewise homotopy theory over X, where X is a fibrewise space over B. In the literature, Dold's original proof can be found in [45] while Hardie and Kamps have given a more conceptual proof in [73]. It is a routine exercise to write out a fibrewise version of the original proof, as in [45], and the more conceptual proof may also be generalized without difficulty. In this survey let us be content with the statement of the fibrewise version of Dold's theorem preceded by that of the auxiliary result which leads up to it.

Proposition 5.3 *Let $p : E \to X$ be a fibrewise fibration, where E and X are fibrewise spaces over B. Let $\theta : E \to E$ be a fibrewise map over X, and suppose that θ, as a fibrewise map over B, is fibrewise homotopic to the identity. Then there exists a fibrewise map $\theta' : E \to E$ over X such that $\theta \circ \theta'$ is fibrewise homotopic to the identity over X.*

Theorem 5.4 *Let X be a fibrewise space over B, and let E and F be fibrewise fibre spaces over X. Let $\phi : E \to F$ be a fibrewise map over X. Suppose that ϕ, as a fibrewise map over B, is a fibrewise homotopy equivalence. Then ϕ is a fibrewise homotopy equivalence over X.*

Corollary 5.5 *Let $p : E \to X$ be a fibrewise fibration, where E and X are fibrewise spaces over B. If p is a fibrewise homotopy equivalence over B then the fibrewise mapping path-space $W_B(p)$ is fibrewise contractible over X.*

Here we regard $W = W_B(p)$ as a fibrewise space over X with projection the fibrewise fibration ρ_1, as in Proposition 5.2. Now $p = \rho_1 \circ \sigma$, where $\sigma : E \to W$ is the standard embedding. Since p and σ are fibrewise homotopy equivalences over B, so is ρ_1. Also p and ρ_1 are fibrewise fibrations and so ρ_1 is a fibrewise homotopy equivalence over X, by Theorem 5.4. In other words, $W_B(p)$ is fibrewise contractible over X, as asserted.

Corollary 5.6 *Let $p : E \to X$ be a fibrewise fibration, where E and X are fibrewise spaces over B. Then the fibrewise path-space $\mathcal{P}_B(X)$ is fibrewise contractible over the fibrewise mapping path-space $W_B(p)$.*

Here we regard $\mathcal{P}_B(X)$ as a fibrewise space over $W_B(p)$ using the projection k as in Proposition 5.1. Now k admits a right inverse, since p is a fibrewise fibration. Hence k is itself a fibrewise fibration (to verify the fibrewise homotopy lifting property one observes that the pairs $(I \times I, (I \times \{0\}) \cup (\{0\} \times I))$ and $(I \times I, I \times \{0\})$ are homeomorphic). Moreover, k is a fibrewise homotopy equivalence over B and so, by Theorem 5.4, a fibrewise homotopy equivalence over $W_B(p)$. This completes the proof.

Sections of a fibrewise fibration can be classified by fibrewise homotopy or, more strictly, by vertical homotopy. In fact there is no difference, as shown in

Proposition 5.7 *Let $p : E \to X$ be a fibrewise fibration, where E and X are fibrewise spaces over B. Let s and s' be fibrewise homotopic sections of E over X. Then s and s' are vertically homotopic.*

We have to show that s and s' are homotopic over X, rather than just over B. So let $h_t : X \to E$ be a fibrewise homotopy of s into s'. Let $k_t : X \to E$ be given by

$$k_t(x) = \begin{cases} h_{2t}(x) & (0 \le t \le \frac{1}{2}), \\ h_1 p h_{2-2t}(x) & (\frac{1}{2} \le t \le 1). \end{cases}$$

Since $p \circ k_t = p \circ k_{1-t}$ the fibrewise map $p \circ k : X \times I \to X$ is fibrewise homotopic, rel $X \times \dot{I}$, to the projection π_2. Therefore k is fibrewise homotopic, rel $X \times \dot{I}$, to a fibrewise map ℓ which constitutes a vertical homotopy of s into s'.

The fibrewise homotopy theorem for fibrewise fibrations

We now come to the fibrewise homotopy theorem for fibrewise fibrations, a result which has no obvious counterpart for fibrewise cofibrations. Again one has a choice between a fibrewise version of the standard proof of the corresponding result in the ordinary theory, as given for example in [44], or a more conceptual proof as given by Hardie and Kamps [73]. Again let us be content with the statement of the theorem itself preceded by that of the auxiliary result from which it follows.

Let X be a fibrewise space over B, and let the cylinder $X \times I$ be regarded as a fibrewise space over B by precomposing with the second projection. Consider a fibrewise space D over $X \times I$. We regard

$$D_t = D \mid (X \times \{t\}) \qquad (0 \le t \le 1)$$

as a fibrewise space over X in the obvious way. The main step in the proof of the fibrewise homotopy theorem is the demonstration that if D is a fibrewise fibre space over $X \times I$ then D_0 is a fibrewise deformation retract of D over X. Similarly, D_1 is a fibrewise deformation retract of D. Hence it follows that D_0 and D_1 have the same fibrewise homotopy type over X.

Theorem 5.8 *Let X be a fibrewise space over B, and let E be a fibrewise fibre space over X. Let $\theta, \phi : X' \to X$ be fibrewise homotopic fibrewise maps, where X' is a fibrewise space over B. Then $\theta^* E$ and $\phi^* E$ have the same fibrewise homotopy type over X'.*

To obtain Theorem 5.8 we apply the auxiliary result with $D = f^* E$, where $f : X' \times I \to X$ is a fibrewise homotopy of θ into ϕ.

Corollary 5.9 *Let X be a fibrewise space over B and let E be a fibrewise fibre space over X. If X is fibrewise contractible over B then E has the same fibrewise homotopy type over B as the fibrewise product $X \times_B T$, for some fibrewise space T over B.*

There is a weaker form of the concept of fibrewise fibration which is also important, as follows. Let X be a fibrewise space over B. By a *weak fibrewise fibre space* over X we mean a fibrewise space E together with a fibrewise map $p : E \to X$ such that E has the same fibrewise homotopy type over X as a fibrewise fibration $p' : E' \to X$. This implies, and is implied by, a weak form of the fibrewise homotopy lifting property: for each fibrewise space A, fibrewise map $f : A \to E$ and fibrewise homotopy $g_t : A \to X$ such that $g_0 = p \circ f$, there exists a fibrewise homotopy $h_t : A \to E$ such that $g_t = p \circ h_t$ and such that h_0 is fibrewise homotopic to f over B. The properties of weak fibrewise fibrations are similar to those of fibrewise fibrations.

An example

The verification of the following result is left as an exercise.

Example 5.10. Let $\phi : (X_1, A_1) \to (X_2, A_2)$ be a fibrewise map of pairs, where (X_i, A_i) is a fibrewise cofibred pair over B $(i = 1, 2)$. Suppose that the fibrewise maps $X_1 \to X_2$ and $A_1 \to A_2$ determined by ϕ are fibrewise homotopy equivalences. Then ϕ is a fibrewise homotopy equivalence of pairs.

6 Numerable coverings

Let us now turn our attention to a series of important theorems due to Dold [45] and tom Dieck [42]. First recall that a *halo* of a subset X' of a space X is a subset V of X, containing X', for which there exists a map $\alpha : X \to I$ with $\alpha = 1$ throughout X' and $\alpha = 0$ away from V. Thus X itself is a halo for every X', since we can take the function to be constant at 1.

Following Dold we say that the fibrewise space X over B has the *section extension property* (SEP) if for each subset B' of B every section of X over B', which can be extended to a halo of B', can be extended to a section of X over B. This condition implies, in particular, that X admits a section, since we can take B' and V to be empty.

Unlike fibrewise contractibility, the section extension property is not natural, in our sense. However, if X has the property then so does any fibrewise space which is fibrewise dominated by X. In particular, X has the property if X is fibrewise contractible.

If the fibrewise space X over B has the section extension property then so does the restriction $X_{B'}$ of X to any numerically defined open set B' of B. By *numerically defined*, here, we mean that B' is the cozero set $\beta^{-1}(0, 1]$ for some map $\beta : X \to I$.

The theorems of Dold

The main theorem of Dold mentioned earlier is similar in spirit to results in the theory of sheaves. We give the statement as follows, but refer to [45] for the proof.

Theorem 6.1 *Let X be a fibrewise space over B. Suppose that there exists a numerable covering of B such that X_V has the SEP over V for each member V of the covering. Then X has the SEP over B.*

Dold uses Theorem 6.1 to prove the following, which is one of the basic results of fibrewise homotopy theory.

Theorem 6.2 *Let $\phi : X \to Y$ be a fibrewise map, where X and Y are fibrewise spaces over B. Suppose that B admits a numerable covering such that the restriction $\phi_V : X_V \to Y_V$ is a fibrewise homotopy equivalence over V for each member V of the covering. Then ϕ is a fibrewise homotopy equivalence over B.*

For consider the fibrewise mapping path-space $W = W_B(\phi)$ of ϕ. Observe that if ϕ is restricted to $\phi_V : X_V \to Y_V$, for any subset V of B, then the fibrewise mapping path-space of ϕ_V is just the restriction to V of the fibrewise mapping path-space W of ϕ itself. By hypothesis ϕ_V is a fibrewise homotopy equivalence over V for each member V of the covering. Therefore $W_V(\phi_V)$, the restriction of W to V, is fibrewise contractible over Y_V, by Corollary 5.5. Hence W_V has the SEP over Y_V. As V runs through the members of the numerable covering of B so Y_V runs through the members of a numerable covering of Y. By Theorem 6.1, therefore, W has the SEP over Y. In particular W admits a section over Y and hence ϕ admits a right inverse ϕ', up to fibrewise homotopy.

Repeating the argument with ϕ' in place of ϕ we obtain a right inverse ϕ'' of ϕ', up to fibrewise homotopy. So ϕ' admits both the left inverse ϕ and the right inverse ϕ'', up to fibrewise homotopy. Hence ϕ' is a fibrewise homotopy equivalence and so ϕ is a fibrewise homotopy equivalence, as asserted.

Returning to the situation where X is a fibrewise space over B, we deduce

Corollary 6.3 *Let $p : E \to X$ and $q : F \to X$ be fibrewise fibrations, where E, F and X are fibrewise spaces over B. Let $\phi : E \to F$ be a fibrewise map such that $q \circ \phi = p$. Suppose that the pull-back*

$$s^*\phi : s^*E \to s^*F$$

is a fibrewise homotopy equivalence over B for each section s of X over B. Also suppose that X admits a numerable fibrewise categorical covering. Then ϕ is a fibrewise homotopy equivalence over X.

For since each member V of the numerable covering is fibrewise categorical it follows from Proposition 5.7 that E_V and F_V are fibrewise trivial over V. Hence ϕ_V is a fibrewise homotopy equivalence over V, taking s in the hypothesis to be given by the fibrewise constant map to which the inclusion is fibrewise homotopic. Now Corollary 6.3 follows at once from Theorem 6.2.

Note that the assumption in Corollary 6.3 is satisfied for all s if it is satisfied for one s in each vertical homotopy class.

Another important application of Theorem 6.2 is

Theorem 6.4 *Let* $p : E \to X$ *be a fibrewise map, where* E *and* X *are fibrewise spaces over* B. *Suppose that the restriction* $p^{-1}V \to V$ *of* p *is a fibrewise fibration for each member* V *of a numerable covering of* X. *Then* p *is a fibrewise fibration.*

This implies, of course, that numerable fibrewise fibre bundles are fibrewise fibrations. The proof of Theorem 6.4 is a straightforward fibrewise version of the proof of the classical result to which it reduces when B is a point (cf. (9.4) of [44], for example, or the globalization theorem of [55]).

A similar result holds for weak fibrewise fibrations.

The theorems of tom Dieck

Some other important results of a similar type are due to tom Dieck [42]. For these it seems necessary that the numerable coverings concerned are closed under finite intersections. In contrast to the results of Dold we have been discussing these are already fully fibrewise homotopy theoretic in character, and so we simply quote them from [42], where proofs are given.

Theorem 6.5 *Let* $\phi : E \to F$ *be a fibrewise map, where* E *and* F *are fibrewise spaces over* B. *Let* $\{U_j\}$ *and* $\{V_j\}$ *be similarly indexed numerable coverings of* E *and* F, *respectively, which are closed under finite intersections. Assume that* $\phi U_j \subseteq V_j$ *for each index* j, *and that each of the fibrewise maps* $U_j \to V_j$ *determined by* ϕ *is a fibrewise homotopy equivalence. Then* ϕ *is a fibrewise homotopy equivalence.*

Theorem 6.6 *Let* $p : E \to X$ *be a fibrewise map, where* E *and* X *are fibrewise spaces over* X. *Let* $\{B_j\}$ *be a numerable covering of* B *and let* $\{E_j\}$ *be a similarly indexed family of subsets of* E, *both families being closed under finite intersections. Assume that* $pE_j \subseteq B_j$ *and that* E_j *is fibrewise contractible over* B_j, *for each index* j. *Then* E *admits a section over* X.

Note that $\{E_j\}$ is not required to be a covering of E. We can deduce another result about fibrewise fibrations, which it is interesting to compare with Theorem 6.4.

Theorem 6.7 *Let* $p : E \to X$ *be a fibrewise map, where* E *and* X *are fibrewise spaces over* B. *Let* $\{E_j\}$ *be a numerable covering of* E *which is closed under finite intersections. Assume that the restriction* $p_j : E_j \to X$ *is a fibrewise fibration for each index* j. *Then* p *is a fibrewise fibration.*

For let $W = W_B(p)$ denote the fibrewise mapping path-space of p. We regard the fibrewise free path-space $\mathcal{P}_B(E)$ of E as a fibrewise space over W, as in Theorem 6.6. Recall that the second projection $r : W \to E$ is a fibrewise homotopy equivalence and that $p \circ r : W \to X$ is a fibrewise fibration, as shown in Proposition 5.2. Now the fibrewise mapping path-spaces W_j of the projections p_j form a numerable covering of W, which is closed under finite intersections. Moreover, $rW_j \subseteq E_j$ for each index j. If p_j is a fibrewise fibration then $\mathcal{P}_B(E_j)$ is fibrewise contractible over W_j for each index j, by Corollary 5.5. Therefore $\mathcal{P}_B(E)$ admits a section over W, by Theorem 6.6, and so p itself is a fibrewise fibration, by Proposition 5.1. The proof we have given is just a fibrewise version of the proof that tom Dieck gives in [42] of the special case when B is a point. A similar result holds for weak fibrewise fibrations.

7 Fibrewise fibre bundles

Principal fibrewise G-spaces

Although we continue to work over the base space B we not only consider fibrewise spaces over B but also fibrewise spaces over those fibrewise spaces, as in Section 6.

For example, let E be a fibrewise G-space, where G is a fibrewise topological group. We may regard E as a fibrewise space over E/G, using the natural projection. Some of the results proved earlier can then be reformulated in a simpler fashion. For example, Proposition 2.4 shows that E is fibrewise open over E/G if G is fibrewise open over B.

In the same spirit, consider a fibrewise map $p : E \to X$, where E and X are fibrewise spaces over B. We may regard E as a fibrewise space over X, as well as over B. Among the fibrewise actions of G on E we single out for attention those where the action of G_b on E_b, for each point b of B, is through fibrewise homeomorphisms of E_b over X_b. This means that p induces a fibrewise map $\rho : E/G \to X$. We describe E as a *principal fibrewise G-space* over X if ρ is a fibrewise homeomorphism. In that case we use ρ to identify E/G with X, as a fibrewise space.

For ρ to be a fibrewise homeomorphism it is necessary, in the first place, that the projection p is surjective. Also when G is fibrewise open it is necessary for p to be open, and when G is also fibrewise compact it is necessary for p to be proper. Sufficient conditions for ρ to be a fibrewise homeomorphism are that the fibrewise action is free and that p is both surjective and either open or closed.

Let us describe a principal fibrewise G-space E over X as *trivial* if E is equivalent to $X \times_B G$, as a fibrewise G-space over X. Local triviality in the same sense is defined similarly. Specifically, a triviality covering of X

is an open covering such that E_V is trivial over V for each member V of the covering. For some purposes, a numerable triviality covering is necessary. When X is Hausdorff and paracompact this is true automatically.

Proposition 7.1 *Let G be a fibrewise topological group and let X be a fibrewise space over B. Let E be a principal fibrewise G-space over X. Suppose that the division function $d : R \to G$ is continuous. If E is sectionable (respectively locally sectionable) over X then E is trivial (respectively locally trivial) as a principal fibrewise G-space over X.*

For let $s : X \to E$ be a section. A fibrewise G-map $\phi : X \times_B G \to E$ over X is given by $\phi(x, g) = s(x).g$. A fibrewise G-map $\psi : E \to R$ is given by $\psi(\xi) = (\xi, sp(\xi))$. Postcomposing with d yields an inverse of ϕ, as required. Similarly in the local case.

Fibrewise G-bundles

Let G be a fibrewise topological group and let X be a fibrewise space over B. By a *principal fibrewise G-bundle* over X we mean a principal fibrewise G-space over X which is locally trivial and for which the division function is continuous. We refer to G, in this situation, as the *fibrewise structural group*. Note that G itself is a (trivial) principal fibrewise G-bundle over B.

For example, let Γ be a fibrewise topological group over B and let G be a subgroup of Γ. Suppose that Γ is locally sectionable over Γ/G. Then Γ, by Proposition 7.1, is a principal fibrewise G-bundle over Γ/G.

Returning to the general case, observe that if E is a trivial principal fibrewise G-bundle over X then the fibrewise mixed product $E \times_G T$ is equivalent to $X \times_B T$, for each fibrewise left G-space T over B. Without the triviality condition this is still true locally and we describe $E \times_G T$ as the *associated fibrewise G-bundle* with fibrewise fibre T. When $T = G$, with G acting on itself by fibrewise multiplication, the associated fibrewise G-bundle can be identified with E. When $T = G$ with G acting on itself by fibrewise conjugation the associated fibrewise G-bundle is not in general principal. In fact, fibrewise conjugation leaves the neutral section of G fixed and so $E \times_G T$ has a section over X in this case.

As before, let E be a principal fibrewise G-bundle over X. For each fibrewise open subgroup H of G we may, as in Proposition 2.10, identify the fibrewise orbit space E/H with the associated fibrewise G-bundle with fibre G/H. In particular, suppose that G is itself a subgroup of a fibrewise topological group Γ. Assuming the existence of local sections, we regard Γ as a principal fibrewise G-bundle over Γ/G. Then Γ/H may be identified with the associated fibrewise G-bundle with fibrewise fibre G/H.

Note that if E is a principal fibrewise G-bundle over X then the pull-back $\lambda^* E$ is a principal fibrewise G-bundle over X' for each fibrewise space X' and fibrewise map $\lambda : X' \to X$. Moreover, if $E \times_G T$ is the associated fibrewise G-bundle over X with fibrewise fibre the fibrewise G-space T then

$$\lambda^*(E \times_G T) = \lambda^* E \times_G T$$

is the associated fibrewise G-bundle over X' with fibrewise fibre T.

Now suppose that we have a fibrewise homomorphism $\alpha : G' \to G$ of fibrewise topological groups. Let E' be a principal fibrewise G'-bundle over the fibrewise space X. Regarding G as a fibrewise G'-space via α, we obtain the principal fibrewise G-bundle $\alpha_* E' = E' \times_{G'} G$ over X, together with a fibrewise G'-map $\phi : E' \to \alpha_* E'$ over X, induced by

$$E' \xrightarrow{\Delta} E' \times_B E' \xrightarrow{1 \times c'} E' \times_B G'.$$

We describe $\alpha_* E'$ as the principal fibrewise G-bundle over X obtained from E' via α.

Conversely, suppose that E is a principal fibrewise G-bundle over X and let $\phi : E' \to E$ be a fibrewise G'-map over X, where G' acts on E via α. Then a fibrewise G-equivalence $\alpha_* E' \to E$ over X is induced by

$$E' \times_B G \xrightarrow{\phi \times 1} E \times_B G \xrightarrow{r} E.$$

If G' is a subgroup of G and α the inclusion we describe E as the principal fibrewise G-bundle over X obtained from E' by extending the fibrewise structural group from G' to G. Although the fibrewise structural group can always be extended in this way, as we have seen, the opposite process of reduction of the fibrewise structural group to a given subgroup is not always possible (for example, reduction to the trivial subgroup is equivalent to trivialization). Moreover, different reductions to a given subgroup G' for the same principal fibrewise G-bundle are not necessarily equivalent, as principal fibrewise G'-bundles.

Proposition 7.2 *Let G be a fibrewise topological group and let X be a fibrewise space over B. Let G' be a subgroup of G such that G is locally sectionable over G/G'. Let E be a principal fibrewise G-bundle over X. Then the fibrewise structural group G of E can be reduced to G' if and only if the associated fibrewise G-bundle E/G' with fibrewise fibre G/G' admits a section.*

In one direction this is almost obvious. Thus if E' is a principal fibrewise G'-bundle over X and $\phi : E' \to E$ is a fibrewise G'-map over X then

$$\phi/G' : E'/G' \to E/G'$$

constitutes a section of the associated fibrewise G-bundle. Conversely, let $s : E/G \to E/G'$ be a section. Regarding E as a principal fibrewise G'-bundle over E/G', consider the induced fibrewise G'-bundle $s^* E$ over E/G. This comes equipped with a fibrewise G'-map $s^* E \to E$ over s, and hence over $E/G = X$ as required.

The Milnor construction

Let G be a fibrewise open group, over a base space B. Following Milnor [113], in the ordinary theory, a countably numerable fibrewise G-bundle can be constructed as follows. We regard the cylinder $A = B \times I$ as a fibrewise space under the first projection, and denote the second projection by $\alpha : A \to I$.

For each point b of B consider sequences

$$(a_1, g_1, a_2, g_2, \ldots), \quad \text{where } a_1, a_2, \ldots \in A_b \text{ and } g_1, g_2, \ldots \in G_b.$$

Restrict attention to those sequences such that $\alpha(a_n) = 0$ for all but a finite number of indices n and such that $\sum \alpha(a_n) = 1$. Impose on this fibrewise set the fibrewise equivalence relation in which two such sequences

$$(a_1, g_1, \ldots) \quad \text{and} \quad (a_1', g_1', \ldots)$$

are equivalent if $a_n = a_n'$ for all n and either $g_n = g_n'$ or $\alpha(a_n) = 0$ for each n. The equivalence class of the sequence $(a_1, g_1, \ldots)$ will be written as $[a_1, g_1, \ldots]$. The fibrewise set of equivalence classes will be denoted by E_G.

Fibrewise functions $a_n : E_G \to A$ and $g_n : (\alpha \circ a_n)^{-1}(0, 1] \to G$ are defined in the obvious way for $n = 1, 2, \ldots$ Let us give E_G the coarsest topology for which all these functions are continuous. Then E_G becomes a fibrewise space, such that for each fibrewise space K a fibrewise function $f : K \to E_G$ is continuous if and only if each function $a_n \circ f$ and $g_n \circ (f \mid (\alpha \circ a_n \circ f)^{-1}(0, 1])$ is continuous.

Now consider the fibrewise action $r_G : E_G \times_B G \to E_G$ of G on E_G given by $r_G([a_1, g_1, a_2, g_2, \ldots], h) = [a_1, g_1 h, a_2, g_2 h, \ldots]$. We denote the fibrewise orbit space E_G/G by X_G and the natural projection by p_G. The fibrewise map a_n is invariant and so induces a fibrewise map $a_n \circ (p_G)^{-1} : X_G \to A$. The open sets $\{a_n \circ p_G^{-1} \circ \alpha^{-1}(0, 1]\}$ form a countably numerable covering of X_G, and E_G is fibrewise G-trivial over each member of the covering. Thus E_G obtains the structure of a countably numerable fibrewise G-bundle over X_G. Note that X_G admits sections, for example the section induced by the fibrewise G-map $G \to E_G$ which sends g into $[1, g, 0, \ldots]$.

The classification theorem

Our aim is to show that the fibrewise G-bundle thus constructed enjoys the universal property for numerable fibrewise G-bundles, classified as above. Specifically, given a fibrewise space X we associate with each fibrewise map $\lambda : X \to X_G$ the induced numerable fibrewise G-bundle $\lambda^* E_G$ over X. It follows by a straightforward fibrewise version of the argument used in the ordinary theory, for example that given by Milnor in [114], that fibrewise homotopic maps induce equivalent fibrewise G-bundles. What remains to be established is first that every numerable fibrewise G-bundle over X is equivalent to $\lambda^* E_G$ for some fibrewise map λ, and second that if $\lambda^* E_G$ is equivalent to $\mu^* E_G$ for some fibrewise maps $\lambda, \mu : X \to X_G$ then λ is fibrewise homotopic to μ.

So let E be a numerable fibrewise G-bundle over X with projection p. We can choose a countable family of open subsets V_n ($n = 1, 2, \ldots$) of X and fibrewise maps $\beta_n : X \to A$ such that E is fibrewise G-trivial over V_n and $(\alpha \circ \beta_n)^{-1}(0, 1] \subseteq V_n$ for each n. Given a local fibrewise G-trivialization

$$\phi_n : E_{V_n} \to V_n \times_B G$$

we define a fibrewise function $g_n : E \to G$ by $g_n = \pi_2 \phi_n$ on E_{V_n} and by $g_n = c$ on $E - E_{V_n}$. Here c, as usual, denotes the fibrewise constant. Although the fibrewise functions g_n may not themselves be continuous they nevertheless define a fibrewise map $f : E \to E_G$, where

$$f = [\beta_1 \circ p, g_1, \beta_2 \circ p, g_2, \ldots].$$

(Here and elsewhere in what follows variables are omitted to ease the notation.) Since f is equivariant, with respect to the fibrewise actions of G on E and E_G, we obtain an induced fibrewise map $\lambda : X \to X_G$, such that E is equivalent to $\lambda^* E_G$. This proves the first assertion.

To prove the second let $\theta, \theta' : E \to E_G$ be fibrewise G-maps, expressed in the form

$$\theta = [a_1, g_1, a_2, g_2, \ldots],$$
$$\theta' = [a'_1, g'_1, a'_2, g'_2, \ldots].$$

We start by showing that θ and θ' are fibrewise G-homotopic to the fibrewise G-maps ϕ and ϕ' given by

$$\phi = [a_1, g_1, 0, c, a_2, g_2, 0, c, \ldots],$$
$$\phi' = [0, c, a'_1, g'_1, 0, c, a'_2, g'_2, \ldots].$$

In fact a fibrewise G-homotopy $H_t : E \to E_G$ of θ into ϕ is given by the expression

$$[(1 - t)a_1, g_1, ta_1, g_1, (1 - t)a_2, g_2, ta_2, g_2, \ldots],$$

and a fibrewise G-homotopy of θ' into ϕ' is given similarly. The next stage is to construct, in infinitely many steps, a fibrewise G-homotopy of ϕ into θ. The first step, indicated by the expression

$$[a_1, g_1, ta_2, g_2, (1 - t)a_2, g_2, ta_3, g_3, (1 - t)a_3, \ldots],$$

ends with the fibrewise G-map indicated by the expression

$$[a_1, g_1, a_2, g_2, 0, c, a_3, g_3, 0, c, \ldots].$$

The second step is defined similarly, and so on, until after the nth step we reach the fibrewise G-map indicated by the expression

$$[a_1, g_1, a_2, g_2, \ldots, a_{n+1}, g_{n+1}, 0, c, a_{n+2}, g_{n+2}, 0, c, \ldots].$$

By juxtaposing the steps in this series of fibrewise G-homotopies we obtain a fibrewise G-homotopy since at each coordinate place in E_G all but a finite number of the steps are stationary. Thus a fibrewise G-homotopy of ϕ into θ is obtained, and similarly, a fibrewise G-homotopy of ϕ' into θ'. Since we have already seen that ϕ is fibrewise G-homotopic to ϕ' we conclude that θ is fibrewise G-homotopic to θ', as required. This completes the proof of the classification theorem. In view of this result we may refer to X_G as the *classifying fibrewise space* of G. Very much the same argument, applied to the non-equivariant case, may be used to show that E_G is fibrewise contractible, and hence that X_G is vertically connected.

In particular, take $X = B \times T$, for some space T. Suppose that B is locally compact regular. Then the set of fibrewise homotopy classes of fibrewise maps of X into X_G is equivalent to the set of homotopy classes of maps of T into $\Gamma(X_G)$, the space of sections of X_G.

To illustrate these ideas consider a topological group K. In the category of K-spaces let E_0 be a space and let G_0 be a topological transformation group of E_0. Thus E_0 and G_0 are both K-spaces, the multiplication of G_0 and the action of G_0 on E_0 are equivariant, and the neutral element of G_0 is fixed. Let P be a principal K-bundle over B. Then the associated bundle $P_\# G_0$ with fibre G_0 is a fibrewise open group. Also the associated bundle $P_\# E_0$ with fibre E_0 is a fibrewise $P_\# G_0$-space, more precisely a principal fibrewise $P_\# G_0$-bundle over the associated bundle $P_\# (E_0/G_0)$ with fibre E_0/G_0. In fact the classifying fibrewise space of $P_\# G_0$ defined as above can be identified with the associated bundle with fibre the classifying space of G_0, in the ordinary sense.

An example

Example 7.3. For example take $K = \mathbb{Z}/2$. Take G_0 to be an Abelian discrete group with $\mathbb{Z}/2$ acting by inversion. The classifying space of G_0 is the Eilenberg–MacLane space $K(G_0, 1)$. For the principal $\mathbb{Z}/2$-bundle P take the sphere S^n, with $\mathbb{Z}/2$ acting by the antipodal transformation and $B = S^n/(\mathbb{Z}/2)$ the real projective n-space. Then the classifying fibrewise space of the fibrewise topological group

$$S^n \times_{\mathbb{Z}/2} G_0$$

with fibre G_0 can be identified with the associated bundle

$$S^n \times_{\mathbb{Z}/2} K(G_0, 1)$$

with fibre $K(G_0, 1)$.

8 Fibrewise mapping-spaces

In this section we consider the problem, first discussed by Thom [131], of constructing an explicit right adjoint to the fibrewise product. Various procedures for doing this have been discussed in the literature, for example Booth and Brown [16, 17] use partial maps while Min and Lee [115] use convergence spaces. However, the method adopted in Section 9 of [81] seems at least as satisfactory as any and so what we give here follows that account quite closely.

The problem is to assign a suitable fibrewise topology to the fibrewise set

$$\mathrm{map}_B(X,Y) = \coprod_{b \in B} \mathrm{map}(X_b, Y_b),$$

where X and Y are fibrewise spaces over B. Although we shall give a fairly full outline of the theory in the general case, for certain technicalities we shall refer the reader to Section 9 of [86], where full details are given. Before dealing with the general case let us consider again the case when X is trivial, which admits of simpler treatment. We begin by describing a variant of the treatment given in Section 1.

The fibrewise compact-open topology

The version of compact-open topology we are going to generalize here is a refinement of the standard one, as follows. Given spaces X and Y, let $\mathrm{map}(X,Y)$ denote the set of maps $\phi : X \to Y$. For $U \subseteq Y$ open, C compact Hausdorff, and $\lambda : C \to X$ a map, let (C, λ, U) denote the subset of map (X,Y) consisting of maps ϕ such that $\phi \lambda C \subseteq U$. We describe such subsets as *compact-open*, and we describe the topology which they generate as the *compact-open topology*. This is a modification of the usual theory, in which C is required to be a subset of X and λ is the inclusion, but it has all the expected properties, as can easily be checked.

Consider first the fibrewise set $\mathrm{map}_B(B \times T, Z)$, where T is a space and Z is a fibrewise space over B. Maps of $\{b\} \times T$ into Z_b can be regarded as maps of T into Z, in the obvious way, and so $\mathrm{map}_B(B \times T, Z)$ can be topologized as a subspace of $\mathrm{map}(T, Z)$, with the compact-open topology. Then for any fibrewise space Y over B a fibrewise map

$$Y \times T = (B \times T) \times_B Y \to Z$$

determines a fibrewise map

$$Y \to \mathrm{map}_B(B \times T, Z),$$

through the standard formula, and the converse holds when T is locally compact and regular. All we need for this is the standard theory of the compact-open topology.

Now let us turn to the general case, where X and Y are fibrewise spaces over B. Given an open set $W \subseteq B$, an open set $U \subseteq Y_W$, a fibrewise compact Hausdorff space K over W, and a fibrewise map $\lambda : K \to X_W$, we denote by $(K, \lambda, U; W)$ the set of maps $\phi : X_b \to Y_b$, where $b \in W$, for which $\phi \lambda K_b \subseteq U$. If $K \subseteq X_W$ and λ is the inclusion then we write $(K, \lambda, U; W)$ as $(K, U; W)$. We describe such a subset $(K, \lambda, U; W)$ of $\mathrm{map}_B(X, Y)$ as *fibrewise compact-open*, and describe the fibrewise topology generated by the fibrewise compact-open subsets as the *fibrewise compact-open topology*. From now on, when we use the term fibrewise mapping-space, this is the fibrewise topology with which it is equipped.

Some caution is necessary when taking pull-backs or even restricting to subspaces of the base. For example, the fibres of the fibrewise mapping-space do not necessarily inherit the compact-open topology. Of course if $\alpha : B' \to B$ is a map, where B' is a space, then a fibrewise bijection

$$\mathrm{map}_{B'}(\alpha^* X, \alpha^* Y) \to \alpha^* \mathrm{map}_B(X, Y)$$

is defined, in the obvious way. The bijection is continuous, since the pull-back $(\alpha^* K, \alpha^* \lambda, \alpha^* U; \alpha^{-1} W)$ of a fibrewise compact-open subset $(K, \lambda, U; W)$ of $\mathrm{map}_B(X, Y)$ is a fibrewise compact-open subset of $\mathrm{map}_{B'}(\alpha^* X, \alpha^* Y)$. Later in this section we will show that this continuous bijection is an equivalence of fibrewise spaces over B' under certain conditions.

It is not difficult to see, however, that when $X = B \times T$, for some space T, the special method of topologizing the fibrewise mapping-space which can be used in this case agrees with the general method. Specifically, consider the injection

$$\sigma : \mathrm{map}_B(B \times T, Y) \to \mathrm{map}(T, Y)$$

given, as before, by transforming each map $\{b\} \times T \to Y_b$, where $b \in B$, into the corresponding map $T \to Y$. We assert that σ constitutes an embedding of the domain with fibrewise compact-open topology in the codomain with compact-open topology.

For let (C, λ, U) be a compact-open subset of $\mathrm{map}(T, Y)$, where $U \subseteq Y$ is open, C is compact Hausdorff and $\lambda : C \to U$ is a map. Then

$$\sigma^{-1}(C, U) = (C \times B, \lambda \circ \pi_1, U; B),$$

where $C \times B$ is fibrewise compact Hausdorff over B.

In the other direction, let $(K, \lambda, U; W)$ be a fibrewise compact-open subset of $\mathrm{map}_B(B \times T, Y)$, so that $W \subseteq B$ is open, $U \subseteq Y_W$ is open, K is fibrewise compact Hausdorff over W and $\lambda : K \to W \times T$ is a fibrewise map. Then $(K, \lambda, U; W)$ is the union of the preimages $\sigma^{-1}(C_b, \lambda_b, U)$, where b runs through the points of W, $C_b \times \{b\} = K_b$ and λ_b is the first component of $\lambda \mid C_b$. Since each subset (C_b, λ_b, U) of $\mathrm{map}(T, Y)$ is compact-open we conclude that σ is an embedding, as stated.

Proposition 8.1 *Let X and Y be fibrewise spaces over B. Then $\mathrm{map}_B(X, B)$ is equivalent to B, as a fibrewise space, and $\mathrm{map}_B(B, Y)$ is equivalent to Y.*

To prove the first assertion we show that the projection $\mathrm{map}_B(X, B) \to B$ is a homeomorphism. Since it is obviously a continuous bijection we only have to show that it is open, and for this it is sufficient to show that the projection of a fibrewise compact-open subset is open. So let $(K, \lambda, U; W)$ be a fibrewise compact-open subset of $\mathrm{map}_B(X, B)$, where $W \subseteq B$ is open, $U \subseteq W$ is open, K is fibrewise compact Hausdorff over W, and $\lambda : K \to X_W$ is a fibrewise map. Then the projection is just $U \cup (W - p\lambda K)$, which is open in B since $p\lambda K$ is closed in W. This proves the first assertion; the proof of the second is equally straightforward.

Clearly if $B' \subseteq B$ then $\mathrm{map}_B(X, Y)|_{B'}$ is equivalent to $\mathrm{map}_{B'}(X', Y')$, as a fibrewise set over B', where $X' = X_{B'}$, $Y = Y_{B'}$. Now the restriction $(X_{B'} \cap K, Y_B \cap V; W \cap B')$ to B' of a fibrewise compact-open set $(K, V; W)$ of $\mathrm{map}_B(X, Y)$ is a fibrewise compact-open set of $\mathrm{map}_{B'}(X', Y')$, hence the identity function

$$\mathrm{map}_{B'}(X', Y') \to \mathrm{map}_B(X, Y)|_{B'}$$

is a continuous bijection. When B' is open in B the function is an equivalence, but this is not true generally.

In particular, the induced topology on the fibre $\mathrm{map}(X_b, Y_b)$ of $\mathrm{map}_B(X, Y)$ may be coarser than the compact-open topology. An exception is when X is fibrewise discrete since in that case a compact subset C of the discrete X_b is necessarily finite, say $C = \{x_1, \dots, x_n\}$. So we can find a neighbourhood W of b and, by using local slices through $x_1, \dots, x_n$, a family $K_1, \dots, K_n$ of subsets of X_W which are fibrewise compact over W and whose union K meets X_b in C. Since K is fibrewise compact over W it follows that the topologies on $\mathrm{map}(X_b, Y_b)$ coincide in this case.

Among the fibrewise compact-open sets of $\mathrm{map}_B(X, Y)$ a special rôle is played by those of the form

$$\langle s, V \rangle = (sW, V; W),$$

where W is open in B, where s is a section of X over W and where V is open in Y. In the case in which X is fibrewise discrete these special fibrewise compact-open sets form a sub-basis for the fibrewise compact-open topology, for a fibrewise topological Y. For let $(K, V; W)$ be a fibrewise compact-open set, where W is open in B, where $K \subseteq X_W$ is fibrewise compact over W, and where V is open in Y. Let $b \in W$ be a given point and assume, to avoid trivialities, that K_b is non-empty. Let $\phi : X_b \to Y_b$ be a continuous function such that $\phi K_b \subseteq V_b$, i.e. $K_b \subseteq \phi^{-1} V_b$. We have $K_b = \{x_1, \dots, x_n\}$, say, since K_b is discrete and compact, therefore finite. Choose a neighbourhood W_i of b, where $W_i \subseteq W$, and a section $s_i : W_i \to X$ such that $s_i(b) = x_i$, for $i = 1, \dots, n$. Note that $U_i = s_i W_i$ is open in X, since X is fibrewise discrete. Since K is fibrewise compact over W the subset

$$W_0 = \{\mu \in W \mid \{s_i(\mu), \ldots, s_n(\mu)\} \supseteq K_\mu\}$$

is open in B. Now $W' = W_0 \cap W_1 \cap \ldots \cap W_n \subseteq W$, is a neighbourhood of b and $\langle s_i', V \rangle$ is a neighbourhood of ϕ for $i = 1, \ldots, n$, where $s_i' = s_i | W'$. Since

$$\langle s_i', V \rangle \cap \ldots \cap \langle s_n', V \rangle \subseteq (K, V; W)$$

this shows that the special fibrewise compact-open sets form a fibrewise sub-basis.

The above observation is due to Lever [98] who uses it to establish a generalization of the fibrewise Tychonoff theorem including, as a special case

Theorem 8.2 *Let X be fibrewise discrete over B. Then $\mathrm{map}_B(X, Y)$ is fibre-wise compact whenever Y is fibrewise compact.*

The proof, which is not easy, can be found in [98] or in Section 9 of [86]. Although the result is obviously of great importance no use of it is made in what follows.

Examples can be given where Y is fibrewise Hausdorff but $\mathrm{map}_B(X, Y)$ is not. We prove

Proposition 8.3 *Let X and Y be fibrewise spaces over B, with X locally sliceable and Y fibrewise Hausdorff. Then $\mathrm{map}_B(X, Y)$ is fibrewise Haus-dorff.*

For consider distinct maps $\phi, \psi : X_b \to Y_b$, where $b \in B$. We have $\phi(x) \neq \psi(x)$ for some point x of X_b. Since X is locally sliceable there exists a neighbourhood W of b and a section $s : W \to X_b$ such that $s(b) = x$. Since Y is fibrewise Hausdorff there exist disjoint neighbourhoods U, V of $\phi(x)$, $\psi(x)$, respectively in Y. Then $(W, s, U; W)$, $(W, s, V; W)$ are disjoint fibre-wise compact-open neighbourhoods of ϕ, ψ, respectively, in $\mathrm{map}_B(X, Y)$.

Functoriality

After these preliminaries we turn to the general theory of fibrewise mapping-spaces, beginning with functoriality. Consider fibrewise spaces X, Y and Z over B. Precomposition with a fibrewise map $\theta : X \to Y$ determines a fibre-wise map

$$\theta^* : \mathrm{map}_B(Y, Z) \to \mathrm{map}_B(X, Z),$$

while postcomposition with a fibrewise map $\phi : Y \to Z$ determines a fibrewise map

$$\phi_* : \mathrm{map}_B(X, Y) \to \mathrm{map}_B(X, Z).$$

We prove two results about these induced fibrewise maps, as follows.

Proposition 8.4 *Let $\theta : X \to Y$ be a proper fibrewise surjection, where X and Y are fibrewise spaces over B. Then the fibrewise map*

$$\theta^* : \mathrm{map}_B(Y, Z) \to \mathrm{map}_B(X, Z)$$

is an embedding for all fibrewise spaces Z.

For let $W \subseteq B$ be open, let $V \subseteq Z_W$ be open, let K be fibrewise compact Hausdorff over W, and let $\mu : K \to Y_W$ be a fibrewise map. The fibrewise product $K \times_{Y_W} X_W$ is also fibrewise compact Hausdorff over W, since the first projection $K \times_{Y_W} X_W \to K$ is proper; we denote the second projection $K \times_{Y_W} X_W \to X_W$ by λ. Since θ^* is injective, because θ is surjective, we have

$$(K, \mu, V; W) = \theta^{*-1}(K \times_{Y_W} X_W, \lambda, V; W).$$

Therefore θ^* is an embedding, as asserted.

Proposition 8.5 *Let $\phi : Y \to Z$ be a fibrewise embedding, where Y and Z are fibrewise spaces over B. Then the fibrewise map*

$$\phi_* : \mathrm{map}_B(X, Y) \to \mathrm{map}_B(X, Z)$$

is an embedding, for all fibrewise spaces Z. If ϕ is closed, furthermore, then ϕ_* is closed provided X is locally sliceable.

For let $(K, \lambda, U; W)$ be a fibrewise compact-open subset of $\mathrm{map}_B(X, Y)$, so that $W \subseteq B$ is open, $U \subseteq Y_W$ is open, K is fibrewise compact Hausdorff over W, and $\lambda : K \to X_W$ is a fibrewise map. Then $U = \phi^{-1}V$ for some open set V of Z_W and so

$$(K, \lambda, U; W) = \phi_*^{-1}(K, \phi_W \circ \lambda, V; W).$$

This proves the first assertion.

To prove the second, let $\alpha : X_b \to Z_b$ $(b \in B)$ belong to the complement of $\phi_*\mathrm{map}_B(X, Y)$ in $\mathrm{map}_B(X, Z)$. Then $\alpha(x) \in U$, for some $x \in X_b$, where $U = Z - \phi Y$ is open. If X is locally sliceable then for some neighbourhood W of b there exists a section $s : W \to X_W$ such that $s(b) = x$. Thus $(W, s, U; W)$ is a fibrewise compact-open neighbourhood of λ which does not meet $\phi_*\mathrm{map}_B(X, Y)$. This completes the proof.

Proposition 8.6 *Let $\{X_j\}$ be a family of fibrewise spaces over B. Then the natural fibrewise map*

$$\mathrm{map}_B(\coprod X_j, Y) \to \prod_B \mathrm{map}_B(X_j, Y)$$

is an equivalence for all fibrewise spaces Y.

Here the ith component of the fibrewise map is

$$\sigma_i^* : \mathrm{map}_B(\textstyle\coprod X_j, Y) \to \mathrm{map}_B(X_i, Y),$$

where $\sigma_i : X_i \to \coprod X_j$ is the standard insertion. The proof of Proposition 8.6 is straightforward.

Our aim is to show that, subject to certain restrictions, the fibrewise mapping-space stands in an adjoint relationship to the fibrewise product. We begin by proving

Proposition 8.7 *Let X, Y and Z be fibrewise spaces over B. If the fibrewise function $h : X \times_B Y \to Z$ is continuous then so is the fibrewise function $k : X \to \mathrm{map}_B(Y, Z)$, where*

$$k(x)(y) = h(x, y) \quad (x \in X_b, \, y \in Y_b, \, b \in B).$$

To establish the continuity of k it is sufficient to show that the preimage of a fibrewise compact-open subset of $\mathrm{map}_B(Y, Z)$ is open in X. So consider the fibrewise compact-open subset $(K, \lambda, V; W)$, where $W \subseteq B$ is open, $V \subseteq Z_W$ is open, K is fibrewise compact Hausdorff over W and $\lambda : K \to Y_W$ is a fibrewise map. Suppose that $k(x) \in (K, \lambda, V; W)$, where $x \in X_b$, $b \in B$. The preimage $(1 \times \lambda^{-1})h^{-1}V$ is a neighbourhood of the preimage $\{x\} \times K_b$ of x under the projection

$$X_W \times_W K \to X_W \times_W W \to X_W.$$

Since the projection is closed we have $U \times_W K \subseteq (1 \times \lambda^{-1})h^{-1}V$ for some neighbourhood U of x in X_W. Then kU is contained in $(K, \lambda, V; W)$. Therefore k is continuous, as asserted.

Our next result requires a technical lemma, as follows.

Lemma 8.8 *Let X and Y be fibrewise spaces over B, with X fibrewise regular. Suppose that the fibrewise topology of Y is generated by a fibrewise sub-basis. Then the fibrewise compact-open topology of $\mathrm{map}_B(X, Y)$ is generated by a fibrewise sub-basis consisting of fibrewise compact-open subsets $(K, \lambda, U; W)$, where $W \subseteq B$ is open, $U \subseteq Y_W$ is fibrewise subbasic, K is fibrewise compact Hausdorff over W, and $\lambda : K \to X_W$ is a fibrewise map.*

We omit the proof since it is a straightforward generalization of the proof of the corresponding result (9.8) of [86]. From this lemma we at once obtain

Proposition 8.9 *Let Y_1, Y_2 be fibrewise spaces over B, and let*

$$Y_1 \xleftarrow{\;\;\pi_1\;\;} Y_1 \times_B Y_2 \xrightarrow{\;\;\pi_2\;\;} Y_2$$

be the standard projections. Then the natural fibrewise map

$$\text{map}_B(X, Y_1 \times_B Y_2) \to \text{map}_B(X, Y_1) \times_B \text{map}_B(X, Y_2)$$

is an equivalence for all fibrewise regular X.

Here the fibrewise map is given by π_{1*} in the first factor, and by π_{2*} in the second. Next we require another technical lemma as follows.

Lemma 8.10 *Let X and Y be fibrewise regular spaces over B. Then for all fibrewise spaces Z the fibrewise compact-open topology of $\text{map}_B(X \times_B Y, Z)$ is generated by fibrewise compact-open subsets $(K \times_W L, \lambda \times \mu, V; W)$, where $W \subseteq B$ is open, $V \subseteq Z_W$ is open, K and L are fibrewise compact Hausdorff over W, and $\lambda : K \to X_W$, $\mu : L \to Y_W$ are fibrewise maps.*

Again we omit the proof since it is straightforward generalization of the corresponding result (9.10) of [86]. From this lemma we at once obtain

Proposition 8.11 *Let X_i, Y_i $(i = 1, 2)$ be fibrewise spaces over B, with X_i fibrewise regular. Then the natural fibrewise function*

$$\text{map}_B(X_1, Y_1) \times_B \text{map}_B(X_2, Y_2) \to \text{map}_B(X_1 \times_B X_2, Y_1 \times_B Y_2)$$

is an embedding.

Here the fibrewise function is given by the fibrewise product functor. Continuity is obvious, while the lemma shows that the condition for an embedding is satisfied.

In particular consider the fibrewise function

$$T_{B\#} : \text{map}_B(X, Y) \to \text{map}_B(T_B X, T_B Y),$$

given by the fibrewise product $\times_B T$ with a given fibrewise regular space T over B. The above result shows that $T_{B\#}$ is an embedding for all fibrewise spaces Y provided X is fibrewise regular. Now let $\Phi_B(X)$ denote the push-out of the cotriad

$$X \times_B T \leftarrow T_0 \times_B X \to T_0$$

and similarly for fibrewise maps, where T_0 is a closed subspace of T. We assert that the fibrewise function

$$\Phi_{B\#} : \text{map}_B(X, Y) \to \text{map}_B(\Phi_B X, \Phi_B Y)$$

is continuous, provided T is fibrewise compact regular and X is fibrewise regular. For example the fibrewise function given by fibrewise suspension is continuous, for fibrewise regular X.

To see this consider the diagram shown below, where $\rho : T_B \to \Phi_B$ is the natural transformation

$$\operatorname{map}_B(X,Y) \xrightarrow{\ T_{B\#}\ } \operatorname{map}_B(T_B X, T_B Y)$$

$$\Phi_{B\#} \downarrow \qquad\qquad\qquad \downarrow \rho_*$$

$$\operatorname{map}_B(\Phi_B X, \Phi_B Y) \xrightarrow[\ \rho^*\ ]{} \operatorname{map}_B(T_B X, \Phi_B Y)$$

Here ρ^* is an embedding, by Proposition 8.4, since $\rho : T_B X \to \Phi_B X$ is a proper surjection. Also $T_{B\#}$ is continuous, by Proposition 8.9, and ρ_* is continuous, from first principles. Therefore $\rho^* \Phi_{B\#} = \rho_* T_{B\#}$ is continuous and so $\Phi_{B\#}$ is an embedding, as asserted.

Fibrewise evaluation

Fibrewise evaluation (i.e. evaluation in each fibre) determines a fibrewise function

$$\operatorname{map}_B(X,Y) \times_B X \to Y$$

for all fibrewise spaces X, Y over B. More generally, fibrewise composition (i.e. composition in each fibre) determines a fibrewise function

$$\operatorname{map}_B(Y,Z) \times_B \operatorname{map}_B(X,Y) \to \operatorname{map}_B(X,Z)$$

for all fibrewise spaces X, Y, Z over B. We prove

Proposition 8.12 *Let Y be fibrewise locally compact regular over B. Then the fibrewise composition function*

$$\operatorname{map}_B(Y,Z) \times_B \operatorname{map}_B(X,Y) \to \operatorname{map}_B(X,Z)$$

is continuous for all fibrewise spaces X and Z.

For let $\theta : X_b \to Y_b$ and $\phi : Y_b \to Z_b$ be maps, where $b \in B$. Let $(K, \lambda, V; W)$ be a fibrewise compact-open neighbourhood of $\phi \circ \theta$ in $\operatorname{map}_B(X,Z)$. Thus W is a neighbourhood of b, $V \subseteq Z_W$ is open, K is fibrewise compact Hausdorff over W, and $\lambda : K \to X_W$ is a fibrewise map. Now

$$\theta(X_b \cap \lambda K) \subseteq \phi^{-1}(Z_b \cap V) = Y_b \cap U$$

for some open $U \subseteq Y_W$. Since Y is fibrewise locally compact regular there exists a neighbourhood $W' \subseteq W$ of b and a neighbourhood N of $\theta(X_b \cap K)$ in $Y_{W'}$ such that the closure $Y'_{W'} \cap \bar{N}$ of N in $Y_{W'}$ is fibrewise compact over W' and contained in U. Fibrewise composition sends the fibrewise product

$$(X_{W'} \cap \bar{N}, V'; W') \times_{W'} (K', \lambda', N; W')$$

into $(K, \lambda, V; W)$, where $K' = K_{W'}$, $\lambda' = \lambda_{W'}$ and $V' = Z_{W'} \cap V$. Since $\theta \in (K', \lambda', N; W')$ and $\phi \in (X_{W'} \cap \bar{N}, V'; W')$ this proves Proposition 8.12. As a special case we obtain

Corollary 8.13 *Let X be fibrewise locally compact regular over B. Then the fibrewise evaluation function*

$$\mathrm{map}_B(X, Y) \times_B X \to Y$$

is continuous for all fibrewise spaces Y.

This leads at once to a converse of Proposition 8.7, subject to the expected restrictions.

Corollary 8.14 *Let X, Y and Z be fibrewise spaces over B, with Y fibrewise locally compact regular. Let $k : X \to \mathrm{map}_B(Y, Z)$ be a fibrewise map. Then the fibrewise function $h : X \times_B Y \to Z$ is continuous, where*

$$h(x, y) = k(x)(y) \quad (x \in X_b, \, y \in Y_b, \, b \in B).$$

To see this it is only necessary to observe that h may be expressed as the composition

$$X \times_B Y \to \mathrm{map}_B(Y, Z) \times_B Y \to Z$$

of $k \times 1_Y$ and the fibrewise evaluation function. When h and k are related as in Proposition 8.7 or Corollary 8.14 we refer to h as the *left adjoint* of k and to k as the *right adjoint* of h. The relationship is placed on a more satisfactory formal basis in our next result, which may be described as the exponential law for our theory.

Proposition 8.15 *Let X, Y and Z be fibrewise spaces over B and let*

$$\xi : \mathrm{map}_B(X \times_B Y, Z) \to \mathrm{map}_B(X, \mathrm{map}_B(Y, Z))$$

be the fibrewise injection defined by taking adjoints in each fibre. If X is fibrewise regular then ξ is continuous. If both X and Y are fibrewise regular then ξ is an open embedding. If, in addition, Y is fibrewise locally compact then ξ is an equivalence.

For by Lemma 8.10, when X is fibrewise regular the fibrewise topology of $\mathrm{map}_B(X, \mathrm{map}_B(Y, Z))$ is generated by fibrewise compact-open subsets of the form $(K, \lambda, (L, \mu, V; W); W)$, where $W \subseteq B$ is open, $V \subseteq Z_W$ is open, K and L are fibrewise compact Hausdorff over W, and $\lambda : K \to X_W$, $\mu : L \to Y_W$ are fibrewise maps. The inverse image of this fibrewise subbasic set is just $(K \times_W L, \lambda \times \mu, V; W)$, which is also fibrewise subbasic, and so ξ is continuous.

By Lemma 8.10 again, when X and Y are fibrewise regular the fibrewise topology of $\mathrm{map}_B(X \times_B Y, Z)$ is generated by fibrewise compact-open subsets of the form $(K \times_W L, \lambda \times \mu, V; W)$, where W, V, K, L, λ, μ are as before. The direct image of this subset is just $(K, \lambda, (L, \mu, V; W); W)$, which is also fibrewise compact-open, and so ξ is an open embedding. The final assertion follows at once from Corollary 8.14.

Another application of Corollary 8.14 is to obtain conditions under which fibrewise mapping-spaces behave naturally with respect to pull-backs. We prove

Proposition 8.16 *Let X and Y be fibrewise spaces over B, with X fibrewise locally compact regular. Then the continuous fibrewise bijection*

$$\alpha_{\#} : \operatorname{map}_{B'}(\alpha^*X, \alpha^*Y) \to \alpha^*\operatorname{map}_B(X, Y)$$

is an equivalence of fibrewise spaces over B' for each space B' and map $\alpha : B' \to B$.

For since the fibrewise evaluation function

$$\operatorname{map}_B(X, Y) \times_B X \to Y$$

is continuous so is its pull-back

$$\alpha^*(\operatorname{map}_B(X, Y) \times_B X) \to \alpha^*Y.$$

Rewriting this as

$$\alpha^*\operatorname{map}_B(X, Y) \times_{B'} \alpha^*X \to \alpha^*Y$$

we take the right adjoint and obtain a fibrewise map

$$\alpha^*\operatorname{map}_B(X, Y) \to \operatorname{map}_{B'}(\alpha^*X, \alpha^*Y)$$

over B', which is inverse to $\alpha_{\#}$. This proves Proposition 8.16.

In particular, take $B = *$ and replace B', X, Y by B, X_0, Y_0, respectively, where X_0, Y_0 are spaces. We deduce that for any space B there is a natural equivalence between $\operatorname{map}_B(X_0 \times B, Y_0 \times B)$ and $\operatorname{map}(X_0, Y_0) \times B$, as fibrewise spaces over B, provided X_0 is locally compact regular.

The space of fibrewise maps

Returning to the general case, let us compare the space $\Gamma(\operatorname{map}_B(X, Y))$ of sections $s : B \to \operatorname{map}_B(X, Y)$ of the fibrewise mapping-space with the space $\operatorname{MAP}_B(X, Y)$ of fibrewise maps $\phi : X \to Y$. Here both the space of sections and the space of fibrewise maps are endowed with the compact-open topology. Consider the function

$$\sigma : \operatorname{MAP}_B(X, Y) \to \Gamma(\operatorname{map}_B(X, Y))$$

which transforms the fibrewise map ϕ into the section s given by $s(b) = \phi_b$ ($b \in B$). Clearly σ is injective.

Proposition 8.17 *Let X, Y be fibrewise spaces over B, and let*

$$\sigma : \text{MAP}_B(X, Y) \to \Gamma(\text{map}_B(X, Y))$$

be the injection defined above. If X is fibrewise locally compact regular then σ is bijective. If in addition B is regular then σ is an equivalence of spaces.

To prove the first assertion, let $s : B \to \text{map}_B(X, Y)$ be a section. Then $s = \sigma(\phi)$, where $\phi : X \to Y$ is the fibrewise map given by the composition

$$X = B \times_B X \to \text{map}_B(X, Y) \times_B X \to Y;$$

here the first stage is $s \times 1$ while the second stage is fibrewise evaluation.

To prove the second assertion, where B is regular, observe that the compact-open topology of $\Gamma(\text{map}_B(X, Y))$ is generated by compact-open subsets of the form $(C, \mu, (K, \lambda, U; W))$, where C is compact Hausdorff, $\mu : C \to B$ is a map, $W \subseteq B$ is open, $U \subseteq Y_W$ is open, K is fibrewise compact Hausdorff over W and $\lambda : K \to Y_W$ is a fibrewise map. The preimage of this subset under σ is the compact-open subset $(C \times_B K, \lambda\pi_2, V)$. Therefore σ is continuous.

Finally, we need to show that σ is open. So let (C, μ, U) be a compact-open subset of $\text{MAP}_B(X, Y)$, where $U \subseteq Y$ is open, C is compact Hausdorff and $\mu : C \to X$ is a map. Then

$$\sigma(C, \mu, U) = (C, p\mu, (C \times B, \mu \circ \pi_1, U; B)),$$

which is fibrewise compact-open. Thus σ is open and therefore an equivalence. This completes the proof.

We conclude with two results about fibrewise fibrations, of which special cases have occurred earlier. The proofs are omitted since they are straightforward exercises in the use of adjoints.

Proposition 8.18 *Let $\phi : E \to F$ be a fibrewise fibration, where E and F are fibrewise spaces over B. Then the postcomposition function*

$$\phi_* : \text{map}_B(Y, E) \to \text{map}_B(Y, F)$$

is a fibrewise fibration, for all fibrewise locally compact regular Y.

Proposition 8.19 *Let $u : A \to X$ be a fibrewise cofibration, where X is fibrewise locally compact regular over B and A is a closed subspace of X. Then the precomposition function*

$$u^* : \text{map}_B(X, E) \to \text{map}_B(A, E)$$

is a fibrewise fibration for all fibrewise spaces E.

Some examples

Example 8.20. Let X be a fibrewise space over B and let $p : E \to X$ be a fibrewise fibration. Then

$$p_* : \mathrm{map}_B(Y, E) \to \mathrm{map}_B(Y, X)$$

is a fibrewise fibration, for all fibrewise compact regular Y.

Example 8.21. Let $u : A \to X$ be a fibrewise cofibration, where X is fibrewise locally compact regular over B and A is a closed subspace of X. Then

$$u^* : \mathrm{map}_B(X, E) \to \mathrm{map}_B(A, E)$$

is a fibrewise fibration for all fibrewise spaces E.

Example 8.22. Let X be a fibrewise space over B. Let E, F be fibrewise fibrations over X. If E is fibrewise locally compact regular over X, then $\mathrm{map}_X(E, F)$ is a fibrewise fibration over X.

Example 8.23. Let $p : E \to X$ be a fibrewise fibration, where E and X are fibrewise spaces over B. Let $A \subseteq X$ be a closed subspace such that the pair (X, A) is fibrewise cofibred. Then the pair (E, E_A) is also fibrewise cofibred.

Example 8.24. The fibrewise homotopy type of $\mathrm{map}_B(X, Y)$ depends only on the fibrewise homotopy types of X and Y, where X, Y are fibrewise spaces over B.

Chapter 2. The Pointed Theory

9 Fibrewise pointed spaces

Basic notions

In this chapter we work over a pointed base space B. A *fibrewise pointed space* over B consists of a space X together with maps

$$B \xrightarrow{\ s\ } X \xrightarrow{\ p\ } B$$

such that $p \circ s = 1_B$. In other words, X is a fibrewise space over B with section s. The alternative terminology *sectioned fibrewise space* is also widely used. Note that the projection is necessarily a quotient map and the section is necessarily an embedding. It is often convenient to regard B as a subspace of X so that the projection retracts X onto B. To simplify the exposition in what follows let us assume, once and for all, that the embedding is closed, as is necessarily the case when X is a Hausdorff space. We regard any subspace of X containing B as a fibrewise pointed space in the obvious way; no other subspaces will be admitted. When s is a cofibration we describe X as *cofibrant*.

We regard B as a fibrewise pointed space over itself using the identity as section and projection. We regard the topological product $B \times T$, for any pointed space T, as a fibrewise pointed space over B using the section given by the basepoint. Of course any map of B into T determines a section of $B \times T$, with the map as first component, and hence a fibrewise pointed space.

Let X be a fibrewise pointed space over B, as above. For each subspace B' of B we regard $X_{B'}$ as a fibrewise pointed space over B' with section $s_{B'}$. In particular, we regard the fibre X_b over the basepoint b of B as a pointed space with basepoint $s(b)$.

Fibrewise pointed spaces over B form a category with the following definition of morphism. Let X and Y be fibrewise pointed spaces over B with sections s and t, respectively. A *fibrewise pointed map* $\phi : X \to Y$ is a fibrewise map which is section-preserving in the sense that $\phi \circ s = t$. If $\phi : X \to Y$ is a fibrewise pointed map over B then the restriction $\phi_{B'} : X_{B'} \to Y_{B'}$ is a fibrewise pointed map over B' for each subspace B' of B. Thus a functor is defined from the category of fibrewise pointed spaces over B to the category of fibrewise pointed spaces over B'.

Equivalences in the category of fibrewise pointed spaces over B are called *fibrewise pointed topological equivalences*. If ϕ, as above, is a fibrewise pointed topological equivalence over B then $\phi_{B'}$ is a fibrewise pointed topological equivalence over B' for each subspace B' of B. In particular ϕ_b is a pointed topological equivalence where b is the basepoint of B.

For each fibrewise pointed space X over B the pull-back $\alpha^* X$ is regarded as a fibrewise pointed space over B', in the obvious way, for each space B' and map $\alpha : B' \to B$, and similarly with fibrewise pointed maps. Thus α^* constitutes a functor from the category of fibrewise pointed spaces over B to the category of fibrewise pointed spaces over B'. When B' is a subspace of B and α the inclusion this is equivalent to the restriction functor described earlier.

Fibrewise collapsing

Let X be a fibrewise space over B and let A be a closed subspace of X. We can define a fibrewise quotient space $X/_B A$ of $X \sqcup_B B$ by identifying points of A with their images under the projection. We refer to $X/_B A$ as the *fibrewise collapse* of A in X. In particular, when $A = X$ the fibrewise collapse reduces to B. If A is fibrewise compact over B the natural projection $X \to X/_B A$ is proper. If in addition X is fibrewise Hausdorff or fibrewise regular then so is $X/_B A$.

Note that if A is a closed subspace of a fibrewise space X over B then the fibrewise collapse $X/_B A$ becomes a fibrewise pointed space with section given by $B = A/_B A \to X/_B A$; the projection $(X, A) \to (X/_B A, B)$ is then a fibrewise relative homeomorphism.

Let X and Y be fibrewise pointed spaces over B with sections s and t, respectively. We regard the fibrewise product $X \times_B Y$ as a fibrewise pointed space with section given by $b \mapsto (s(b), t(b))$. The subspace

$$(X \times_B B) \cup (B \times_B Y) \subseteq X \times_B Y$$

is denoted by $X \vee_B Y$ and called the *fibrewise pointed coproduct* (or fibrewise wedge product). The fibrewise collapse

$$X \wedge_B Y = (X \times_B Y)/_B(X \vee_B Y)$$

is called the *fibrewise smash product*. Of course these constructions are functorial in nature.

Note that $X \times_B Y$ is fibrewise compact, and hence $X \wedge_B Y$ is fibrewise compact, whenever X and Y are fibrewise compact. Also if A is a fibrewise closed subspace of X the natural projection

$$(X \wedge_B Y)/_B(A \wedge_B Y) \to (X/_B A) \wedge_B Y$$

is a fibrewise pointed topological equivalence.

There exists a similar equivalence

$$(X \wedge_B Z) \vee_B (Y \wedge_B Z) \to (X \vee_B Y) \wedge_B Z$$

for all fibrewise pointed spaces X, Y, Z over B. Moreover, provided any two of the three are fibrewise compact Hausdorff there is also an equivalence

$$(X \wedge_B Y) \wedge_B Z \to X \wedge_B (Y \wedge_B Z).$$

We may refer to these as the *distributive law* and the *associative law* for the fibrewise smash product.

Given a pointed space F a functor Ψ_B is defined which transforms each fibrewise pointed space X over B into the fibrewise smash product

$$(B \times F) \wedge_B X,$$

and similarly for fibrewise pointed maps. Suppose that $F = D/E$, the pointed space obtained from a space D by collapsing a closed subspace E. The corresponding functor Φ_B is defined as in Section 1, and $\Psi_B(X)$ is equivalent to the fibrewise collapse $\Phi_B(X)/_B \Phi_B(B)$. We refer to Ψ_B in this case as the *reduction* of Φ_B, and write $\Psi_B = \Phi_B^B$. In particular, the reduction C_B^B of the fibrewise cone is given by $F = I$, with basepoint $\{0\}$, and the reduction Σ_B^B of the fibrewise suspension is given by $E = I/\dot{I}$, with basepoint $\dot{I}/\dot{I}$.

There is an important relationship between the fibrewise join of Section 1 and the fibrewise smash product, as follows. Using the coarse topology the pair $(C_B(X_0 *_B X_1), X_0 *_B X_1)$ is fibrewise homeomorphic to the pair

$$(C_B X_0 \times_B C_B X_1, C_B X_0 \times_B X_1 \cup X_0 \times_B C_B X_1).$$

Hence the fibrewise collapse

$$C_B(X_0 *_B X_1)/_B X_0 *_B X_1 = \Sigma_B(X_0 *_B X_1)$$

is fibrewise pointed homeomorphic to the fibrewise collapse

$$(C_B X_0 \times_B C_B X_1)/_B (C_B X_0 \times_B X_1 \cup X_0 \times_B C_B X_1) = (\Sigma_B X_0) \wedge_B (\Sigma_B X_1).$$

When X_0 and X_1 are fibrewise compact Hausdorff, therefore, we conclude that $\Sigma_B(X_0 *_B X_1)$ and $(\Sigma_B X_0) \wedge_B (\Sigma_B X_1)$ are equivalent, as fibrewise pointed spaces, whichever topology is used.

Bundles of pointed spaces

A fibrewise pointed space X over B is said to be *trivial* if X is fibrewise pointed homeomorphic to $B \times T$ for some pointed space T, and then a fibrewise pointed homeomorphism $\phi : X \to B \times T$ is called a *trivialization* of X. A fibrewise pointed space X over B is said to be *locally trivial* if there exists an open covering of B such that X_V is trivial over V for each member V of the covering. A locally trivial fibrewise pointed space is the simplest form of fibrewise pointed (or sectioned) fibre bundle or bundle of pointed spaces. A more sophisticated form involves a structural group G. A principal G-bundle P over B determines a functor $P_\#$ from the category of pointed G-spaces to the category of fibrewise pointed fibre bundles over B. Specifically, $P_\#$ transforms each pointed G-space A into the associated bundle $P \times_G A$ with fibre A and section $P \times_B \{a_0\}$, where a_0 is the basepoint, and similarly with pointed G-maps.

If X is a fibrewise pointed fibre bundle over B then $\alpha^* X$ is a fibrewise pointed fibre bundle over B' for each space B' and map $\alpha : B' \to B$. The triviality covering in the case of B' is just the pull-back of the triviality covering for B.

We may refer to a fibrewise pointed fibre bundle, as above, as a *bundle of pointed spaces*. The question naturally arises as to whether a bundle of (non-pointed) spaces which admits a section is then a bundle of pointed spaces. This is true when the fibre is a manifold, as shown in [31] by

Proposition 9.1 *Let B be a space and let X be a fibre bundle over B with a topological manifold A, without boundary, as fibre. If X admits a section then X (with this section) is locally trivial as a fibrewise pointed space, so that $X - sB$ is a fibre bundle over B with fibre A minus a point.*

In other words, a bundle of spaces which admits a section is equivalent as a fibrewise pointed space, to a bundle of pointed spaces. One easily reduces to the case in which $B = A$, X is the fibrewise space $A \times A$ with the projection onto the first factor, and s is the diagonal map. The proof in that case is given in Part II, Remark 11.22.

Adjoints

Finally, a word about adjoints. As we have seen in Section 1 the fibrewise mapping-space

$$\mathrm{map}_B(X, Z) = \coprod_{b \in B} \mathrm{map}(X_b, Z_b)$$

is defined, at least when $X = B \times T$ for some space T. If Z is fibrewise pointed with section $u : B \to Z$ there is an induced embedding

$$u_* : \mathrm{map}_B(X, B) \to \mathrm{map}_B(X, Z).$$

Now $\mathrm{map}_B(X, B)$ reduces to B and so we may regard $\mathrm{map}_B(X, Z)$ as a fibrewise pointed space with section u_*. When X, as well as Z, is fibrewise pointed we may consider the subspace $\mathrm{map}_B^B(X, Z)$ of pointed maps. In particular, take $X = B \times T$, where T is pointed. Then for any fibrewise pointed space Y over B a fibrewise pointed map

$$(B \times T) \wedge_B Y \to Z$$

determines a fibrewise pointed map

$$Y \to \mathrm{map}_B^B(B \times T, Z),$$

through the standard formula, and the converse holds when T is compact Hausdorff.

This relationship holds, in particular, when $T = I$, and shows that fibre-wise pointed maps of $C_B(Y)$ into Z correspond precisely to fibrewise pointed maps of Y into $P_B(Z)$, the fibrewise path-space. It also holds when T is the circle $I/\dot{I}$, and shows that fibrewise pointed maps of $\Sigma_B(Y)$ into Z correspond precisely to fibrewise pointed maps of Y into $\Omega_B(Z)$, the fibrewise loop-space.

Given a fibrewise space X over B we regard the coproduct $X \sqcup B$ as a fibrewise pointed space with section given by the second insertion, and similarly with fibrewise maps. Thus a functor is defined from the category of fibrewise spaces to the category of fibrewise pointed spaces, representing the former category as a full subcategory of the latter category.

10 Fibrewise one-point (Alexandroff) compactification

To illustrate some of the ideas we have been discussing, we give a fibrewise version of the theory of one-point (or Alexandroff) compactification. This construction, which is functorial in character, has formal properties which render it of considerable interest to fibrewise homotopy theorists. Fibrewise versions of other types of compactification can also be considered but these are less relevant to fibrewise homotopy theory.

The construction

Given a fibrewise space X over B, a fibrewise pointed space X_B^+ can be constructed as follows, and shown to be fibrewise compact. As a fibrewise pointed set X_B^+ is just the coproduct $X \sqcup B$ with section s. The fibrewise topology of X_B^+ is generated by (i) the open sets of X (so that X is embedded as an open subspace) and (ii) the subsets of the form $(X_W - K) \sqcup sW$, where W is open in B and $K \subseteq X_W$ is fibrewise compact over W.

To show that X_B^+ is fibrewise compact, let $\mathcal{U}$ be a covering of the fibre over the point b of B by open sets of X_B^+. Since $s(b)$ is covered, in particular, there exists a member U of $\mathcal{U}$ which contains a subset of the form $(X_W - K) \sqcup sW$, where W is a neighbourhood of b and $K \subseteq X_W$ is fibrewise compact over W. Now K_b is covered by members of $\mathcal{U}$. Hence W contains a neighbourhood V of b such that K_V is covered by a finite subfamily of $\mathcal{U}$. By adding U, if necessary, we obtain a finite subfamily of $\mathcal{U}$ which covers the restriction of X_B^+ to V. Therefore X_B^+ is fibrewise compact.

In view of this result we refer to X_B^+ as the *fibrewise one-point (or Alexandroff) compactification* of X. When X itself is fibrewise compact the construction reduces to the coproduct $X \sqcup B$. It is important to know when X_B^+ is fibrewise Hausdorff, and so we prove

Proposition 10.1 *Let X be fibrewise locally compact Hausdorff over B. Then the fibrewise compactification X_B^+ is fibrewise Hausdorff.*

For since X is open in X_B^+ it is sufficient to show that each point x of X_b can be separated from $s(b)$, for each point of B. Since X is fibrewise locally compact there exists a neighbourhood W of b in B and a neighbourhood U of x in X_W such that $X_W \cap \bar{U}$ is fibrewise compact over W. Then U and $(X_W - X_W \cap \bar{U}) \sqcup sW$ are disjoint neighbourhoods of x and $s(b)$, respectively. Hence X_B^+ is fibrewise Hausdorff.

Let X be a fibrewise space over B. Given a space B' and map $\alpha : B' \to B$, we can form the pull-back $\alpha^* X$ of X, as a fibrewise space over B', and then form the fibrewise compactification $(\alpha^* X)_{B'}^+$. Or we can form the pull-back $\alpha^*(X_B^+)$ of the fibrewise compactification of X. The canonical fibrewise function determines a continuous fibrewise pointed bijection

$$\theta : (\alpha^* X)_{B'}^+ \to \alpha^*(X_B^+).$$

Suppose that X is fibrewise locally compact Hausdorff. Then X_B^+ is fibrewise Hausdorff over B and so $\alpha^*(X_B^+)$ is fibrewise Hausdorff over B'. Since $(\alpha^* X)_{B'}^+$ is fibrewise compact over B' we obtain that θ is an equivalence of fibrewise pointed spaces over B'. In particular $(B \times T)_B^+$ is equivalent to $B \times T^+$, as a fibrewise pointed space over B, for locally compact Hausdorff T. For example, $(B \times \mathbb{R}^n)_B^+$ is equivalent to $B \times (\mathbb{R}^n)^+$ and hence to $B \times S^n$.

Functoriality

Let us turn now to the question of functoriality. To each fibrewise map $\phi : X \to Y$, where X and Y are fibrewise spaces over B, there corresponds a fibrewise pointed function $\phi^+ : X_B^+ \to Y_B^+$, given by ϕ on X and fibrewise constant on $X_B^+ - X$. Suppose that ϕ is proper. Then the preimage under ϕ^+ of the fibrewise subbasic open set $(Y_W - L) \sqcup tW$ of Y_B^+, where W is open in B and L is fibrewise compact over W, is the fibrewise subbasic open set $(X_W - \phi^{-1}L) \sqcup sW$ of X_B^+. It follows that ϕ^+ is continuous. Thus we may regard fibrewise one-point compactification as a functor from the category of fibrewise spaces and proper fibrewise maps to the category of fibrewise pointed spaces and fibrewise pointed maps.

Proposition 10.2 *Let X be fibrewise locally compact Hausdorff over B, and let A be a fibrewise compact subspace of X. Then there exists a natural equivalence*

$$(X/_B A)^+_B \to X^+_B /_B A^+_B$$

of fibrewise pointed spaces.

Since A is fibrewise compact the natural projection $X \to X/_B A$ is proper and so determines a fibrewise pointed map

$$\theta : X^+_B \to (X/_B A)^+_B.$$

Since θ is fibrewise constant on A^+_B we have an induced fibrewise pointed map

$$\theta' : X^+_B /_B A^+_B \to (X/_B A)^+_B$$

which is clearly bijective. The domain here is fibrewise compact, since X^+_B is fibrewise compact. Also $X/_B A$ is fibrewise locally compact Hausdorff and so $(X/_B A)$ is fibrewise Hausdorff. Therefore θ' is closed and so an equivalence of fibrewise pointed spaces.

Proposition 10.3 *Let X and Y be fibrewise locally compact Hausdorff over B. Then there exist natural equivalences*

$$X^+_B \vee_B Y^+_B \to (X \sqcup_B Y)^+_B, \quad X^+_B \wedge_B Y^+_B \to (X \times_B Y)^+_B$$

of fibrewise pointed spaces.

To prove the first part of Proposition 10.3 consider the standard insertions

$$X \to X \sqcup_B Y \leftarrow Y.$$

These are both proper and so they induce fibrewise pointed maps

$$X^+_B \to (X \sqcup_B Y)^+_B \leftarrow Y^+_B.$$

The fibrewise coproduct

$$X^+_B \vee_B Y^+_B \to (X \sqcup_B Y)^+_B$$

of these fibrewise pointed maps is a continuous bijection and hence is an equivalence of fibrewise pointed spaces, since $X^+_B \vee_B Y^+_B$ is fibrewise compact and $(X \sqcup_B Y)^+_B$ is fibrewise Hausdorff.

To prove the second part, consider the fibrewise surjection

$$\xi : X^+_B \times_B Y^+_B \to (X \times_B Y)^+_B,$$

which is given by the identity on $X \times_B Y$ and by the fibrewise constant on the complement $X^+_B \vee Y^+_B$. The domain of ξ is fibrewise compact Hausdorff, since X^+_B and Y^+_B are fibrewise compact Hausdorff. Also the codomain of ξ is fibrewise compact Hausdorff, since $X \times_B Y$ is fibrewise locally compact Hausdorff. Now the preimages of the open sets of $X \times_B Y$ in the codomain are open sets of $X \times_B Y$ in the domain, therefore open in the domain. Also if W is open in B and L is a fibrewise compact subset of $X_W \times_W Y_W$ in the codomain then $\xi^{-1}L$ is a fibrewise compact subset of $X_W \times_W Y_W$ in the domain. Therefore ξ is continuous and so induces a continuous fibrewise bijection

$$\eta : X_B^+ \wedge Y_B^+ \to (X \times_B Y)_B^+.$$

But the domain of η is fibrewise compact and the codomain is fibrewise Hausdorff, hence η is an equivalence.

Finally, let us consider fibrewise one-point compactification from the equivariant point of view. Let X be fibrewise Hausdorff over B and let G be a fibrewise compact Hausdorff fibrewise group acting fibrewise on X. Now $G \times_B X$ is open in $G \times_B X_B^+$, since X is open in X_B^+, also the action is proper, since G is fibrewise compact. It follows that we can extend the fibrewise action of G to X_B^+ so that points of the canonical section are left fixed. We prove

Proposition 10.4 *Let X be fibrewise locally compact Hausdorff over B. Let G be a fibrewise compact Hausdorff group acting fibrewise on X. Then $(X_B^+)/G$ is equivalent to $(X/G)_B^+$, as a fibrewise pointed space.*

For since G is fibrewise compact the natural projection $\pi : X \to X/G$ is proper and so determines a fibrewise pointed map

$$\pi^+ : X_B^+ \to (X/G)_B^+.$$

Also π^+ is invariant with respect to the action, since π is invariant, and so π^+ induces a fibrewise map

$$\rho : X_B^+/G \to (X/G)_B^+.$$

By inspection ρ is bijective. Now X_B^+ is fibrewise compact, by construction, and so X_B^+/G is fibrewise compact. Also X/G is fibrewise locally compact Hausdorff. Therefore $(X/G)_B^+$ is fibrewise Hausdorff and so ρ is an equivalence.

11 Fibrewise pointed homotopy

Basic notions

Fibrewise pointed homotopy is an equivalence relation between fibrewise pointed maps. Specifically, let $\theta, \phi : X \to Y$ be fibrewise pointed maps, where X and Y are fibrewise pointed spaces over B. A *fibrewise pointed homotopy* of θ into ϕ is a homotopy $f_t : X \to Y$ of θ into ϕ which is fibrewise pointed for all $t \in I = [0, 1]$.

If there exists a fibrewise pointed homotopy of θ into ϕ we say that θ is *fibrewise pointed homotopic* to ϕ and write $\theta \simeq^B_B \phi$. In this way an equivalence relation is defined on the set of fibrewise pointed maps of X into Y, and the pointed set of equivalence classes is denoted by $\pi^B_B[X; Y]$. Formally, π^B_B constitutes a binary functor from the category of fibrewise pointed spaces to the category of pointed sets, contravariant in the first entry and covariant in the second.

The operation of composition for fibrewise pointed maps induces a function

$$\pi^B_B[Y; Z] \times \pi^B_B[X; Y] \to \pi^B_B[X; Z],$$

for any fibrewise pointed spaces X, Y, Z over B. Moreover, there are natural equivalences between $\pi^B_B[X \vee_B Y; Z]$ and $\pi^B_B[X; Z] \times \pi^B_B[Y; Z]$, and between $\pi^B_B[X; Y \times_B Z]$ and $\pi^B_B[X; Y] \times \pi^B_B[X; Z]$.

The fibrewise pointed map $\phi : X \to Y$ is called a *fibrewise pointed homotopy equivalence* if there exists a fibrewise pointed map $\psi : Y \to X$ such that

$$\psi \circ \phi \simeq^B_B 1_X, \ \phi \circ \psi \simeq^B_B 1_Y.$$

Thus an equivalence relation is defined; the equivalence classes are called *fibrewise pointed homotopy types*.

It should be appreciated that the fibrewise pointed homotopy type of a fibrewise space which admits a section depends very much on the choice of section. For example the torus, fibred over the circle using the first projection, has an infinite number of fibrewise pointed homotopy types, depending on the choice of section.

Let $p : E \to X$ be a fibrewise pointed map, where E and X are fibrewise pointed spaces over B. Sections of E over X, in this context, are required to be fibrewise pointed maps, and similarly with vertical homotopies.

Initially, the pointed theory is similar to the non-pointed theory, as in Section 3. For example, fibrewise pointed maps $\phi : X \to Y$, where X and Y are fibrewise pointed spaces over B, correspond precisely to sections of the fibrewise product $X \times_B Y$, regarded as a fibrewise pointed space over X. Similarly, fibrewise pointed homotopy classes of fibrewise pointed maps correspond to vertical homotopy classes of sections.

A fibrewise pointed homotopy into the fibrewise constant is called a *fibrewise pointed null-homotopy*. A fibrewise pointed space is said to be *fibrewise pointed contractible* if it has the same fibrewise pointed homotopy type as the base space, in other words if the identity is fibrewise pointed null-homotopic.

For example, consider the reduced fibrewise cone $C^B_B(X)$ on the fibrewise pointed space X. The fibrewise contraction of $C_B(X)$ given in Section 4 induces a fibrewise pointed contraction of $C^B_B(X)$. Similarly, the fibrewise path-space $P_B(X)$ is fibrewise pointed contractible. A subset U of the fibrewise pointed space X is said to be *fibrewise pointed categorical* if the inclusion $U \to X$ is fibrewise pointed null-homotopic.

Let $\phi : X \to Y$ and $\psi : Y \to X$ be fibrewise pointed maps such that $\psi \circ \phi \simeq^B_B 1_X$. Then ψ is said to be a *left inverse* of ϕ, up to fibrewise pointed homotopy, and ϕ to be a *right inverse* of ψ, up to fibrewise pointed homotopy. Note that if ϕ admits both a left inverse ψ and a right inverse ψ', up to fibrewise pointed homotopy, then $\psi \simeq^B_B \psi'$ and so ϕ is a fibrewise pointed homotopy equivalence.

Suppose that X and Y are fibrewise spaces over B, with closed subspaces $X' \subseteq X$ and $Y' \subseteq Y$. Fibrewise maps $(X, X') \to (Y, Y')$ of the pair induce fibrewise pointed maps $X/_B X' \to Y/_B Y'$, after fibrewise collapsing, and similarly with fibrewise homotopies. Hence fibrewise homotopy equivalences of the pair induce fibrewise pointed homotopy equivalences.

Of course the associated bundle functor $P_\#$ defined in Section 9 transforms pointed G-homotopy classes of pointed G-maps into fibrewise pointed homotopy classes of fibrewise pointed maps, for each principal G-bundle P over B.

Fibrewise multiplications

Given a fibrewise pointed space X over B, a fibrewise pointed map $m : X \times_B X \to X$ is called a *fibrewise multiplication*. We describe m as *fibrewise homotopy-commutative* if

$$m \simeq^B_B m \circ t : X \times_B X \to X,$$

where $t : X \times_B X \to X \times_B X$ switches factors. We describe m as *fibrewise homotopy-associative* if

$$m \circ (m \times 1) \simeq^B_B m \circ (1 \times m) : X \times_B X \times_B X \to X.$$

By a *fibrewise Hopf structure* on X we mean a fibrewise multiplication m such that

$$m \circ (1 \times c) \circ \Delta \simeq^B_B 1_X \simeq^B_B m \circ (c \times 1) \circ \Delta,$$

where c denotes the fibrewise constant map, as shown below

$$X \xrightarrow{\;\Delta\;} X \times_B X \underset{c \times 1}{\overset{1 \times c}{\longrightarrow}} X \times_B X \xrightarrow{\;m\;} X.$$

Given such a structure we describe X as a *fibrewise Hopf space*. Of course, the associated bundle functor $P_\#$ mentioned earlier transforms Hopf G-spaces into fibrewise Hopf spaces over B for each principal G-bundle P over B.

Sectionable fibrewise spaces may admit fibrewise Hopf structure with one choice of section but not with another. Thus consider the product $S \times S$, for some (pointed) S. If S is a Hopf space, wth basepoint e, then $S \times S$ is a fibrewise Hopf space with axial section $S \times \{e\}$, as we have just observed. Suppose, however, that we replace the axial section by the diagonal. Then a fibrewise Hopf structure on $S \times S$ would determine a map of $S \times S \times S$ into S which satisfies the Hopf condition on $S \times S \times \{e\}$ and maps the diagonal of $S \times S \times S$ identically onto S. When S is a sphere (say $S = S^1$) this is a homological impossibility.

A fibrewise homotopy right inverse for a fibrewise multiplication m on X is a fibrewise pointed map $u : X \to X$ such that the composition

$$X \xrightarrow{\Delta} X \times_B X \xrightarrow{1 \times u} X \times_B X \xrightarrow{m} X$$

is fibrewise pointed null-homotopic. Fibrewise homotopy left inverses are defined similarly. When m is fibrewise homotopy-associative a fibrewise homotopy right inverse is also a fibrewise homotopy left inverse, and the term fibrewise homotopy inverse may be used.

A fibrewise homotopy-associative fibrewise Hopf space for which the fibrewise multiplication admits a fibrewise homotopy inverse is called a *fibrewise group-like space*. For example the topological product $B \times T$ is fibrewise group-like for each group-like space T. Again, the fibrewise loop-space $\Omega_B(Z)$ of a fibrewise pointed space Z is fibrewise group-like.

A fibrewise multiplication on the fibrewise pointed space Y over B determines a multiplication on the pointed set $\pi_B^B[X; Y]$ for all fibrewise pointed spaces X. If the former is fibrewise homotopy-commutative then the latter is commutative, and similarly with the other conditions we have mentioned. Thus $\pi_B^B[X; Y]$ is a group if Y is fibrewise group-like.

Fibrewise comultiplications

This is an area where the formal duality in homotopy theory operates satisfactorily. Thus, given a fibrewise pointed space X over B, a fibrewise pointed map $m : X \to X \vee_B X$ is called a *fibrewise comultiplication*. We describe m as *fibrewise homotopy-commutative* if

$$m \simeq_B^B t \circ m : X \to X \vee_B X,$$

where $t : X \vee_B X \to X \vee_B X$ switches factors. We describe m as *fibrewise homotopy-associative* if

$$(m \vee 1) \circ m \simeq_B^B (1 \vee m) \circ m : X \to X \vee_B X \vee_B X.$$

By a *fibrewise coHopf structure* on X we mean a fibrewise comultiplication m such that

$$\nabla \circ (1 \vee c) \circ m \simeq_B^B 1_X \simeq_B^B \nabla \circ (c \vee 1) \circ m,$$

where

$$X \xrightarrow{m} X \vee_B X \xrightarrow[c \vee 1]{1 \vee c} X \vee_B X \xrightarrow{\nabla} X.$$

Given such a structure we describe X as a *fibrewise coHopf space*. Of course, the associated bundle functor $P_\#$ mentioned earlier transforms coHopf G-spaces into fibrewise coHopf spaces for each principal G-bundle P over B. Sunderland [130] has studied the problem of the existence of fibrewise coHopf structures in the case of fibrewise pointed sphere-bundles over a given base.

A fibrewise homotopy right inverse for a fibrewise comultiplication m is a fibrewise pointed map $u : X \to X$ such that the composition

$$X \xrightarrow{\ m\ } X \vee_B X \xrightarrow{1 \vee u} X \vee_B X \xrightarrow{\ \nabla\ } X$$

is fibrewise pointed null-homotopic. Fibrewise homotopy left inverses are defined similarly. When m is fibrewise homotopy-associative a fibrewise homotopy right inverse is always a fibrewise homotopy left inverse, and the term fibrewise homotopy inverse may be used.

A fibrewise homotopy-associative fibrewise coHopf space for which the fibrewise comultiplication admits a fibrewise homotopy inverse is called a *fibrewise cogroup-like space*. For example, the topological product $B \times T$ is fibrewise cogroup-like for each cogroup-like space T.

A fibrewise comultiplication on the fibrewise pointed space X over B determines a multiplication on the pointed set $\pi_B^B[X; Y]$ for all fibrewise pointed spaces Y. If the former is fibrewise homotopy-commutative then the latter is commutative, and similarly with the other conditions we have mentioned. Thus $\pi_B^B[X; Y]$ is a group if X is fibrewise cogroup-like.

If X is a fibrewise coHopf space and Y is a fibrewise Hopf space then the multiplication on $\pi_B^B[X; Y]$ determined by the fibrewise comultiplication on X coincides with the multiplication determined by the fibrewise multiplication on Y. Furthermore, the multiplication is both commutative and associative.

By the distributive law for the fibrewise smash product, a fibrewise comultiplication on X determines a fibrewise comultiplication on $X \wedge_B Y$ for all fibrewise pointed spaces Y. If the former is fibrewise homotopy-commutative then so is the latter, and similarly with the other conditions. Thus $X \wedge_B Y$ is fibrewise cogroup-like if X is fibrewise cogroup-like.

For example, take $X = S^1 \times B$, which is fibrewise cogroup-like since S^1 is cogroup-like. We see that the reduced fibrewise suspension $\Sigma_B^B(Y)$ is fibrewise cogroup-like for all fibrewise pointed spaces Y.

A fibrewise group-like space G may be described as a fibrewise topological group up to fibrewise pointed homotopy. In a similar manner we may describe a fibrewise pointed map

$$r : E \times_B G \to E,$$

for any fibrewise pointed space E, as a fibrewise pointed action up to fibrewise pointed homotopy if the following two conditions are satisfied. First

$$r \circ (r \times 1) \simeq_B^B r \circ (1 \times m) : E \times_B G \times_B G \to E,$$

where m denotes the fibrewise multiplication on G. Second

$$r \circ (1 \times c) \circ \Delta : E \to E,$$

where $c : E \to G$ is fibrewise constant. For example, let Z be a fibrewise pointed space. Juxtaposition of fibrewise paths defines a fibrewise action, up to fibrewise pointed homotopy, of the fibrewise loop-space $\Omega_B(Z)$ on the fibrewise path-space $P_B(Z)$.

Fibrewise fibre bundles over a (reduced) fibrewise suspension can, of course, be classified by the general method, but there is also a direct approach as follows. Consider a fibrewise open group G. Let Y be a fibrewise pointed space with projection $q : Y \to B$. Each section $\sigma : B \to G$ transforms a fibrewise pointed map $\phi : Y \to G$ into a fibrewise pointed map $\phi' : Y \to G$, where

$$\phi'(y) = \sigma q(y).\phi(y).(\sigma q(y))^{-1}.$$

In this way the group $\pi_B[B; G]$ of vertical homotopy classes of sections of G acts on the group $\pi_B^B[Y; G]$ of fibrewise pointed homotopy classes of fibrewise pointed maps of Y into G.

Now given ϕ as above, we can construct a principal fibrewise G-bundle $E(\phi)$ over the reduced fibrewise suspension $X = \Sigma_B^B(Y)$ by taking two copies of the fibrewise product $C_B^B(Y) \times_B G$ and identifying $Y \times_B G$ in the first copy with $Y \times_B G$ in the second copy through the relation

$$(y, g) \sim (y, g.\phi(y)) \qquad (g \in G_b, \, y \in Y_b, \, b \in B).$$

Here $C_B^B(Y)$ denotes the reduced fibrewise cone on Y with Y embedded as a subspace in the usual way. The principal fibrewise G-bundle $E(\phi)$ over X thus constructed is obviously numerable. It is not difficult to show that every numerable fibrewise G-bundle over X is equivalent to $E(\phi)$ for some ϕ. Moreover, if ϕ and ϕ' are related as above then $E(\phi)$ is equivalent to $E(\phi')$, also the equivalence class of $E(\phi)$ depends only on the fibrewise pointed homotopy class of ϕ. Finally, if $E(\phi)$ is equivalent to $E(\psi)$, for some fibrewise pointed map ψ, then there exists some section σ such that ψ is fibrewise pointed homotopic to ϕ', where ϕ' is obtained from ϕ by transforming under the action of σ. We conclude, therefore, that the equivalence classes of numerable principal fibrewise G-bundles over $\Sigma_B^B(Y)$ correspond precisely to the factor set of $\pi_B^B[Y; G]$ with respect to the action of $\pi_B[B; G]$. When G is vertically connected, of course, the correspondence is with $\pi_B^B[Y; G]$ itself.

12 Fibrewise pointed cofibrations

Basic notions

Not much adjustment is necessary to adapt the basic theory of Section 4 to the fibrewise pointed case. Thus let A be a fibrewise pointed space over B. By a *fibrewise pointed cofibre space* under A we mean a fibrewise pointed space X together with a fibrewise pointed map $u : A \to X$ with the following fibrewise pointed homotopy extension property. Let E be a fibrewise pointed space, let $f : X \to E$ be a fibrewise pointed map, and let $g_t : A \to E$ be a fibrewise pointed homotopy of $f \circ u$. Then there exists a fibrewise pointed homotopy $h_t : X \to E$ of f such that $g_t = h_t \circ u$.

Instead of describing X as a fibrewise pointed cofibre space under A we may say that u is a *fibrewise pointed cofibration*. An important special case is when A is a subspace of X and u is the inclusion. In that case we describe (X, A) as a *fibrewise pointed cofibred pair* when the above condition is satisfied. Note that (X, B) and (X, X) are fibrewise pointed cofibred pairs.

Of course, the fibrewise pointed map $u : A \to X$ is a fibrewise pointed cofibration if u is a fibrewise cofibration. In fact the converse holds, by Proposition 4.1, when A is a closed subspace of X and u the inclusion.

If X is a fibrewise pointed cofibre space under A the push-out $\pi_* X$ is defined, as a fibrewise pointed space over B, where π is the projection. We refer to $\pi_* X$ as the *fibrewise pointed cofibre* of the fibrewise cofibre space. When $A \subseteq X$ and u is the inclusion the fibrewise pointed cofibre is just the fibrewise collapse $X/_B A$.

The Puppe sequence

The remainder of this section is devoted to an outline of the fibrewise version of the well-known exact sequence of D. Puppe. This concerns sequences

$$X_1 \xrightarrow{f_1} X_2 \xrightarrow{f_2} X_3 \longrightarrow \cdots$$

of fibrewise pointed spaces and fibrewise pointed maps, over B. In this context we describe such a sequence as *exact* if the induced sequence

$$\pi_B^B[X_1;\, E] \xleftarrow{f_1^*} \pi_B^B[X_2;\, E] \xleftarrow{f_2^*} \pi_B^B[X_3;\, E] \longleftarrow \cdots$$

of pointed sets and pointed functions is exact, for all fibrewise pointed spaces E.

Given a fibrewise pointed map $\phi : X \to Y$, where X and Y are fibrewise pointed spaces over B, the *reduced fibrewise mapping cone* $C_B^B(\phi)$ of ϕ is defined to be the push-out of the cotriad

$$C_B^B(X) \xleftarrow{\sigma_1} X \xrightarrow{\phi} Y.$$

Now $C_B^B(\phi)$ comes equipped with a fibrewise embedding

$$\phi' : Y \to C_B^B(\phi),$$

and we easily see that the sequence

$$\pi_B^B[X;\, E] \xleftarrow{\phi^*} \pi_B^B[Y;\, E] \xleftarrow{\phi'^*} \pi_B^B[C_B^B(\phi);\, E]$$

of pointed sets is exact, for all fibrewise pointed spaces E. Obviously, the procedure can be iterated so as to obtain exact sequences of unlimited length, but that in itself is not particularly interesting.

To understand the situation better, consider the case of a fibrewise pointed cofibration $u : A \to X$, where A and X are fibrewise pointed spaces over B. We assert that the natural projection

$$C_B^B(u) \to C_B^B(u)/_B C_B^B(A) = X/_B A$$

is a fibrewise pointed homotopy equivalence.

For consider the fibrewise pointed null-homotopy $g_t : C_B^B(A) \to C_B^B(u)$ of the inclusion given by $g_t(a, s) = (a, s(1 - t))$, where $s, t \in I$ and $a \in A$. Since u is a fibrewise pointed cofibration there exists a fibrewise pointed homotopy $h_t : X \to C_B^B(u)$ of the inclusion. Now g_t and h_t together form a fibrewise pointed homotopy of the identity on $C_B^B(u)$ which deforms $C_B^B(A)$ over itself into the fibrewise constant. Hence we obtain a fibrewise pointed homotopy inverse of the natural projection, as required. Notice, incidentally, that the fibrewise pointed homotopy equivalence thus constructed transforms u' into the natural projection $X \to X/_B A$, where u' is derived from u in the way that ϕ' is derived from ϕ.

Returning to the general case, where $\phi : X \to Y$, we next show that the embedding $\phi' : Y \to C_B^B(\phi)$ is a fibrewise pointed cofibration. In fact the embedding $X \to C_B^B(X)$ is a fibrewise pointed cofibration, from first principles, and so the conclusion follows from the observation that the push-out of a fibrewise pointed cofibration is again a fibrewise pointed cofibration.

By combining these last two results we see that the reduced fibrewise mapping cone $C_B^B(\phi')$ is equivalent to the reduced fibrewise suspension

$$\Sigma_B^B(X) = C_B^B(\phi)/_B Y = C_B^B(\phi')/_B C_B^B(\phi),$$

up to fibrewise pointed homotopy equivalence. In the process, moreover, $(\phi')'$ is transformed into a fibrewise pointed map

$$\phi'' : C_B^B(\phi) \to \Sigma_B^B(X).$$

Repeating the procedure we find that $C_B^B((\phi')')$ is equivalent to the reduced fibrewise suspension $\Sigma_B^B(Y)$, in the same sense. In the process, moreover, $((\phi')')'$ is transformed into the reduced fibrewise suspension

$$\Sigma_B^B(\phi) : \Sigma_B^B(X) \to \Sigma_B^B(Y)$$

of ϕ, precomposed with the fibrewise reflection in which $(x, t) \mapsto (x, 1 - t)$. This last does not affect the exactness property and so we arrive at an exact sequence of the form

$$X \xrightarrow[\phi]{} Y \longrightarrow C_B^B(\phi) \longrightarrow \Sigma_B^B(X) \xrightarrow[\Sigma_B^B(\phi)]{} \Sigma_B^B(Y) \longrightarrow \cdots$$

When the given fibrewise pointed map ϕ is varied by a fibrewise pointed homotopy the exact sequence varies similarly. In particular, if ϕ is fibrewise pointed null-homotopic the sequence has the same fibrewise pointed homotopy type (in an obvious sense) as in the case of the fibrewise constant map, where it reduces to

$$X \to Y \to Y \vee_B \Sigma_B^B(X) \to \Sigma_B^B(X) \to \cdots$$

A special case

For example, take $B = S^n$ ($n \geq 1$). We regard $S^n \vee S^i$ ($i = 1, 2, \ldots$) as a fibrewise pointed space over S^n with projection the identity on S^n and the constant on S^i, and with section the first insertion of the coproduct. Clearly, fibrewise pointed maps of $S^n \vee S^q$ into $S^n \vee S^r$ correspond precisely to pointed maps of S^q into $S^n \vee S^r$, and similarly with homotopies. Note that the correspondence is additive, in the sense that the fibrewise track sum of fibrewise pointed maps corresponds to the track sum of the corresponding pointed maps. Now a pointed map $f : S^q \to S^r$ determines a pointed map $f' : S^q \to S^n \vee S^r$, by composition with the second insertion of the coproduct, and hence a fibrewise pointed map $f'' : S^n \vee S^q \to S^n \vee S^r$. If we replace f by its reduced suspension $S^{q+1} \to S^{r+1}$, in this process, then clearly f'' is replaced by its reduced fibrewise suspension $S^n \vee S^{q+1} \to S^n \vee S^{r+1}$.

The following remark, however, is not quite so obvious. Take $q = n + r - 1$ and consider the fibrewise pointed map

$$g : S^n \vee S^{n+r-1} \to S^n \vee S^r$$

which is given by the Whitehead product of the identity ι_n on S^n with the identity ι_r on S^r. We assert that the reduced fibrewise suspension

$$S^n \vee S^{n+r} \to S^n \vee S^{r+1}$$

of g is given by the Whitehead product of the identity ι_n on S^n with the identity ι_{r+1} on S^{r+1}. This can most easily be seen by observing that $S^n \times S^r$ can be interpreted either as the reduced fibrewise mapping cone of g or the reduced ordinary mapping cone of the first Whitehead product, while the reduced fibrewise suspension $S^n \times S^{r+1}$ of $S^n \times S^r$ can be interpreted either as the reduced fibrewise mapping cone of the reduced fibrewise suspension of g or the reduced ordinary mapping cone of the second Whitehead product.

More generally, consider the reduced fibrewise mapping cone

$$Z = e^{n+r} \cup (S^n \vee S^r),$$

where the map g is of homotopy class

$$\iota_r \circ \alpha + k[\iota_n, \iota_r],$$

for $\alpha \in \pi_{n+r-1}(S^r)$ and $k \in \mathbb{Z}$. By combining our two observations we see that the reduced fibrewise suspension

$$\Sigma_{S^n}^{S^n}(Z) = e^{n+r+1} \cup (S^n \vee S^{r+1})$$

can be constructed similarly using a map of homotopy class

$$\iota_{r+1} \circ \Sigma_* \alpha + k[\iota_n, \iota_{r+1}],$$

where Σ_* denotes the suspension homomorphism.

The conclusion we arrive at, then, is that there is an exact sequence

$$S^n \vee S^{n+r-1} \to S^n \vee S^r \to Z \to S^n \vee S^{n+r} \to S^n \vee S^{r+1} \to \Sigma_{S^n}^{S^n}(Z) \to \cdots$$

Hence if E is a fibrewise pointed space with fibrewise fibre F the sequence

$$\pi_{n+r-1}(F) \leftarrow \pi_r(F) \leftarrow \pi_{S^n}^{S^n}(Z, E) \leftarrow \pi_{n+r}(F) \leftarrow \pi_{r+1}(F) \leftarrow \cdots$$

is exact, where fibrewise homotopy groups have been replaced by ordinary homotopy groups as explained above. Further details of the structure of this exact sequence, with some calculations, can be found in [81].

13 Fibrewise pointed fibrations

Basic notions

Let X be a fibrewise pointed space over B. By a *fibrewise pointed fibre space* over X we mean a fibrewise pointed space E together with a fibrewise pointed map $p : E \to X$ with the following fibrewise pointed homotopy lifting property. Let A be a fibrewise pointed space, let $f : A \to E$ be a fibrewise pointed map, and let $g_t : A \to X$ be a fibrewise pointed homotopy such that $g_0 = p \circ f$. Then there exists a fibrewise pointed homotopy $h_t : A \to E$ of f such that $g_t = p \circ h_t$. For example, the fibrewise product $X \times_B T$ is a fibrewise pointed fibre space over X for each fibrewise pointed space T. In particular, take $X = B$; then every fibrewise pointed space over B is a fibrewise pointed fibre space.

Returning to the general case, instead of describing E as a fibrewise pointed space over X we may describe p as a *fibrewise pointed fibration*. The *fibrewise fibre* is the pull-back s^*E, where $s : B \to X$ is the section of X.

The reader may feel that the introduction of the term fibrewise pointed fibration is superfluous since it follows from Proposition 13.1 that a fibrewise pointed map is a fibrewise pointed fibration if and only if it is a fibrewise fibration. Nevertheless it seems to improve the look of the theory, if not the substance, if the term is available for use.

One can expect to find fibrewise pointed counterparts to the other results of Section 5, such as the theorems of Dold. First we have

Proposition 13.1 *Let $p : E \to X$ be a fibrewise pointed fibration, where E and X are fibrewise pointed spaces over B. Let $\theta : E \to E$ be a fibrewise pointed map over X and suppose that θ, as a fibrewise pointed map over B, is fibrewise pointed homotopic to the identity. Then there exists a fibrewise* pointed map $\theta' : E \to E$ over X such that $\theta \circ \theta'$ is fibrewise pointed homotopic to the identity.

Theorem 13.2 *Let X be a fibrewise pointed space over B, and let E and F be fibrewise pointed fibre spaces over X. Let $\phi : E \to F$ be a fibrewise pointed map over X. Suppose that ϕ, as a fibrewise pointed map over B, is a fibrewise pointed homotopy equivalence. Then ϕ is a fibrewise pointed homotopy equivalence over X.*

In the fibrewise pointed theory the terms section and vertical homotopy are used in the fibrewise pointed sense, as in

Proposition 13.3 *Let $p : E \to X$ be a fibrewise pointed fibration, where E and X are fibrewise pointed spaces over B. Let s and s' be fibrewise pointed homotopic sections of E over X. Then s and s' are vertically homotopic.*

The proofs are straightforward adaptations of the proofs of the corresponding results in Section 5. Similarly we have the fibrewise pointed counterpart of the key result about induced fibrewise fibrations, as follows.

Theorem 13.4 *Let $p : E \to X$ be a fibrewise pointed fibration, where E and X are fibrewise pointed spaces over B. Let $\theta, \phi : X' \to X$ be fibrewise pointed homotopic fibrewise pointed maps, where X' is a fibrewise pointed space over B. Then $\theta^* E$ and $\phi^* E$ have the same fibrewise pointed homotopy type over X'.*

Corollary 13.5 *Let $p : E \to X$ be a fibrewise pointed fibration, where E and X are fibrewise pointed spaces over B. If X is fibrewise pointed contractible over B then E has the same fibrewise pointed homotopy type over X as the fibrewise product $X \times_B T$ for some fibrewise pointed space T over B.*

The Nomura sequence

As Eckmann and Hilton [56] have pointed out, this is an area where the concept of duality between pointed fibrations and pointed cofibrations operates successfully, and this is also true in the fibrewise theory. For example, let us dualize the fibrewise version of the exact sequence of Section 12, in other words obtain a fibrewise version of the Nomura sequence.

Suppose that we have a sequence

$$\cdots \longrightarrow X_3 \xrightarrow{\ f_2\ } X_2 \xrightarrow{\ f_1\ } X_1$$

of fibrewise pointed spaces and fibrewise pointed maps. In this context let us describe such a sequence as *exact* if the induced sequence

$$\cdots \to \pi_B^B[A; X_3] \xrightarrow{\ f_{2*}\ } \pi_B^B[A; X_2] \xrightarrow{\ f_{1*}\ } \pi_B^B[A; X_1]$$

of pointed sets and pointed functions is exact, for all fibrewise pointed spaces A.

For any fibrewise pointed space X over B the *fibrewise path-space* $P_B X$ is defined as

$$P_B(X) = \mathrm{map}_B^B(B \times I, X).$$

Thus the fibre of $P_B X$ over the point b of B is just the space of paths in the fibre X_b originating at the basepoint given by the section. The fibrewise pointed map $P_B X \to X$ given by the ends of the paths is a fibrewise pointed fibration. Note that if X is a fibrewise pointed bundle over B with fibre X_0 then $P_B X$ is a fibrewise pointed bundle over B with fibre $P X_0$, the ordinary space of based paths.

Given a fibrewise pointed map $\phi : Y \to X$, where X and Y are fibrewise pointed spaces, the *fibrewise mapping fibre* (or *fibrewise homotopy-fibre*) $F_B(\phi)$ of ϕ is defined as the pull-back $\phi^* P_B(X)$. In other words, it is the fibrewise fibre of the fibrewise fibration $W_B(\phi) \to X$ (Proposition 5.2). The fibrewise mapping fibre is a fibrewise pointed space and comes equipped with a fibrewise pointed map

$$\phi' : F_B(\phi) \to Y.$$

It is easy to see that the sequence

$$\pi_B^B[A; F_B(\phi)] \xrightarrow{\phi'_*} \pi_B^B[A; Y] \xrightarrow{\phi_*} \pi_B^B[A; X]$$

of pointed sets and pointed functions is exact, for all fibrewise pointed spaces A. Obviously, the procedure can be iterated so as to obtain exact sequences of unlimited length but that in itself is not particularly interesting.

To understand the situation better consider the case of a fibrewise pointed fibration $p : E \to X$, where E and X are fibrewise pointed spaces over B. The fibrewise fibre F is included as a subspace of the fibrewise mapping fibre $F_B(p)$. We assert that the inclusion is a fibrewise pointed homotopy equivalence. For consider the fibrewise pointed null-homotopy $g_t : F_B(p) \to P_B(X)$ of the second projection, given by

$$g_t(\xi, \lambda)(s) = \lambda(s(1-t)).$$

Since p is a fibrewise pointed fibration there exists a fibrewise pointed homotopy $h_t : F_B(p) \to E$ of the first projection. Now g_t and h_t together form a fibrewise pointed deformation of $F_B(p)$ into F, keeping F fixed. Therefore $F_B(p)$ and F have the same fibrewise pointed homotopy type as asserted. In fact we have described a rather specific construction for the fibrewise pointed homotopy equivalence, and it is this construction which we shall be using in what follows.

Returning to the general case, where $\phi : Y \to X$, we next show that the fibrewise pointed map $\phi' : F_B(\phi) \to Y$ defined earlier is a fibrewise pointed fibration. In fact since the evaluation $P_B(X) \to X$ is a fibrewise pointed fibration this follows at once from the Cartesian property of pull-backs.

By combining these last two results we see that the fibrewise mapping fibre $F_B(\phi')$ has the same fibrewise pointed homotopy type as the fibrewise loop-space $\Omega_B(Y)$. With the specific construction described above the process transforms $(\phi')'$ into a fibrewise pointed map

$$\phi'' : \Omega_B(X) \to F_B(\phi).$$

Repeating the procedure we find that $F_B((\phi')')$ has the same fibrewise pointed homotopy type as $\Omega_B(X)$, and in the process $((\phi')')'$ is transformed into

$$\Omega_B(\phi) : \Omega_B(Y) \to \Omega_B(X),$$

apart from a reflection $t \mapsto 1-t$ in the domain of the fibrewise loops. This last does not affect the exactness property and so we arrive at an exact sequence of the form

$$\cdots \to \Omega_B(Y) \xrightarrow[\Omega_B(\phi)]{} \Omega_B(X) \longrightarrow F_B(\phi) \longrightarrow Y \xrightarrow[\phi]{} X.$$

Now let us return to the situation where E is a fibrewise pointed fibre space over X. Then a fibrewise action (up to fibrewise pointed homotopy) of the fibrewise loop-space $\Omega_B(X)$ on the fibrewise fibre F can be constructed as follows. Let f denote the composition

$$F \times_B \Omega_B(X) \xrightarrow[\pi_1]{} F \xrightarrow[j]{} E,$$

where j is the inclusion, and let g_t denote the composition

$$F \times_B \Omega_B(X) \xrightarrow[\pi_2]{} \Omega_B(X) \xrightarrow[\rho_t]{} X,$$

where ρ_t is the evaluation. Then $p \circ f = g_0$, the fibrewise constant, and so we may lift g_t to a fibrewise pointed homotopy

$$h_t : F \times_B \Omega_B(X) \to E$$

of f into a fibrewise pointed map r, as required. It is not difficult to show that the fibrewise pointed homotopy class of r is independent of the choice of lifting. Moreover, r satisfies the requirements for a fibrewise action, up to fibrewise pointed homotopy, so that for each fibrewise pointed space A the group $\pi_B^B[A; \Omega_B(X)]$ acts on the pointed set $\pi_B^B[A; F]$ in a way which depends only on the fibrewise pointed fibration.

Using the action thus defined we can say rather more about the properties of the exact sequence in the case of the fibrewise pointed fibration $p : E \to X$. In the first place the hitherto unidentified pointed function in the sequence is induced by the fibrewise pointed map $s : \Omega_B(X) \to F$, where s is given in terms of r by the composition

$$\Omega_B(X) \xrightarrow[\sigma_2]{} F \times_B \Omega_B(X) \xrightarrow[r]{} F.$$

Specifically, the exact sequence takes the form

$$\cdots \to \pi_B^B[A;\ \Omega_B(E)] \xrightarrow[\Omega(p)_*]{} \pi_B^B[A;\ \Omega_B(X)] \xrightarrow[s_*]{}$$

$$\pi_B^B[A;\ F] \xrightarrow[j_*]{} \pi_B^B[A;\ E] \xrightarrow[p_*]{} \pi_B^B[A;\ X].$$

The terms of the sequence are groups, linked by homomorphisms, until we reach the last three. Even there certain algebraic properties still hold, as stated in

Proposition 13.6 *In the above sequence:*
(i) $s_*\beta_1 = s_*\beta_2$, *where* $\beta_1, \beta_2 \in \pi_B^B[A;\ \Omega_B(X)]$, *if and only if* $\beta_1 - \beta_2 \in (\Omega p)_* \pi_B^B[A;\ \Omega_B(E)]$;
(ii) $j_*\xi_1 = j_*\xi_2$, *where* $\xi_1, \xi_2 \in \pi_B^B[A;\ F]$, *if and only if* $\xi_1 = \xi_2.\beta$ *for some* $\beta \in \pi_B^B[A;\ \Omega_B(X)]$.

Here the dot refers to the action of $\pi_B^B[A;\ \Omega_B(X)]$ on $\pi_B^B[A;\ F]$ described above. For the proofs of (i) and (ii), which are quite elementary, we just dualize the proofs of the corresponding results for cofibrations, as given in [119], and then make a fibrewise version in a routine fashion.

An application

Proposition 13.7 *Let* $p : E \to X$ *be a fibrewise pointed fibration such that the fibrewise fibre* F *is fibrewise pointed contractible in* E. *Then* $F \times_B \Omega_B(E)$ *has the same fibrewise pointed homotopy type as* $\Omega_B(X)$.

It follows, of course, that F is a fibrewise Hopf space; in fact this will emerge at an early stage in the proof, which is in several steps.

First let $j_t : F \to E$ be a fibrewise pointed homotopy of the fibrewise constant c into the inclusion j. Then a fibrewise pointed map $d : F \to \Omega_B(X)$ is given by the right adjoint of $p \circ j_t$. We assert that the composition

$$F \times_B \Omega_B(E) \xrightarrow[d \times \Omega(p)]{} \Omega_B(X) \times_B \Omega_B(X) \xrightarrow{m} \Omega_B(X)$$

is a fibrewise pointed homotopy equivalence. Here m, of course, is the fibrewise multiplication given by juxtaposition of loops.

Before proving this let us show that s is a left inverse of d, up to fibrewise pointed homotopy. For let $\nu_t : F \to E$ be the composition

$$F \xrightarrow[\sigma_2 \circ d]{} F \times_B \Omega_B(X) \xrightarrow[h_t]{} E,$$

where h_t is the fibrewise pointed homotopy used in the definition of r. Then $\nu_0 = j_0$ and $p \circ \nu_t = g_t \circ \sigma_2 \circ d = p \circ j_t$. Now j_1 induces the identity on F while ν_1 induces $s \circ d$. Hence $s \circ d$ is fibrewise pointed homotopic to the identity, from the fibrewise pointed homotopy lifting condition. At this stage we can already deduce that F is a fibrewise Hopf space, since $\Omega_B(X)$ is a fibrewise Hopf space.

For the next step we consider, for various fibrewise pointed A, the exact sequence

$$\cdots \to \pi_B^B[A;\ \Omega_B(E)] \xrightarrow{(\Omega p)_*} \pi_B^B[A;\ \Omega_B(X)] \xrightarrow{s_*}$$

$$\pi_B^B[A;\ F] \xrightarrow{j_*} \pi_B^B[A;\ E] \xrightarrow{p_*} \pi_B^B[A;\ X].$$

First take A to be $\Omega_B(X)$. Since $s_*\{1\} = s_*\{d \circ s\}$, by what we have just proved, there exists by Proposition 13.6(i) a fibrewise pointed map $k : \Omega_B(X) \to \Omega_B(E)$ such that the identity on $\Omega_B(X)$ is fibrewise pointed homotopic to $(d \circ s).(\Omega(p) \circ k)$. Here, as before, the dot denotes the operation determined by r. Now let $\ell : \Omega_B(X) \to F \times_B \Omega_B(E)$ be the fibrewise pointed map with components $\pi_1 \circ \ell = s$ and $\pi_2 \circ \ell = k$. We assert that ℓ is an inverse of $m \circ (d \times \Omega(p))$, up to fibrewise pointed homotopy.

Clearly, ℓ is an inverse on the right, since

$$m \circ (d \times \Omega(p)) \circ \ell = (d \circ s).(\Omega(p) \circ k),$$

which is fibrewise pointed homotopic to the identity. So it remains to be shown that

$$\ell \circ m \circ (d \times \Omega(p)) : F \times_B \Omega_B(E) \to F \times_B \Omega_B(E)$$

is fibrewise pointed homotopic to the identity. Since $\pi_1 \circ \ell = s$ and $\pi_2 \circ \ell = k$, by definition of ℓ, what we have to prove is first that

$$s \circ m \circ (d \times \Omega(p)) : F \times_B \Omega_B(E) \to F$$

is equivalent to π_1, up to fibrewise pointed homotopy, and secondly that $k \circ m \circ (d \times \Omega(p)) : F \times_B \Omega_B(E) \to \Omega_B(E)$ is equivalent to π_2, in the same sense.

To establish the first relation observe that $m \circ (d \times \Omega(p))$ is fibrewise pointed homotopic to $(d \circ \pi_1).(\Omega(p) \circ \pi_2)$. Thus the elements

$$\{m \circ (d \times \Omega(p))\}, \{d \circ \pi_1\} \in \pi_B^B[F \times_B \Omega_B(E);\ \Omega_B(X)]$$

differ by the action of an element of

$$(\Omega p)_* \pi_B^B[F \times_B \Omega_B(E);\ \Omega_B(E)].$$

By Proposition 13.6(ii), therefore, the elements have the same image under

$$s_* : \pi_B^B[F \times_B \Omega_B(E);\ \Omega_B(X)] \to \pi_B^B[F \times_B \Omega_B(E);\ F].$$

In other words $s \circ m \circ (d \times \Omega(p))$ is fibrewise pointed homotopic to $s \circ d \circ \pi_1$ and hence to π_1, as asserted.

Now s is a left inverse of d, up to fibrewise pointed homotopy, as we have seen, and so $\Omega(s)$ is a left inverse of $\Omega(d)$, in the same sense. Hence the homomorphism

$$(\Omega s)_* : \pi_B^B[A;\ \Omega_B(\Omega_B(X))] \to \pi_B^B[A;\ \Omega_B(F)]$$

is surjective, for all A, and so by exactness the homomorphism

$$(\Omega p)_* : \pi_B^B[A; \ \Omega_B(E)] \to \pi_B^B[A; \ \Omega_B(X)]$$

is injective. Taking A to be $F \times_B \Omega_B(E)$ we see that the second relation will follow if we can show that $\Omega(p) \circ k \circ m \circ (d \times \Omega(p))$ is fibrewise pointed homotopic to $\Omega(p) \circ \pi_2$. In fact $m \circ (d \times \Omega(p))$ is fibrewise pointed homotopic to

$$(d \circ s \circ m \circ (d \times \Omega(p))).(\Omega(p) \circ k \circ m \circ (d \times \Omega(p)))$$

from the relation by which k is defined, and hence to

$$(d \circ \pi_1).(\Omega(p) \circ k \circ m \circ (d \times \Omega(p))),$$

by the relation we have just proved. But $m \circ (d \times \Omega(p))$ is fibrewise pointed homotopic to $(d \circ \pi_1).(\Omega(p) \circ \pi_2)$, and so we conclude that $\Omega(p) \circ k \circ m \circ (d \times \Omega(p))$ is fibrewise pointed homotopic to $\Omega(p) \circ \pi_2$, as required. This completes the proof of Proposition 13.7.

The argument given here is essentially just a fibrewise version of that given, with more detail, by Eckmann and Hilton [56] in the ordinary theory. As they observe, the dual result is also noteworthy. In the fibrewise version of the dual we begin with a fibrewise pointed space A and a fibrewise pointed cofibre space X under A with fibrewise cofibre Y. If the natural projection $X \to Y$ is fibrewise pointed null-homotopic the conclusion is that $\Sigma_B^B(A)$ has the same fibrewise pointed homotopy type as $Y \vee_B \Sigma_B^B(X)$, in particular Y is a fibrewise coHopf space.

14 Numerable coverings (continued)

Some pointed versions of earlier results

Recall that in the case of a fibrewise pointed space the members of a covering are required to contain the section. That implies, of course, that a point-finite, *a fortiori* locally finite, covering has to be finite. Hence numerable coverings are necessarily finite, in the fibrewise pointed case.

As we have already seen there are fibrewise pointed versions of many of our earlier results. Here we reconsider those stated in Section 6 from this point of view. First we take Dold's theorem (Theorem 6.2) of which the fibrewise pointed version is

Theorem 14.1 *Let $\phi : X \to Y$ be a fibrewise pointed map, where X and Y are fibrewise pointed spaces over B. Suppose that B admits a numerable covering such that the restriction $\phi_V : X_V \to Y_V$ is a fibrewise pointed homotopy equivalence over V for each member V of the covering. Then ϕ is a fibrewise pointed homotopy equivalence over B.*

Essentially this modification of Dold's theorem is due to Eggar [58] (see also [59] and [60]). Although Eggar, in his version, assumes that B is paracompact there is a result of Mather [101] which enables this restriction to be dispensed with. (The relevant result is quoted, more conveniently, in [46], p. 354.)

Corollary 14.2 *Let $p : E \to X$ and $q : F \to X$ be fibrewise pointed fibrations, where X is a fibrewise pointed space over B. Let $\phi : E \to F$ be a fibrewise pointed map such that $q \circ \phi = p$. Suppose that the pull-back*

$$s^* \phi : s^* E \to s^* F$$

is a fibrewise pointed homotopy equivalence over B, where s is the section of X. Also suppose that X admits a numerable fibrewise pointed categorical covering. Then ϕ is a fibrewise pointed homotopy equivalence over X.

Since each member V of the covering is fibrewise pointed categorical it follows from Corollary 13.5 that E_V and F_V are fibrewise pointed trivial over V. Hence ϕ_V is a fibrewise pointed homotopy equivalence over V, since the inclusion is fibrewise pointed homotopic to the fibrewise constant. Now Corollary 14.2 follows at once from Theorem 14.1.

Corollary 14.3 *Let E be a fibrewise pointed space and let $p : E \to X$ be a fibrewise pointed map. Suppose that the restriction $p^{-1}V \to V$ is a fibrewise pointed fibration for each member V of a numerable covering of X. Then p is a fibrewise pointed fibration.*

This implies, of course, that numerable fibrewise pointed fibre bundles are fibrewise pointed fibrations. The deduction of Corollary 14.3 from Theorem 14.1 is straightforward.

We now turn to some of the results of tom Dieck [42], which we state without proofs.

Theorem 14.4 *Let $\phi : E \to F$ be a fibrewise pointed map, where E and F are fibrewise pointed spaces over B. Let $\{U_j\}$ and $\{V_j\}$ be similarly indexed numerable coverings of E and F, which are closed under intersections. Assume that $\phi U_j \subseteq V_j$ for each index j, and that each of the fibrewise pointed maps $U_j \to V_j$ determined by ϕ is a fibrewise pointed homotopy equivalence. Then ϕ is a fibrewise pointed homotopy equivalence.*

Theorem 14.5 *Let $p : E \to X$ be a fibrewise pointed map, where E and X are fibrewise pointed spaces over B. Let $\{X_j\}$ be a numerable covering of X and let $\{E_j\}$ be a similarly indexed family of subsets of E, both families being closed under intersections. Assume that $pE_j \subseteq X_j$ and that E_j is fibrewise pointed contractible over X_j, for each index j. Then E admits a section over X.*

From Theorem 14.5 we can deduce another result about fibrewise pointed fibrations, which it is interesting to compare with Corollary 14.3.

Theorem 14.6 *Let $p : E \to X$ be a fibrewise pointed map, where E and X are fibrewise pointed spaces over B. Let $\{E_j\}$ be a numerable covering of E which is closed under intersections. Assume that the restriction $p_j : E_j \to X$ is a fibrewise pointed fibration for each index j. Then p is a fibrewise pointed fibration.*

Similar results hold for weak fibrewise pointed fibrations.

15 Fibrewise pointed mapping-spaces

The adjoint of the fibrewise smash product

In this section our aim is to show that, subject to certain conditions, the fibrewise pointed mapping-space provides an adjoint to the fibrewise smash product. The results we state here are all more or less straightforward consequences of the corresponding results for the non-pointed theory proved earlier in Section 8 and so proofs can for the most part be omitted. The restrictions which appear are, of course, inconvenient. To avoid them a fibrewise version of Steenrod's 'convenient category' may be used. The necessary machinery may be found in [91]; we do not reproduce it here.

Let X and Y be fibrewise pointed spaces over B. Assume, as before, that the sections are closed. By the *fibrewise pointed mapping-space* $\operatorname{map}_B^B(X, Y)$ we mean the subspace of the fibrewise mapping-space $\operatorname{map}_B(X, Y)$ consisting of pointed maps of the fibres. Here the basepoints in the fibres are determined by the sections in the usual way. The section t of Y determines the section t_* of $\operatorname{map}_B^B(X, Y)$ through

$$B = \operatorname{map}_B(X, B) \xrightarrow{t_*} \operatorname{map}_B(X, Y).$$

Note that t_* is closed, by Proposition 8.5, when X is locally sliceable.

So far as naturality is concerned we simply observe, as this stage, that a continuous fibrewise bijection

$$\alpha_\# : \operatorname{map}_{B'}^{B'}(\alpha^* X, \alpha^* Y) \to \alpha^* \operatorname{map}_B^B(X, Y)$$

is defined for each space B' and map $\alpha : B' \to B$.

From Proposition 8.1 we have at once

Proposition 15.1 *Let X and Y be fibrewise pointed spaces over B. Then the fibrewise pointed mapping-spaces $\operatorname{map}_B^B(X, B)$ and $\operatorname{map}_B^B(B, Y)$ are equivalent to B, as fibrewise pointed spaces.*

Let X, Y and Z be fibrewise pointed spaces over B. Precomposition with a fibrewise pointed map $\theta : X \to Y$ determines a fibrewise pointed map

$$\theta^* : \operatorname{map}_B^B(Y, Z) \to \operatorname{map}_B^B(X, Z),$$

while postcomposition with a fibrewise pointed map $\phi : Y \to Z$ determines a fibrewise pointed map

$$\phi_* : \operatorname{map}_B^B(X, Y) \to \operatorname{map}_B^B(X, Z).$$

From Propositions 8.4 and 8.5 we obtain

Proposition 15.2 *Let* $\theta : X \to Y$ *be a proper fibrewise pointed surjection, where* X *and* Y *are fibrewise pointed spaces over* B. *Then the fibrewise pointed map*

$$\theta^* : \operatorname{map}_B^B(Y, Z) \to \operatorname{map}_B^B(X, Z)$$

is an embedding for all fibrewise pointed Z.

Proposition 15.3 *Let* $\phi : Y \to Z$ *be a fibrewise pointed embedding, where* Y *and* Z *are fibrewise pointed spaces over* B. *Then the fibrewise pointed map*

$$\phi_* : \operatorname{map}_B^B(X, Y) \to \operatorname{map}_B^B(X, Z)$$

is an embedding for all fibrewise pointed spaces X. *If in addition* ϕ *is closed then* ϕ_* *is closed provided* X *is locally sliceable.*

Given a finite family $\{X_j\}$ of fibrewise pointed spaces over B the natural projection

$$\coprod_B X_j \to \bigvee_B X_j$$

of the fibrewise sum onto the fibrewise wedge product is a fibrewise pointed map. Hence and from Proposition 8.6 we obtain

Proposition 15.4 *Let* $\{X_j\}$ *be a finite family of fibrewise pointed spaces over* B. *Then the natural fibrewise pointed map*

$$\operatorname{map}_B^B(\bigvee_B X_j, Y) \to \prod_B \operatorname{map}_B^B(X_j, Y)$$

is an equivalence for all fibrewise pointed spaces Y.

Here the ith component of the fibrewise pointed map is induced by σ_i^*, where $\sigma_i : X_i \to \bigvee_B X_j$ is the standard insertion.

Adjoints

We now begin our account of the adjoint relationship between the fibrewise pointed mapping-space and the fibrewise smash product, with

Proposition 15.5 *Let X, Y and Z be fibrewise pointed spaces over B. If the fibrewise pointed function $h : X \wedge_B Y \to Z$ is continuous then so is the fibrewise pointed function $k : X \to \operatorname{map}_B^B(Y, Z)$, where*

$$k(x)(y) = h(x, y) \quad (x \in X_b, \, y \in Y_b, \, b \in B).$$

This follows at once from Proposition 8.7. We go on to prove

Proposition 15.6 *Let X_i and Y_i $(i = 1, 2)$ be fibrewise pointed spaces over B, with X_i fibrewise compact regular. Then the fibrewise pointed injection*

$$\operatorname{map}_B^B(X_1, Y_1) \wedge_B \operatorname{map}_B^B(X_2, Y_2) \to \operatorname{map}_B^B(X_1 \wedge_1 X_2, Y_1 \wedge_B Y_2),$$

given by the fibrewise smash product, is continuous.

For consider the diagram shown below, where ρ is the generic fibrewise quotient map, where ξ is given by the fibrewise product, and where η is given by the fibrewise smash product.

$$\operatorname{map}_B^B(X_1, Y_1) \times_B \operatorname{map}_B^B(X_2, Y_2) \quad \overset{\xi}{\longrightarrow} \quad \operatorname{map}_B^B(X_1 \times_B X_2, Y_1 \times_B Y_2)$$

$$\downarrow \rho_*$$

$$\operatorname{map}_B^B(X_1 \times_B X_2, Y_1 \wedge_B Y_2)$$

$$\downarrow \rho^*$$

$$\rho \downarrow$$

$$\operatorname{map}_B^B(X_1, Y_1) \wedge_B \operatorname{map}_B^B(X_2, Y_2) \quad \underset{\eta}{\longrightarrow} \quad \operatorname{map}_B^B(X_1 \wedge_B X_2, Y_1 \wedge_B Y_2)$$

By Proposition 8.11 ξ is continuous, also ρ_* and ρ^* on the right are continuous, so that $\rho^* \circ \rho_* \circ \xi$ is continuous. Hence η is continuous, since ρ is a fibrewise quotient map.

Fibrewise evaluation determines a fibrewise pointed function

$$\operatorname{map}_B^B(X, Y) \wedge_B X \to Y$$

for all fibrewise pointed spaces X, Y over B. More generally, fibrewise composition determines a fibrewise pointed function

$$\operatorname{map}_B^B(Y, Z) \wedge_B \operatorname{map}_B^B(X, Y) \to \operatorname{map}_B^B(X, Z)$$

for all fibrewise pointed spaces X, Y, Z over B. From Proposition 8.12 we obtain

Proposition 15.7 *Let* Y *be a fibrewise locally compact regular fibrewise pointed space over* B. *Then the fibrewise composition function*

$$\mathrm{map}_B^B(Y,Z) \wedge_B \mathrm{map}_B^B(X,Y) \to \mathrm{map}_B^B(X,Z)$$

is continuous for all fibrewise pointed spaces X *and* Z.

Corollary 15.8 *Let* X *and* Y *be fibrewise pointed spaces over* B, *with* X *fibrewise locally compact regular. Then the fibrewise evaluation function*

$$\mathrm{map}_B^B(X,Y) \wedge_B X \to Y$$

is continuous.

Furthermore, from Corollary 8.14 we obtain

Proposition 15.9 *Let* X, Y *and* Z *be fibrewise pointed spaces over* B, *with* Y *fibrewise locally compact regular. Suppose that the fibrewise pointed function* $k : X \to \mathrm{map}_B^B(Y,Z)$ *is continuous. Then so is the fibrewise pointed function* $h : X \wedge_B Y \to Z$, *where*

$$h(x,y) = k(x)(y) \quad (x \in X_b,\ y \in Y_b,\ b \in B).$$

When h and k are related as in Proposition 15.5 or 15.9 we refer to h as the *left adjoint* of k, and to k as the *right adjoint* of h.

In particular, take $X = B \times T$, where T is pointed. Then for any fibrewise pointed space Y over B a fibrewise pointed map

$$(B \times T) \wedge_B Y \to Z$$

determines a fibrewise pointed map

$$Y \to \mathrm{map}_B(B \times T, Z),$$

through the standard formula, and the converse holds when T is locally compact regular.

This relationship holds, in particular, when $T = \dot{I}$, and shows that fibrewise pointed maps of $C_B^B(Y)$ into Z correspond precisely to fibrewise pointed maps of Y into $P_B(Z)$, the fibrewise path-space. It also holds when T is the circle $I/\dot{I}$, and shows that fibrewise pointed maps of $\Sigma_B^B(Y)$ into Z correspond precisely to fibrewise pointed maps of Y into $\Omega_B(Z)$, the fibrewise loop-space.

The exponential law

Returning to the general case we obtain from Proposition 8.15, the exponential law for fibrewise mapping-spaces, the corresponding result for fibrewise pointed mapping-spaces as follows.

Proposition 15.10 *Let X, Y and Z be fibrewise pointed spaces over B, with X and Y fibrewise compact regular. Then the fibrewise pointed function*

$$\mathrm{map}_B^B(X \wedge_B Y, Z) \to \mathrm{map}_B^B(X, \mathrm{map}_B^B(Y, Z)),$$

defined by taking adjoints, is an equivalence of fibrewise pointed spaces.

Note that fibrewise compact is the condition here, rather than fibrewise locally compact, because in using Proposition 8.4 we need the natural projection from $X \times_B Y$ to $X \wedge_B Y$ to be proper.

Proposition 15.11 *Let X, Y be fibrewise pointed spaces over B, with X fibrewise locally compact regular. Then the continuous fibrewise pointed bijection*

$$\alpha_\# : \mathrm{map}_{B'}^{B'}(\alpha^* X, \alpha^* Y) \to \alpha^* \mathrm{map}_B^B(X, Y)$$

is an equivalence of fibrewise pointed spaces for each space B' and map $\alpha : B' \to B$.

This follows at once from Proposition 8.16. In particular, replace B by $*$ and replace B', X, Y by B, X_0, Y_0, respectively, where X_0, Y_0 are pointed spaces. We deduce that for any space B there is an equivalence, as fibrewise pointed spaces over B, between $\mathrm{map}_B^B(X_0 \times B, Y_0 \times B)$ and $\mathrm{map}^*(X_0, Y_0) \times B$, provided X_0 is locally compact regular.

The space of fibrewise pointed maps

Returning to the general case, let us compare the pointed space $\mathrm{MAP}_B^B(X, Y)$ of fibrewise pointed maps $\phi : X \to Y$ with the pointed space $\Gamma(\mathrm{map}_B^B(X, Y))$ of sections $s : B \to \mathrm{map}_B^B(X, Y)$. Here both the pointed space of sections and the pointed space of fibrewise pointed maps are endowed with the compact-open topology. Consider the pointed function

$$\sigma : \mathrm{MAP}_B^B(X, Y) \to \Gamma(\mathrm{map}_B^B(X, Y))$$

which transforms the fibrewise pointed map ϕ into the pointed section s given by $s(b) = \phi_b$ ($b \in B$). Clearly σ is injective. From Proposition 8.17 we obtain

Proposition 15.12 *Let X, Y be fibrewise pointed spaces over B, and let*

$$\sigma : \mathrm{MAP}_B^B(X, Y) \to \Gamma(\mathrm{map}_B^B(X, Y))$$

be the injection defined above. If X is fibrewise locally compact regular then σ is bijective. If, in addition, B is regular then σ is an equivalence of pointed spaces.

Given fibrewise pointed spaces X and Y, the fibrewise pointed mapping-space $\operatorname{map}_B^B(X, Y)$ obtains fibrewise Hopf structure if either (i) X is fibre-wise coHopf or (ii) Y is fibrewise Hopf and X is fibrewise regular. If both (i) and (ii) hold the fibrewise Hopf structures on $\operatorname{map}_B^B(X, Y)$ which arise are equivalent, in the sense of fibrewise pointed homotopy, and are fibrewise homotopy-commutative.

Two examples

Example 15.13. Let $\phi : E \to F$ be a fibrewise pointed fibration, where E and F are fibrewise pointed spaces over B. Then the postcomposition function

$$\phi_* : \operatorname{map}_B^B(Y, E) \to \operatorname{map}_B^B(Y, F)$$

is a fibrewise pointed fibration, for all fibrewise pointed compact regular Y.

Example 15.14. Let $u : A \to X$ be a fibrewise pointed cofibration, where X is fibrewise pointed compact regular over B and A is a closed subspace of X. Then the precomposition function

$$u^* : \operatorname{map}_B(X, E) \to \operatorname{map}_B(A, E)$$

is a fibrewise pointed fibration for all fibrewise pointed spaces E.

16 Fibrewise well-pointed and fibrewise non-degenerate spaces

Fibrewise well-pointed spaces

Let us describe a fibrewise pointed space over B as *fibrewise well-pointed* (or *well-sectioned*) if the section is a fibrewise cofibration. For example, B is always fibrewise well-pointed, as a fibrewise pointed space over itself. For another example, let (X, X') be a fibrewise cofibred pair over B; then the fibrewise collapse $X/_B X'$ is fibrewise well-pointed, by Corollary 4.2.

Note that the associated bundle functor $P_\#$ mentioned earlier transforms equivariant well-pointed G-spaces into fibrewise well-pointed spaces over B, for each principal G-bundle P over B.

Clearly, the fibrewise pointed coproduct of fibrewise well-pointed spaces is fibrewise well-pointed. Also it follows from Corollary 4.2 that the fibrewise product and fibrewise smash product of fibrewise well-pointed spaces is fibrewise well-pointed.

As a direct consequence of the fibrewise homotopy extension property we have the following useful result. Let $\phi : X \to Y$ be a fibrewise map such that $\phi \mid s \simeq_B t$, where X and Y are fibrewise pointed spaces over B with sections s and t, respectively. Suppose that X is fibrewise well-pointed. Then ϕ is fibrewise homotopic to a fibrewise pointed map. In particular, take $Y = B$; if the section of X is a fibrewise retract up to fibrewise homotopy then it is a fibrewise retract.

Proposition 16.1 *Let X be a fibrewise well-pointed space over B. Let $\theta : X \to X$ be a fibrewise pointed map which is fibrewise homotopic to the identity. Then there exists a fibrewise pointed map $\theta' : X \to X$ such that $\theta' \circ \theta$ is fibrewise pointed homotopic to the identity.*

In the following argument we embed B in X by means of the section, thus releasing the letters s and t to denote parameters. So let f_t be a fibrewise homotopy of θ into 1_X. Then $f_t \mid B$ is a fibrewise homotopy of the inclusion into itself. By the fibrewise homotopy extension property there exists a fibrewise homotopy $g_t : X \to X$ of 1_X such that $f_t \mid B = g_t \mid B$. We assert that the condition in Proposition 16.1 is satisfied with $\theta' = g_1$.

For consider the juxtaposition k_s of $g_{1-s} \circ \theta$ and f_s, as a fibrewise homotopy of $\theta' \circ \theta$ into 1_X. Now $f_t \mid B = g_t \mid B$ and hence $H_{(s,0)} = k_s \mid B$, where

$$H(b,s,t) = \begin{cases} g(b, 1 - 2s(1-t)) & (0 \le s \le \tfrac{1}{2}), \\ f(b, 1 - 2(1-s) - (1-t)) & (\tfrac{1}{2} \le s \le 1). \end{cases}$$

Again using the fibrewise homotopy extension property we extend H to a fibrewise map $K : X \times I \times I \to X$ such that $K_{(s,0)}$ is fibrewise pointed homotopic to k_s, for each s. Then

$$k_0 \simeq_B^B K_{(0,0)} \simeq_B^B K_{(0,1)} \simeq_B^B K_{(1,0)} \simeq_B^B k_1$$

and so $\theta' \circ \theta \simeq_B^B 1_X$, as asserted.

The main use of Proposition 16.1 is to prove the useful

Theorem 16.2 *Let $\phi : X \to Y$ be a fibrewise pointed map, where X and Y are fibrewise well-pointed spaces over B. If ϕ is a fibrewise homotopy equivalence then ϕ is a fibrewise pointed homotopy equivalence.*

For let $\psi : Y \to X$ be an inverse of ϕ, up to fibrewise homotopy. Since $\psi \circ t = \psi \circ \phi \circ s \simeq_B s$ there exists a fibrewise pointed map $\psi' : Y \to X$ such that $\psi \simeq_B \psi'$. Since $\psi' \circ \phi \circ s = t$ and since $\psi' \circ \phi \simeq_B 1_X$ there exists, by Proposition 16.1, a fibrewise pointed map $\psi'' : X \to Y$ such that $\psi'' \circ \psi' \circ \phi$

is fibrewise pointed homotopic to the identity. Thus ϕ admits a left inverse $\phi' = \psi'' \circ \psi'$ up to fibrewise pointed homotopy.

Now ϕ' is a fibrewise homotopy equivalence, since ϕ is a fibrewise homotopy equivalence, and so the same argument, applied to ϕ' instead of ϕ, shows that ϕ' admits a left inverse ϕ'' up to fibrewise pointed homotopy. Thus ϕ' admits both a right inverse ϕ and a left inverse ϕ'' up to fibrewise pointed homotopy. Hence ϕ' is a fibrewise pointed homotopy equivalence and so ϕ itself is a fibrewise pointed homotopy equivalence, as asserted.

The proof of our next result is similar but easier and so will be omitted.

Proposition 16.3 *Let $p : E \to X$ be a fibrewise pointed map, where E and X are fibrewise pointed spaces over B. Suppose that E is fibrewise well-pointed and that p is a fibrewise fibration. Then p is a fibrewise pointed fibration.*

Fibrewise non-degenerate spaces

The class of fibrewise well-pointed spaces has many good properties but is too restrictive for some purposes. Often, however, it can be replaced by a wider class, defined as follows.

Consider a fibrewise pointed space X over B with section $s : B \to X$. In this context we denote by $\check{X}_B$ the fibrewise mapping cylinder $M_B(s)$ of s, regarded as a fibrewise pointed space with section the insertion σ_1. Note that the inclusion $\sigma : X \to \check{X}_B$ is a fibrewise map, in fact a fibrewise homotopy equivalence, but not a fibrewise pointed map. The natural projection $\rho : \check{X}_B \to X$, which fibrewise collapses $M_B(1_B) = B \times I$, is a fibrewise pointed map as well as a fibrewise homotopy equivalence. Let us describe X as *fibrewise non-degenerate* if ρ is a fibrewise pointed homotopy equivalence.

Note that $\check{X}_B$ itself is always fibrewise non-degenerate, so that every fibrewise pointed space has the same fibrewise homotopy type as a fibrewise non-degenerate space. Furthermore, every fibrewise well-pointed space is fibrewise non-degenerate, by Theorem 16.2.

Fibrewise non-degenerate spaces over B can be characterized as follows. By a *fibrewise Puppe structure* on a fibrewise pointed space X we mean a pair (α, U), where U is a fibrewise categorical neighbourhood of B in X and $\alpha : X \to I$ is a map such that $\alpha = 1$ throughout B and $\alpha = 0$ away from U. For example, take $X = \Sigma_B(Y)$, where Y is a fibrewise space over B. We take $\alpha : X \to I$ to be given by the suspension parameter and we take $U = \alpha^{-1}(0, 1]$, where X is regarded as a fibrewise pointed space with section $\alpha^{-1}(1)$.

Proposition 16.4 *Let X be a fibrewise pointed space over B. Then X is fibrewise non-degenerate if and only if X admits a fibrewise Puppe stucture.*

For suppose that $\rho : \check{X}_B \to X$ is a fibrewise pointed homotopy equivalence. Take U to be the preimage of the open cylinder $B \times (0, 1] \subseteq \check{X}_B$ under a fibrewise pointed homotopy inverse $\rho' : X \to \check{X}_B$ of ρ. Now $\rho' \mid U$ is fibrewise pointed null-homotopic. Also $\rho \circ \rho' \mid U$ is fibrewise pointed homotopic to the inclusion. Hence the inclusion $U \to X$ is fibrewise pointed null-homotopic. Take $\alpha : X \to I$ to be the composition

$$X \xrightarrow{\rho'} \check{X}_B \longrightarrow \check{X}_B/_B X = B \times I \xrightarrow{\pi_1} I,$$

where the middle stage is fibrewise collapse. Then $\alpha = 1$ throughout B and $\alpha = 0$ away from U, so that (α, U) is a fibrewise Puppe structure.

Conversely, let (α, U) be a fibrewise Puppe structure on X. With no real loss of generality we may suppose that U is a closed neighbourhood of B, since otherwise we can replace (α, U) by the fibrewise Puppe structure (α', U'), where $U' = \alpha^{-1}[0, \frac{1}{2}]$ and $\alpha' = \min(2\alpha, 1)$. Choose a fibrewise pointed null-homotopy $f : U \times I \to X$ of the inclusion, and consider the fibrewise pointed homotopy $f' : U \times I \to \check{X}_B$ given by

$$f'(x, t) = \begin{cases} f(x, 2t) & (0 \leq t \leq \frac{1}{2}) \\ (p(x), 2t - 1) & (\frac{1}{2} \leq t \leq 1). \end{cases}$$

A fibrewise pointed null-homotopy g_t of the identity on $\check{X}_B$ is given on the subspace $X \subseteq \check{X}_B$ by

$$g_t(x) = \begin{cases} x & (x \notin U), \\ f'(x, t.\alpha(x)) & (x \in U), \end{cases}$$

and on $B \times I \subseteq \check{X}_B$ by

$$g_t(b, s) = \begin{cases} (b, s) & (0 \leq t \leq \frac{1}{2}) \\ (b, 1 - (1 - s)(2 - 2t)) & (\frac{1}{2} \leq t \leq 1). \end{cases}$$

Therefore X is fibrewise non-degenerate and the proof is complete.

We use fibrewise Puppe structures to prove

Theorem 16.5 *If X and Y are fibrewise non-degenerate spaces over B then so is the fibrewise product $X \times_B Y$.*

For let (α, U) and (β, V) be fibrewise Puppe structures on X and Y, respectively. Then (γ, W) is a fibrewise Puppe structure on $X \times_B Y$, where $W = U \times_B V$ and $\gamma : X \times_B Y \to I$ is given by $\gamma(x, y) = \alpha(x).\beta(y)$.

It can be shown, as in Section 22 of [86], that the fibrewise smash product of fibrewise non-degenerate spaces is also fibrewise non-degenerate.

If X is a fibrewise pointed space over B then for each subspace V of B we may identify the restriction to V of the fibrewise mapping cylinder of the section $s : B \to X$ with the fibrewise mapping cylinder of the restriction $s_V : V \to X_V$. Hence and from Proposition 7.2 we obtain

Proposition 16.6 *Let X be a fibrewise pointed space over B. Suppose that B admits a numerable covering such that X_V is fibrewise non-degenerate over V for each member V of the covering. Then X is fibrewise non-degenerate.*

Proposition 16.7 *Let X and Y be vertically connected fibrewise pointed spaces over B. Suppose that X and Y have the same fibrewise homotopy type. Also suppose that X and Y are fibrewise non-degenerate. Then X and Y have the same fibrewise pointed homotopy type.*

As usual we denote by a dot the binary operation (juxtaposition) in the groupoid of vertical homotopies, and we denote by e the neutral element (stationary homotopy) at any section.

Let s denote the section of X and t the section of Y. Let $\theta : X \to Y$ be a fibrewise homotopy equivalence with fibrewise homotopy inverse $\phi : Y \to X$. With any vertical homotopy $u : B \times I \to X$ of $\phi \circ t$ into s we associate the fibrewise pointed map $\check{\phi} : \check{Y}_B \to \check{X}_B$ given by ϕ on Y and by $u.e$ on $B \times I$. With any vertical homotopy $v : B \times I \to Y$ of $\theta \circ s$ into t we associate the fibrewise pointed map $\check{\theta} : \check{X}_B \to \check{Y}_B$ given by θ on X and by $v.e$ on $B \times I$.

Choose a fibrewise homotopy of $\phi \circ \theta$ into 1_X. Precomposing this with s we obtain a vertical homotopy H of some section of X into s. Choose any v, as above, and then choose u so that $(\phi \circ v).u$ is equivalent to H, by a vertical homotopy rel $(B \times \dot{I})$. Then the fibrewise homotopy of $\phi \circ \theta$ into the identity on X can be extended to a fibrewise pointed homotopy of $\check{\phi} \circ \check{\theta}$ into the identity on $\check{X}_B$.

Now choose a fibrewise homotopy of $\theta \circ \phi$ into 1_Y. By precomposing this with t we obtain a vertical homotopy K of some section of Y into t. With u as before we choose a vertical homotopy w of $\theta \circ s$ into t so that $(\theta \circ u).w$ is equivalent to K. Then the fibrewise homotopy of $\theta \circ \phi$ into the identity on Y extends to a fibrewise homotopy of $\check{\psi} \circ \check{\phi}$ into the identity on $\check{Y}_B$, where $\check{\psi} : \check{X}_B \to \check{Y}_B$ is the fibrewise pointed map given by θ on X and by $w.e$ on $B \times I$.

Therefore ϕ admits both a left inverse $\check{\psi}$ and a right inverse $\check{\theta}$ up to fibrewise pointed homotopy, and so is a fibrewise pointed homotopy equivalence. Thus $\check{X}_B$ and $\check{Y}_B$ have the same fibrewise pointed homotopy type. But $\check{X}_B$ has the same fibrewise pointed homotopy type as X and $\check{Y}_B$ has the same fibrewise pointed homotopy type as Y, by the assumption of fibrewise non-degeneracy. Therefore X and Y have the same fibrewise pointed homotopy type, as asserted.

When a fibrewise space admits a section and so can be regarded as a fibrewise pointed space we must expect the fibrewise pointed homotopy type to depend on the choice of section, in general. However, if $s_0, s_1 : B \to X$ are vertically homotopic sections of the fibrewise space X then an argument similar to, but simpler than, that used to prove Proposition 16.4 shows that the fibrewise mapping cylinders $M_B(s_0)$ and $M_B(s_1)$ have the same fibrewise pointed homotopy type. Suppose, therefore, that the fibrewise pointed space X_i $(i = 0, 1)$ obtained from X by using s_i as section is fibrewise non-degenerate. Then X_0 and X_1 have the same fibrewise pointed homotopy type. This means, for example, that if X_0 admits fibrewise Hopf structure then so does X_1. Similarly with fibrewise coHopf structure. More generally, X_0 and X_1 have the same fibrewise pointed category (see Section 19).

17 Fibrewise complexes

The category of pairs

Fibrewise homotopy theory may be regarded as a branch of the homotopy theory of the category Top(2) of pairs of spaces and maps, as discussed by Eckmann and Hilton [57] and others. Unfortunately the term *pair* is potentially confusing in our situation, where it is preferable to describe Top(2) as the category of *spaces over spaces* and *maps over maps*, the latter being classified by *homotopies over homotopies*.

Various treatments of the homotopy theory of the category Top(2) may be found in the literature. Perhaps that of tom Dieck, Kamps and Puppe [44] is the most appropriate for our purposes. However, it is convenient to adopt a modification of their terminology and notation, as follows.

The objects of Top(2), of course, are the morphisms of Top. Thus an object consists of a base space B and a space X over B with projection p, say. The morphisms of Top(2) are commutative diagrams of morphisms of Top. Thus if X is a space over B with projection p and X' is a space over B' with projection p' then a morphism from X to X' consists of a map $f : B \to B'$ and a map $F : X \to X'$ such that $p' \circ F = f \circ p$. We may refer to F as a map over f.

Maps over maps are classified by homotopies over homotopies, as follows. Let $f_i : B \to B'$ be a map $(i = 0, 1)$ and let $F_i : X \to X'$ be a map over f_i. Let $f_t : B \to B'$ be a homotopy of f_0 into f_1 and let $F_t : X \to X'$ be a homotopy of F_0 into F_1 such that $p' \circ F_t = f_t \circ p$ for all t. We may refer to F_t as a homotopy over f_t. Homotopy equivalences over homotopy equivalences, etc., are defined in a similar manner.

The category Top_B of spaces over a given base space B may be regarded as contained in the category Top(2) as the subcategory of spaces over B and maps over the identity of B. However, the morphisms of Top_B are classified by homotopies over the stationary homotopy of the identity, which is generally a finer classification than that by homotopies over self-homotopies of the identity. It is the former classification which is appropriate here.

Fibrewise complexes

The notion of CW complex, introduced by J.H.C. Whitehead [135], is covered
in all the standard textbooks, for example in Chapter 7 of [127]. To keep this
section short we do not consider infinite complexes and so the letters CW
(standing for closure finite, weak topology) can be omitted.

Recall that a cellular decomposition of a Hausdorff space B consists, in
each dimension n, of a finite collection of maps $\theta : D^n \to B$, satisfying certain
conditions. The image of the closed n-ball D^n under the characteristic map
θ is called the closed n-cell. That of $D^n - S^{n-1}$ is called the open n-cell,
and that of S^{n-1} is called the boundary of the n-cell (the terminology does
not refer to the topology of B). The conditions are that θ maps $D^n - S^{n-1}$
homeomorphically onto the open n-cell, and that the boundary of the n-cell
is the union of open m-cells for $m < n$. Also the whole collection of open
cells forms a decomposition of B, so that every point of B is contained in
precisely one open cell. When these conditions are satisfied we describe B as
a *complex*.

Choose a cellular decomposition of B, and let K be a fibrewise Hausdorff
space over B. A cellular block decomposition of K consists of a decomposition
of K into open cellular blocks. Specifically, over each closed n-cell of B,
with characteristic map $\theta : D^n \to B$, there exists a finite collection of maps
$\Theta : D^n \times T \to K$, over the map θ, where T is compact Hausdorff (the factors
T may vary with Θ), satisfying certain conditions. The image of $D^n \times T$
under Θ is called the closed n-cellular block, that of $(D^n - S^{n-1}) \times T$ the
open n-cellular block, and that of $S^{n-1} \times T$ the boundary of the n-cellular
block (the terminology does not refer to the topology of K). The conditions
are that Θ maps $(D^n - S^{n-1}) \times T$ homeomorphically onto the open n-cellular
block, and that the boundary of the n-cellular block is the union of open m-
cellular blocks for $m < n$. Also the whole collection of open cellular blocks,
over the chosen cellular decomposition of B, forms a decomposition of K, so
that every point of K is contained in precisely one open cellular block. When
these conditions are satisfied we describe K as a *fibrewise complex* over the
complex B. There is no requirement for K to be a complex in the ordinary
sense.

Clearly, any fibre bundle with compact Hausdorff fibre over the complex
B can be regarded as a fibrewise complex over B. Examples of fibrewise
complexes which are not fibre bundles arise in the theory of transformation
groups. Specifically, if K is a G-complex, where G is a compact group, then
the orbit space K/G is a complex and K is a fibrewise complex over K/G,
as described in [19], the factors of the cellular blocks being orbits of different
types.

Thus consider the much-studied family of $O(n)$-manifolds W_k^{2n-1} (see I.7
of [19]) for which the orbit space is the 2-disc D^2. We can construct W_k^{2n-1}
by adjoining the 2-cellular block $D^2 \times V_{n,2}$ to S^{n-1} by means of the map

$\psi_k : S^1 \times V_{n,2} \to S^{n-1}$, where $\psi_k((\cos\theta, \sin\theta), (u,v)) = (u\cos k\theta, v\sin k\theta)$. Here the factor $V_{n,2}$ is the Stiefel manifold of orthonormal pairs (u,v) in $\mathbb{R}^n$.

Special fibrewise complexes

Fibrewise complexes in which all the factors in the cellular blocks are complexes in the ordinary sense play a special rôle in the theory and so we will refer to them as *special* fibrewise complexes. For example, the family of $O(n)$-complexes W_k^{2n-1} we have just described are special fibrewise complexes over D^2. Also sphere-bundles over complexes are special. If K is a special fibrewise complex over B the dimension $\dim K$ of K is defined to be the maximum dimension of the cellular blocks in the decomposition of K. For example $2n-1$ is the dimension of W_k^{2n-1}. Also if K is a q-sphere bundle over the complex B then the dimension of K is $q + \dim B$.

Let B be a complex and let K be a fibrewise complex over B. We describe a subspace L of K as a *subcomplex* of K if L is the union of open cellular blocks of K subject to the condition that the boundary of each of the cellular blocks of L is also in L. This ensures that L itself is a fibrewise complex over B. If K is special then $\dim(K - L)$ is defined to be the maximum dimension of the cellular blocks in the decomposition of K which do not belong to L. When K is a sphere-bundle over B with section corresponding to a reduction of the structural group then the section forms a subcomplex of K assuming K is regarded as a fibrewise complex in the obvious way.

Returning to the general situation, let us denote by K^n, where $n \geq 0$, the subcomplex formed by m-cellular blocks for $m \leq n$. Then $L^n = L \cap K^n$ when L is a subcomplex of K.

It would be convenient if the inclusion of a subcomplex in a fibrewise complex satisfied the condition for a fibrewise cofibration. While this may not be so we can demonstrate a weaker result in this direction which is still useful.

Proposition 17.1 *Let B be a complex and let X be a fibre space over B. Let K be a fibrewise complex over B and let L be a subcomplex of K. Let $f : K \to X$ be a fibrewise map, and let $g_t : L \to X$ be a fibrewise homotopy of $f \mid L$. Then there exists a fibrewise homotopy $h_t : K \to X$ of f such that $g_t = h_t \mid L$.*

Here the term *fibre space*, as distinct from fibrewise space, means that the homotopy lifting property holds. The first step in the proof of Proposition 17.1 is to establish the following.

Lemma 17.2 *Let B be a complex and let $\theta : D^n \to B$ be a map. Let X be a fibre space over B. Let*

$$\phi : (\{0\} \times D^n \cup I \times S^{n-1}) \times T \to X$$

be a map over θ, where T is a complex. Then ϕ can be extended to a map

$$\psi : I \times D^n \times T \to X$$

over θ.

By taking adjoints we obtain from ϕ a map

$$\hat{\phi} : \{0\} \times D^n \cup I \times S^{n-1} \to \mathrm{map}(T, X)$$

over θ, where the codomain is the space of maps with compact-open topology. Since X is a fibre space over B so is $\mathrm{map}(T, X)$. Hence the induced fibre space $\theta^*\mathrm{map}(T, X)$ over D^n is equivalent to the product $D^n \times \mathrm{map}(T, X_0)$, where X_0 is the fibre of X.

Now $\hat{\phi}$ determines a section

$$s : \{0\} \times D^n \cup I \times S^{n-1} \to \theta^*\mathrm{map}(T, X),$$

equivalently a section

$$s' : \{0\} \times D^n \cup I \times S^{n-1} \to D^n \times \mathrm{map}(T, X_0).$$

Consider the second projection

$$s'' : \{0\} \times D^n \cup I \times S^{n-1} \to \mathrm{map}(T, X_0).$$

Since the inclusion $S^{n-1} \to D^n$ is a cofibration we can extend s'' over $I \times D^n$. Therefore s' can be extended to a section over $I \times D^n$, and hence s can be extended to a section over $I \times D^n$. Therefore $\hat{\phi}$ can be extended to a map

$$\hat{\psi} : I \times D^n \to \mathrm{map}(T, X)$$

over θ and finally, taking the adjoint, ϕ can be extended to a map

$$\psi : I \times D^n \times T \to X$$

over θ, as asserted.

Having established this we can now prove Proposition 17.1 in the special case where K is obtained from L by adjoining the single n-cellular block $D^n \times T$. All that needs to be done is to precompose with the characteristic map of the block, apply Lemma 17.2, and then precompose again with the inverse of the characteristic map. We may then proceed by iteration to the case where K is obtained from L by adjoining a succession of n-cellular blocks, for given n.

In the general case we make an induction on dimension as follows. Assume, for $n \geq 1$, that there exists a fibrewise homotopy $f_t^{n-1} : K^{n-1} \to X$ of $f \mid K^{n-1}$ such that $f_t^{n-1} \mid L^{n-1} = g_t \mid L^{n-1}$, as is clearly true when $n = 1$. Use the special case to extend f_t^{n-1} to a fibrewise homotopy $f_t^n : K^n \to X$ of $f \mid K^n$ such that $f_t^n \mid L^n = g_t \mid L^n$. This deals with the inductive step and so, since $K^n = K$ for sufficiently large n, proves Proposition 17.1. In particular, taking $X = \{0\} \times K \cup I \times L$ we obtain

Corollary 17.3 *Let B be a complex. Let K be a fibre complex over B and let L be a fibre subcomplex of K. Then the inclusion $L \to K$ is a fibrewise cofibration.*

Here we use the term *fibre complex* to mean a fibre space which is also a fibrewise complex. Next we prove

Proposition 17.4 *Let B be a complex. Let K be a special fibrewise complex over B and let L be a subcomplex of K. Let X be a fibre space over B and let Y be a subspace of X which is also a fibre space over B. Suppose that the pair (X, Y) is d-connected and that $\dim(K - L) \le d$. Then any fibrewise map*

$$f : (K, L) \to (X, Y)$$

is fibrewise homotopic, relative to L, to a fibrewise map of K into Y.

The proof proceeds on similar lines to that of Proposition 17.1. The first step is to establish

Lemma 17.5 *Let X be a fibre space over the complex B and let Y be a subspace of X which is also a fibre space over B. Suppose that the pair (X, Y) is d-connected. Let $\theta : D^n \to B$ be a map and let*

$$\phi : (D^n \times T, S^{n-1} \times T) \to (X, Y)$$

be a map over θ, where T is a complex such that $n + \dim T \le d$. Then relative to $S^{n-1} \times T$, ϕ is homotopic over θ to a map of $D^n \times T$ into Y.

The adjoint of the given map ϕ is a map

$$\hat{\phi} : (D^n, S^{n-1}) \to (\text{map}(T, X), \text{map}(T, Y)).$$

Now $\hat{\phi}$, like ϕ, is over $\theta : D^n \to B$.

As in the proof of Lemma 17.2, we see that $\theta^* \text{map}(T, X)$ is equivalent to $D^n \times \text{map}(T, X_0)$ and $\theta^* \text{map}(T, Y)$ is equivalent to $D^n \times \text{map}(T, Y_0)$, where X_0 and Y_0 are the fibres of X and Y, respectively. Now $\hat{\phi}$ determines a section

$$s : (D^n, S^{n-1}) \to (\theta^* \text{map}(T, X), (\theta \mid S^{n-1})^* \text{map}(T, Y)),$$

equivalently a section

$$s' : (D^n, S^{n-1}) \to (D^n \times \text{map}(T, X_0), S^{n-1} \times \text{map}(T, Y_0)).$$

Consider the second projection of s'

$$s'' : (D^n, S^{n-1}) \to (\mathrm{map}(T, X_0), \mathrm{map}(T, Y_0)).$$

By standard theory (see (7.6.13) of [127], for example) s'' is homotopic rel S^{n-1} to a map of D^n into $\mathrm{map}(T, Y_0)$, since the pair $(\mathrm{map}(T, X_0), \mathrm{map}(T, Y_0))$ is $(d - \dim T)$-connected. Therefore s' is vertically homotopic rel S^{n-1} to a section into $S^{n-1} \times \mathrm{map}(T, Y_0)$, and hence s is vertically homotopic rel S^{n-1} to a section into $(\theta \mid S^{n-1})^* \mathrm{map}(T, Y)$. Finally, $\hat{\phi}$ is homotopic, over ψ and relative to S^{n-1}, to a map of D^n into $\mathrm{map}(T, Y)$ and then, taking the adjoint, the original map ϕ is homotopic over θ and relative to $S^{n-1} \times T$, to a map of $D^n \times T$ into Y. This proves the lemma.

Having established this we can now prove the special case of Proposition 17.4 where K is obtained from L by adjoining the single n-cellular block $D^n \times T$. All that needs to be done is to precompose with the characteristic map of the block, apply Lemma 17.5, and then precompose again with the inverse of the characteristic map. We may then proceed by iteration to the case where K is obtained from L by adjoining a succession of n-cellular blocks, for given n.

In the general case we make an induction on dimension, as follows. Assume, for $n \geq 1$, that there exists a fibrewise homotopy $f_t^{n-1} : K^{n-1} \to X$ of $f \mid K^{n-1}$, relative to L^{n-1}, such that $f_1^{n-1} K^{n-1} \subseteq Y$, as is clearly true when $n = 1$. Use the fibrewise homotopy extension property, as in Proposition 17.1, to extend f_t^{n-1} to a fibrewise homotopy $h_t : K^n \to X$ of $f \mid K^n$ relative to L^n. Using Lemma 17.5, since $h_1 K^{n-1} \subseteq Y$ there exists a fibrewise homotopy $k_t : K^n \to X$ of h_1, relative to $K^{n-1} \cup L^n$, such that $k_1 K^n \subseteq Y$. By juxtaposition of k_t and h_t we obtain a fibrewise homotopy $f_t^n : K^n \to X$ of $f \mid K^n$, relative to L^n, such that $f_1^n K^n \subseteq Y$. This deals with the inductive step and so, since $K^n = K$ for sufficiently large n, proves Proposition 17.4.

Applications

Most of the applications of Proposition 17.4 can be derived from special cases of the following

Proposition 17.6 *Let B be a complex and let K be a special fibrewise pointed complex over B. Let $u : E \to F$ be a k-connected fibrewise pointed map, where E and F are fibrewise pointed fibre spaces over B. Then the induced function*

$$u_* : \pi_B^B[K; E] \to \pi_B^B[K; F]$$

is injective when $\dim K < k$, surjective when $\dim K \leq k$.

By replacing F by the fibrewise mapping cylinder of u we may suppose, without real loss of generality that $E \subseteq F$. Surjectivity in Proposition 17.6 follows at once from Proposition 17.4, applied to the pair (K, B), while injectivity follows from Proposition 17.4 applied to the pair $(K \times I, K \times \{0\})$. Of course, there is a relative version of this result, proved in the same way.

Having reached this stage we can now improve a number of results in the literature by replacing assumptions that fibrewise spaces are complexes, which is contrary to the spirit of fibrewise homotopy theory, by assumptions that they are special fibrewise complexes. Here we give just one illustration of this out of many possibilities, the fibrewise Freudenthal theorem.

Proofs of this fundamental result, under somewhat different hypotheses, have been given by Becker [8] and James [78]. However, Proposition 17.1 enables us to prove the result in the following form.

Theorem 17.7 *Let B be a complex and let K be a special fibrewise pointed complex over B. Let E be a fibrewise pointed fibre space over B with $(m-1)$-connected fibre. Then the fibrewise suspension*

$$\Sigma_* : \pi_B^B[K; E] \to \pi_B^B[\Sigma_B^B K; \Sigma_B^B E]$$

is injective for $\dim K < 2m - 1$, surjective for $\dim K \le 2m - 1$.

To deduce Theorem 17.7 from Proposition 17.6 we note that $\Sigma_B^B E$ is a fibre space over B, by (6.37) of [79], since E is a fibre space over B, and so the fibrewise loop-space $\Omega_B \Sigma_B^B E$ is a fibre space over B, by (6.32) of [85]. Since the fibre of E is $(m-1)$-connected the classical Freudenthal suspension theorem shows that the adjoint

$$u : E \to \Omega_B \Sigma_B^B E$$

of the identity is $(2m-1)$-connected, and so Theorem 17.7 follows at once from Proposition 17.6. Proceeding in the same way as in [79], we deduce

Corollary 17.8 *Let B be a complex. Let K be a fibrewise pointed k-sphere bundle and let L be a fibrewise pointed l-sphere bundle over B. Then for each fibrewise pointed sphere-bundle N over B the fibrewise smash product*

$$N_\# : \pi_B^B[K; L] \to \pi_B^B[N \wedge_B K; N \wedge_B L]$$

is injective when $\dim B < 2l - k - 1$, surjective when $\dim B \le 2l - k - 1$.

Another treatment of the fibrewise Freudenthal theorem, under different assumptions, will be given in Part II, Section 3. The theory presented here originally appeared in [92].

18 Fibrewise Whitehead products

Definition of the product

Although other approaches are possible, perhaps the best way to establish the properties of the fibrewise Whitehead product is by studying the behaviour of the fibrewise product under fibrewise suspension, as in

Proposition 18.1 *Let X and Y be fibrewise non-degenerate over B. Then $\Sigma_B^B(X \times_B Y)$ has the same fibrewise pointed homotopy type as the fibrewise pointed coproduct*

$$\Sigma_B^B(X) \vee_B \Sigma_B^B(Y) \vee_B \Sigma_B^B(X \wedge_B Y).$$

In fact a fibrewise pointed homotopy equivalence may be constructed by taking the fibrewise track sum, in some order, of the reduced fibrewise suspensions of the natural projection

$$X \times_B Y \to X \wedge_B Y$$

and the projections of the fibrewise product into its factors. The proof of Proposition 18.1 is fairly lengthy: details are given in Sections 21 and 22 of [86]. In fact Proposition 18.1 can be iterated so that, more generally, we obtain

Proposition 18.2 *Let $X_1, \ldots, X_n$ be fibrewise non-degenerate spaces over B. Then*

$$\Sigma_B^B(X_1 \times_B \ldots \times_B X_n)$$

has the same fibrewise pointed homotopy type as the fibrewise pointed coproduct

$$\bigvee_N \Sigma_B^B \bigwedge_{i \in N} {}_B X_i,$$

where N runs through all non-empty subsets of the integers 1 to n.

In fact a fibrewise pointed homotopy equivalence may be constructed by taking the fibrewise track sum, in some order, of the reduced fibrewise suspensions of the natural projections

$$X_1 \times_B \ldots \times_B X_n \to \bigwedge_{i \in N} {}_B X_i,$$

where X_i is mapped by the identity when $i \in N$, the projection p_i otherwise. In this way we obtain a monomorphism

$$\pi_B^B[\Sigma_B^B(\bigwedge_{i \in N} {}_B X_i); E] \to \pi_B^B[\Sigma_B^B(X_1 \times_B \ldots \times_B X_n); E],$$

for any fibrewise pointed space E, of which the image is a normal subgroup. We may therefore regard each of the groups

$$\pi_B^B[\Sigma_B^B(\bigwedge_{i \in N}{}_B X_i); E]$$

as a normal subgroup of

$$\pi_B^B[\Sigma_B^B(X_1 \times_B \ldots \times_B X_n); E].$$

Note that if the fibrewise cogroup-like structure of $\Sigma_B^B(X_i)$ is fibrewise homotopy-commutative for some $i \in N$ then the subgroup in question is commutative since, as we have seen, $\Sigma_B^B(\bigwedge_{B j \in N} X_j)$ has the same fibrewise pointed homotopy type as

$$\Sigma_B^B(X_i) \wedge_B \bigwedge_{j \in N, \, j \neq i}{}_B X_j,$$

and moreover the fibrewise pointed homotopy equivalence can be chosen so as to preserve the fibrewise cogroup-like structure.

Each of the standard projections

$$\pi_i : \prod_B X_i \to X_i \qquad (i = 1, \ldots, n)$$

admits a right inverse. Hence the reduced fibrewise suspension of π_i admits a right inverse and so embeds

$$\pi_B^B[\Sigma_B^B(X_i); E] \qquad (i = 1, \ldots, n)$$

as a normal subgroup of the group

$$\pi_B^B[\Sigma_B^B(\prod_B X_i); E],$$

for each fibrewise pointed space E. There are, however, other normal subgroups.

For example, take $n = 2$. In that case the reduced fibrewise suspension of the natural projection

$$X_1 \times_B X_2 \to X_1 \wedge_B X_2$$

is one of the fibrewise pointed maps used to split $\Sigma_B^B(X_1 \times_B X_2)$, as in Proposition 18.1. Hence the image of

$$\pi_B^B[\Sigma_B^B(X_1 \wedge_B X_2); E]$$

under the corresponding induced homomorphism is a normal subgroup of the group

$$\pi_B^B[\Sigma_B^B(X_1 \times_B X_2); E].$$

Again, take $n = 3$. In that case the reduced fibrewise suspension of the natural projection

$$X_1 \times_B X_2 \times_B X_3 \to X_1 \wedge_B X_2 \wedge_B X_3$$

is one of the fibrewise pointed maps used to split $\Sigma_B^B(X_1 \times_B X_2 \times_B X_3)$, as in Proposition 18.2. Hence the image of

$$\pi_B^B[\Sigma_B^B(X_1 \wedge_B X_2 \wedge_B X_3); E]$$

under the corresponding induced homomorphism is a normal subgroup of the group

$$\pi_B^B[\Sigma_B^B(X_1 \times_B X_2 \times_B X_3); E].$$

Of course, the group also contains normal subgroups of the type described in the previous paragraph, such as

$$\pi_B^B[\Sigma_B^B((X_1 \times_B X_2) \wedge_B X_3); E].$$

Returning to the case $n = 2$, let $\alpha_i \in \pi_B^B[\Sigma_B^B(X_i); E]$ $(i = 1, 2)$. The commutator $[\alpha_1, \alpha_2]$ is then defined in the group $\pi_B^B[\Sigma_B^B(X_1 \times_B X_2); E]$, and lies in the kernel of the homomorphism

$$\pi_B^B[\Sigma_B^B(X_1 \times_B X_2); E] \to \pi_B^B[\Sigma_B^B(X_1 \vee_B X_2); E]$$
$$= \pi_B^B[\Sigma_B^B(X_1); E] \times \pi_B^B[\Sigma_B^B(X_2); E]$$

induced by the reduced fibrewise suspension of the inclusion

$$X_1 \vee_B X_2 \to X_1 \times_B X_2.$$

Now the kernel, by exactness of the fibrewise Puppe sequence, is the normal subgroup

$$\pi_B^B[\Sigma_B^B(X_1 \wedge_B X_2); E].$$

We refer to the commutator, regarded as an element of this normal subgroup, as the *fibrewise Whitehead product* of α_1 and α_2. We have at once that

$$[\alpha_2, \alpha_1] = -(\Sigma_B^B t)^*[\alpha_1, \alpha_2], \tag{18.3}$$

where $t : X_1 \wedge_B X_2 \to X_2 \wedge_B X_1$ is the switching equivalence.

Clearly, if E is a fibrewise Hopf space the fibrewise Whitehead product is trivial, since the group

$$\pi_B^B[\Sigma_B^B(X_1 \times_B X_2); E]$$

is commutative.

We assert that the fibrewise Whitehead product vanishes under reduced fibrewise suspension, at least when X_1 and X_2 are fibrewise compact regular. For since the fibrewise loop-space $\Omega_B \Sigma_B^B(E)$ is fibrewise Hopf the fibrewise Whitehead product

$$\pi_B^B[\Sigma_B^B(X_1);\ \Omega_B\Sigma_B^B(E)] \times \pi_B^B[\Sigma_B^B(X_2);\ \Omega_B\Sigma_B^B(E)]$$

$$\downarrow$$

$$\pi_B^B[\Sigma_B^B(X_1 \wedge_B X_2);\ \Omega_B\Sigma_B^B(E)]$$

is trivial. If we now precompose with the product of the homomorphisms induced by the adjoint $E \to \Omega_B\Sigma_B^B(E)$ of the identity on E, and postcompose with the standard isomorphism

$$\pi_B^B[\Sigma_B^B(X_1 \wedge_B X_2);\ \Omega_B\Sigma_B^B(E)] \to \pi_B^B[\Sigma_B^B \circ \Sigma_B^B(X_1 \wedge_B X_2);\ \Sigma_B^B(E)]$$

we obtain the fibrewise suspension of the fibrewise Whitehead product

$$\pi_B^B[\Sigma_B^B(X_1);\ E] \times \pi_B^B[\Sigma_B^B(X_2);\ E] \to \pi_B^B[\Sigma_B^B(X_1 \wedge_B X_2);\ E].$$

This proves the assertion.

In group theory, we recall, a given element k of a group G determines an automorphism of each normal subgroup H of G, by conjugation. We denote the automorphism thus: $h \mapsto h^k$ ($h \in H$). At this stage it becomes more convenient to express the group operations in additive rather than multiplicative notation although the groups in question are not, in general, commutative.

Bilinearity

There are various ways to show that the fibrewise Whitehead product is bilinear, under appropriate conditions, but the following argument also provides an illustration of the method we shall be using to establish the Jacobi identity. We prove

Proposition 18.4 *Let X_i ($i = 1, 2$) be a fibrewise non-degenerate space over B such that $\Sigma_B^B(X_i)$ is fibrewise homotopy-commutative. If*

$$\alpha_i, \alpha_i' \in \pi_B^B[\Sigma_B^B(X_i);\ E],$$

where E is a fibrewise pointed space, then

$$[\alpha_1 + \alpha_1', \alpha_2] = [\alpha_1, \alpha_2] + [\alpha_1', \alpha_2],$$
$$[\alpha_1, \alpha_2 + \alpha_2'] = [\alpha_1, \alpha_2] + [\alpha_1, \alpha_2'].$$

To establish the first relation, we begin with the identity

$$[\alpha_1 + \alpha_1', \alpha_2] = [\alpha_2', \alpha_1]^{\alpha_1} + [\alpha_1, \alpha_2],$$

which holds in any group. The elements $[\alpha_1', \alpha_2]^{\alpha_1}$ and $[\alpha_1', \alpha_2]$ differ by the iterated commutator $[[\alpha_1', \alpha_2], -\alpha_1]^{\alpha_1}$. By exactness the latter element lies in the normal subgroup

$$\pi_B^B[\Sigma_B^B(X_1 \wedge_B X_2);\ E] \subseteq \pi_B^B[\Sigma_B^B(X_1 \times_B X_2);\ E].$$

Since $\Sigma_B^B(X_1)$ is fibrewise homotopy-commutative, so is

$$(\Sigma_B^B(X_1)) \wedge_B X_2 = \Sigma_B^B(X_1 \wedge_B X_2).$$

Hence the element in question is zero and so we may drop the index α_1, in the above identity, obtaining the first relation in Proposition 18.4. The second part may be proved similarly or else deduced from the first part with the help of the commutation law (18.3).

The Jacobi identity

To conclude our discussion we turn to the Jacobi identity, as follows.

Proposition 18.5 *Let X_i ($i = 1, 2, 3$) be a fibrewise non-degenerate space over B such that $\Sigma_B^B(X_i)$ is fibrewise homotopy-commutative. Let*

$$\alpha_i \in \pi_B^B[\Sigma_B^B(X_i); E],$$

where E is a fibrewise pointed space. Then

$$[[\alpha_1, \alpha_2], \alpha_3] + (\Sigma_B^B \tau)^*[[\alpha_2, \alpha_3], \alpha_1] + (\Sigma_B^B \tau^2)^*[[\alpha_3, \alpha_1], \alpha_2] = 0,$$

where τ is the appropriate cyclic permutation of the factors in the fibrewise smash products.

To establish the identity we begin with the relation

$$[[-\alpha_2, \alpha_1], \alpha_3]^{\alpha_2} + [[-\alpha_3, \alpha_2], \alpha_1]^{\alpha_3} + [[-\alpha_1, \alpha_3], \alpha_2]^{\alpha_1} = 0,$$

which holds in any group. The elements $[[-\alpha_2, \alpha_1], \alpha_3]^{\alpha_2}$ and $[[\alpha_1, \alpha_2], \alpha_3]$ differ by the iterated commutator $[[\alpha_2, \alpha_1], [-\alpha_3, \alpha_1]]^{\alpha_3}$. By exactness $[\alpha_2, \alpha_1]$ and $[-\alpha_3, \alpha_1]$ both lie in the normal subgroup

$$\pi_B^B[\Sigma_B^B(X_1 \wedge_B (X_2 \times_B X_3)); E] \subseteq \pi_B^B[\Sigma_B^B(X_1 \times_B X_2 \times_B X_3); E].$$

Since $\Sigma_B^B(X_1)$ is fibrewise homotopy-commutative, so is

$$\Sigma_B^B(X_1 \wedge_B (X_1 \times_B X_2)).$$

Hence the normal subgroup is commutative and the element in question vanishes, so that we may replace the first term in the general relation by $[[\alpha_1, \alpha_2], \alpha_3]$. Treating the second and third terms similarly the relation reduces to

$$[[\alpha_1, \alpha_2], \alpha_3] + [[\alpha_2, \alpha_3], \alpha_1] + [[\alpha_3, \alpha_1], \alpha_2] = 0.$$

This relation holds in the group

$$\pi_B^B[\Sigma_B^B(X_1 \times_B X_2 \times_B X_3); E].$$

When we translate it into terms of fibrewise Whitehead products in the group

$$\pi_B^B[\Sigma_B^B(X_1 \wedge_B X_2 \wedge_B X_3); E]$$

the automorphisms in Proposition 18.5 appear and we obtain the Jacobi identity as stated.

Fibrewise Whitehead products are not easy to calculate; various interesting problems suggest themselves. For example, consider the fibrewise Whitehead square

$$[1_{\Sigma_B^B(X)}, 1_{\Sigma_B^B(X)}] \in \pi_B^B[\Sigma_B^B(X \wedge_B X); \Sigma_B^B(X)],$$

of the identity on $\Sigma_B^B(X)$, which is the obstruction to the existence of a fibrewise Hopf structure on the fibrewise pointed space X. If X is a fibrewise pointed q-sphere bundle over B, with q odd, and B is of finite numerable category cat $B = n$, the order of the fibrewise Whitehead square can be shown (see [80]) to be a divisor of 2^n. When the bundle is trivial the order is 1 for $q = 1, 3$ or 7, otherwise the order is 2. Examples have been given [81] of sphere-bundles over spheres where the fibrewise Whitehead square is of order 4. Berrick [11] has studied fibrewise Samelson products, which are related.

Chapter 3. Applications

19 Numerical invariants

Although the survey [83] of category, in the sense of Lusternik–Schnirelmann, has been somewhat outdated by subsequent research, it may still serve as a convenient reference for the basic definitions and standard results in the ordinary theory. The following is a brief summary.

Category and pointed category

Recall that a subset V of a space X is said to be *categorical* if V is contractible in X. The *category* cat X of X is defined to be the least number of categorical open sets required to cover X. When no such number exists the category is said to be infinite.

Although this is not always made explicit, much of the literature is more concerned with the pointed version of category, as follows. Given a pointed space X, a subset V of X (necessarily containing the basepoint) is said to be *pointed categorical* if V is pointed contractible in X. The *pointed category* cat* X of X is defined to be the least number of pointed categorical open sets required to cover X. When no such number exists the pointed category is said to be infinite. If we disregard the basepoint then cat X is defined and cannot exceed cat* X. In fact equality holds when X is path-connected, under reasonably general conditions.

In 1956 G.W. Whitehead [134] gave a characterization of pointed category which many homotopy theorists found it convenient to adopt as their definition. When necessary we refer to this as pointed category in the new sense, to distinguish it from pointed category in the previous sense.

Whitehead's characterization is as follows. Let x_0 be the basepoint of the pointed space X. In the r-fold topological product $\prod^r X$ $(r = 1, 2, \ldots)$, consider the 'fat-wedge' subspace

$$T^r(X, x_0) = \pi_1^{-1}(x_0) \cup \ldots \cup \pi_r^{-1}(x_0),$$

where π_i is the ith projection $(i = 1, \ldots, r)$. Whitehead proved

Proposition 19.1 *Suppose that the basepoint x_0 of the pointed space X admits a pointed categorical neighbourhood. If the diagonal map $\Delta : X \to \prod^r X$*

can be compressed into $T^r(X, x_0)$ *by a pointed homotopy, for some* $r \geq 1$, *then* $\text{cat}^* X \leq r$.

Proposition 19.2 *Let X be a normal pointed space. If $\text{cat}^* X \leq r$, for $r \geq 1$, then the diagonal map $\Delta : X \to \prod^r X$ can be compressed into $T^r(X, x_0)$ by a pointed homotopy.*

Berstein and Hilton [13] adopted the new definition of category, based on Whitehead's characterization, and compared this invariant with another, defined as follows. Recall that the r-fold smash product $\bigwedge^r X$ of X is obtained from $\prod^r X$ by collapsing $T^r(X, x_0)$. We denote by $\Delta' : X \to \bigwedge^r X$ the projection of the diagonal map into the smash product.

Definition 19.3 The *weak pointed category* $\text{wcat}^* X$ of the pointed space X is the least number r such that $\Delta' : X \to \bigwedge^r X$ is pointed null-homotopic.

If no such number exists the weak pointed category is said to be infinite. Of course, $\text{cat}^* X \geq \text{wcat}^* X$; equality holds under certain conditions. Examples are given in [13] and [64] where the two invariants are not the same.

Lower bounds for weak pointed category, and hence for pointed category, can be given by using cohomology. Thus consider the reduced cohomology ring $H^*(X, x_0)$ of the pointed space X, with arbitrary coefficients. If $\text{wcat}^* X$ is defined then the ring is nilpotent and the index $\text{nil}\, H^*(X, x_0)$ of nilpotency cannot exceed r. Results of this type have a long history but this particular form of the result will be found at the end of [13].

Using pointed category in the new sense, it is obvious that $\text{cat}^* X \leq 2$ if and only if X admits coHopf structure. This suggests saying that X admits weak coHopf structure when $\text{wcat}^* X \leq 2$, without attempting to give a meaning to the concept of weak coHopf structure as such.

Occasionally, in what follows, we shall need to refer to category in the equivariant sense. Equivariant category has a literature of its own but the appropriate definition for our purposes is not the one usually adopted. Specifically, let X be a G-space, where G is a topological group. An invariant subset U of X is said to be *G-categorical* if the inclusion $U \to X$ is G-null-homotopic (to a point, which is necessarily a fixed point). The *G-category*, G-cat X, of X is defined to be the least number of G-categorical open sets required to cover X. If no such number exists the G-category is said to be infinite. There is, of course, a pointed version of the definition as well.

Fibrewise category

After these preliminaries we are now ready to define fibrewise category, as follows. Recall that a subset U of a fibrewise space X over B is said to be *fibrewise categorical* if the inclusion $U \to X$ is fibrewise null-homotopic.

The *fibrewise category* $\mathrm{cat}_B X$ of X is defined to be the least number of fibrewise categorical open sets required to cover X. If no such number exists the fibrewise category is said to be infinite.

Note that for any space B' and map $\alpha : B' \to B$ we have

$$\mathrm{cat}_{B'} \, \alpha^* X \le \mathrm{cat}_B X.$$

In particular, $\mathrm{cat}_B X$ is bounded below by the category of the fibres of X.

Of course, $\mathrm{cat}_B X = 1$ if and only if X is fibrewise contractible. Also $\mathrm{cat}_B X \le 2$ if X is a fibrewise suspension since X is the union of two open fibrewise cones.

One further remark before we proceed to the fibrewise pointed case. Let P be a principal G-bundle over B, where G is a topological group. Given a G-space Y we define $P_{\#} Y$, as before, to be the associated bundle with fibre Y, and similarly with G-maps and G-homotopies. Now if U is a G-categorical open set of Y then $P_{\#} U$ is a fibrewise categorical open set of $P_{\#} Y$. It follows that

$$\mathrm{cat}_B P_{\#} Y \le G\text{-cat}\, Y. \tag{19.4}$$

Fibrewise pointed category

Let us turn now to the fibrewise pointed theory, where the base space B is pointed. Recall that a subset U of the fibrewise pointed space X (necessarily containing the section) is said to be *fibrewise pointed categorical* if the inclusion $U \to X$ is fibrewise pointed null-homotopic. The *fibrewise pointed category* $\mathrm{cat}_B^B X$ of X is defined to be the least number of fibrewise pointed categorical open sets required to cover X. If no such number exists the fibrewise pointed category is said to be infinite.

Note that for any pointed space B' and pointed map $\alpha : B' \to B$ we have

$$\mathrm{cat}_{B'}^{B'} \, \alpha^* X \le \mathrm{cat}_B^B X.$$

In particular, $\mathrm{cat}_B^B X$ is bounded below by the pointed category of the fibres of X.

Of course, $\mathrm{cat}_B^B X = 1$ if and only if X is fibrewise pointed contractible. Also $\mathrm{cat}_B^B X \le 2$ if X is the reduced fibrewise suspension of a fibrewise pointed space.

If we disregard the section then the fibrewise category $\mathrm{cat}_B X$ of X is defined and cannot exceed $\mathrm{cat}_B^B X$. The relation between these invariants will be considered further below.

For any fibrewise pointed space X over B the r-fold fibrewise product $\prod_B^r X$ is defined ($r = 1, 2, \ldots$), and contains the union $T_B^r(X, B)$ of the preimages $\pi_i^{-1}(B)$ ($i = 1, \ldots, r$) of the section. We may refer to $T_B^r(X, B)$ as the *fibrewise fat wedge*. Note that $\prod_B^r X$ contains the diagonal ΔX of X while $T_B^r(X, B)$ contains the diagonal ΔB of B. In other words the pair

$$\prod_B^r(X, B) = (\prod_B^r X, T_B^r(X, B))$$

contains the diagonal $\Delta(X,B) = (\Delta X, \Delta B)$ of the pair (X,B). By (6.1) and (6.2) of [65] we have fibrewise versions of the Whitehead characterization of pointed category, as follows.

Proposition 19.5 *Suppose that the section B of the fibrewise pointed space X admits a fibrewise pointed categorical neighbourhood. If the diagonal map $\Delta : X \to \prod_B^r X$ can be compressed into $T_B^r(X,B)$ by a fibrewise pointed homotopy, for some $r \geq 1$, then $\mathrm{cat}_B^B X \leq r$.*

Proposition 19.6 *Let X be a normal fibrewise pointed space over B. If $\mathrm{cat}_B^B X \leq r$, for $r \geq 1$, then the diagonal map $\Delta : X \to \prod_B^r X$ can be compressed into $T_B^r(X,B)$ by a fibrewise pointed homotopy.*

One way in which fibrewise pointed categorical neighbourhoods of the section can arise is as follows. Let P, as before, be a principal G-bundle over B, where G is a topological group. Let Y be a pointed G-space such that the (closed) basepoint y_0 admits a pointed G-categorical neighbourhood U in Y. Then the section $P_{\#} y_0$ admits the fibrewise pointed categorical neighbourhood $P_{\#} U$ in $P_{\#} Y$.

In view of Propositions 19.5 and 19.6 we may choose to define fibrewise pointed category through the above characterization, distinguishing it when necessary by adding the phrase 'in the new sense'. Note that $\mathrm{cat}_B^B X \leq 2$, in this sense, if and only if X admits fibrewise coHopf structure.

Recall that the r-fold fibrewise smash product $\bigwedge_B^r X$ is obtained from $\prod_B^r X$ by fibrewise collapsing $T_B^r(X,B)$. We denote by $\Delta' : X \to \bigwedge_B^r X$ the projection of the diagonal into the fibrewise smash product.

Definition 19.7 The *weak fibrewise pointed category* $\mathrm{wcat}_B^B X$ of the fibrewise pointed space X is the least number r such that $\Delta' : X \to \bigwedge_B^r X$ is fibrewise pointed null-homotopic.

If no such number exists the weak fibrewise pointed category is said to be infinite. Of course

$$\mathrm{wcat}_B^B X \leq \mathrm{cat}_B^B X, \qquad (19.8)$$

in the new sense. The relation between the two invariants will be considered below.

We say that X admits *weak fibrewise coHopf structure* when $\mathrm{wcat}_B^B X \leq 2$, without attempting to give a meaning to the concept of weak fibrewise coHopf structure as such.

Note that for any pointed space B' and pointed map $\alpha : B' \to B$ we have

$$\mathrm{wcat}_{B'}^{B'} \alpha^* X \leq \mathrm{wcat}_B^B X.$$

In particular, $\mathrm{wcat}_B^B X$ is bounded below by the weak pointed category of the fibres of X.

There is a useful functor which sends each fibrewise pointed space X into the mapping cone $C(s)$ of the section, and similarly for fibrewise pointed maps and fibrewise pointed homotopies. Then

$$\mathrm{cat}_B^B X \geq \mathrm{cat}^* C(s) \leq \mathrm{cat}^* B + 1; \tag{19.9}$$

the first inequality resulting from use of the functor, the second being due to Berstein and Ganea [12]. Similarly

$$\mathrm{wcat}_B^B X \geq \mathrm{wcat}^* C(s) \leq \mathrm{wcat}^* B + 1. \tag{19.10}$$

When the section s is a cofibration we may replace $C(s)$ by the pointed space X/B obtained from X by collapsing B. The index of nilpotency nil $H^*(X, B)$ of the cohomology of the pair (X, B), with arbitrary coefficients, is a lower bound for $\mathrm{wcat}_B^B X$ and hence for $\mathrm{cat}_B^B X$.

Polar category

In what follows we shall need another invariant, the polar category. This is similar to the well-known sectional category, discussed in Section 8 of [83], and so let us start with a few notes on that subject. We work over a base space B without basepoint. Given a fibrewise space X over B we describe a subset W of B as *section categorical* if X_W admits a section over W. The *sectional category* secat X of X is defined to be the least number of section categorical open sets required to cover B. If no such number exists the sectional category is said to be infinite. Note that

$$\mathrm{secat}\, X \leq \mathrm{cat}\, B \tag{19.11}$$

when X is fibrant.

Consider the r-fold fibrewise join

$$X^{(r)} = X *_B * \ldots *_B X$$

of X with itself, in the Milnor topology. Provided B is paracompact, Schwarz has shown that secat $X \leq r$ if and only if $X^{(r)}$ admits a section. This result, for which [64] is a convenient reference, leads at once to an upper bound for sectional category, as in (8.2) of [83]:

Proposition 19.12 *Let B be a finite complex and let X be a fibre bundle over B with $(q-1)$-connected fibre, where $q \geq 1$. Then*

$$\mathrm{secat}\, X < (q+1)^{-1}(\dim B + 2).$$

For lower bounds we turn to cohomology again and consider the homomorphism

$$p^* : H^*(B) \to H^*(X)$$

induced by the projection. We find that

$$\operatorname{secat} X \geq \operatorname{nil} \ker p^*. \tag{19.13}$$

In particular, suppose that X is an (orthogonal) $(q-1)$-sphere bundle over B, associated with a euclidean q-plane bundle ξ. The Euler class $W_q(\xi)$ is defined with integral coefficients when ξ is orientable, with mod 2 coefficients in any case. Either directly or via Schwarz' theorem, we find that $\operatorname{secat} X \leq r$ implies $(W_q(\xi))^r = 0$.

When X is fibrewise pointed the sectional category itself is without interest. For our purposes, however, another numerical invariant plays an important rôle. Let us say that X is *polarized* if every section of $X - B$ is vertically homotopic in X to the standard section.

Proposition 19.14 *Let B be a finite complex and let E be a $(q-1)$-sphere bundle over B. Suppose that $\dim B < 2q - 2$. If the fibrewise suspension $\Sigma_B E$ is polarized then E admits a section.*

For consider the fibrewise space $\Lambda_B \Sigma_B E$ of fibrewise interpolar paths in $\Sigma_B E$, i.e. of paths in each fibre from the south pole to the north. Since E is a fibre bundle over B with fibre S^{q-1} it follows that $\Lambda_B \Sigma_B E$ is a fibre bundle over B with fibre ΛS^q, the space of interpolar paths in S^q. We observe that ΛS^q is homeomorphic to ΩS^q, the space of loops. We embed E in $\Lambda_B \Sigma_B E$ using the adjoint of the identity on $\Sigma_B E$ and consider the pair $(\Lambda_B \Sigma_B E, E)$ as a fibre bundle over B with fibre the pair $(\Lambda S^q, S^{q-1})$. Since the fibre is $(2q-3)$-connected, by the Freudenthal theorem, the conclusion of Proposition 19.14 follows by obstruction theory.

By way of illustration, take $B = S^n$ and let E be obtained by the clutching construction from an element $\alpha \in \pi_{n-1} SO(q)$. The necessary and sufficient condition for E to admit a section is that $\rho_* \alpha = 0$ in $\pi_{n-1}(S^{q-1})$, where ρ is the evaluation map. The corresponding condition for $\Lambda_B \Sigma_B E$ to admit a section is that $\Sigma_* \rho_* \alpha = 0$ in $\pi_n(S^q)$, where Σ_* denotes the suspension operator, and this is therefore the condition for $\Sigma_B E$ to be polarized. For an example where this condition is satisfied but E does not admit a section, take $n = 4$ and $q = 3$. Take α to be twice a generator of the infinite cyclic group $\pi_3 SO(3)$. Then $\rho_* \alpha \neq 0$, since ρ_* is an isomorphism, but $\Sigma_* \rho_* \alpha = 0$, since $\pi_4(S^3) = Z_2$.

Proposition 19.15 *Let B be a space and let E be a fibrewise space over B. Suppose that the fibrewise suspension $\Sigma_B E$ is polarized. Then $\Sigma_B E$ is fibrewise coHopf.*

Here we regard $\Sigma_B E$ as a fibrewise pointed space with the section s where the suspension parameter $t = 0$. Then if $h_u : \Sigma_B E \to \Sigma_B E$ is a fibrewise homotopy of $s'p$ into sp, where s' is the section where $t = 1$, then a fibrewise coHopf structure

$$m : \Sigma_B E \to \Sigma_B E \vee_B \Sigma_B E$$

is given by

$$m(x,t) = \begin{cases} ((x, 4t), (x, 0)) & (0 \le t \le \tfrac{1}{4}) \\ (h_{2-4t}(x, t), (x, 0)) & (\tfrac{1}{4} \le t \le \tfrac{1}{2}) \\ ((x, 0), (x, 4t - 2)) & (\tfrac{1}{2} \le t \le \tfrac{3}{4}) \\ ((x, 0), h_{4-4t}(x, t)) & (\tfrac{3}{4} \le t \le 1). \end{cases}$$

After these preliminaries we are ready to define the polar category. Given a fibrewise pointed space X over B, we say that a subset W of B is *polar categorical* if X_W is polar. Then we define the *polar category* polcat X of X to be the least number of polar categorical open sets required to cover B. If no such number exists the polar category is said to be infinite.

If $X = \Sigma_B E$, for some fibrewise space E, then we have polcat $X =$ secat $\Lambda_B \Sigma_B E$, hence

$$\text{polcat } \Sigma_B E \le \text{secat } E \tag{19.16}$$

In the other direction we prove

Proposition 19.17 *Let E be a $(q-1)$-sphere bundle over the finite complex B. Suppose that $\dim B \le 2qr - 3$, for some r, and that polcat $\Sigma_B E \le r$. Then secat $E \le r$.*

As in the proof of Schwarz' theorem the condition polcat $\Sigma_B E \le r$ implies that the polar sections of $\Sigma_B E^{(r)}$ are vertically homotopic. Then by Proposition 19.14 the condition $\dim B \le 2qr - 3$ implies that $E^{(r)}$ itself admits a section. Now apply Schwarz' theorem.

Corollary 19.18 *Let ξ be a euclidean q-plane bundle over the finite complex B. Suppose that polcat $\Sigma_B E \le r$, where $E = S(\xi)$ is the associated $(q-1)$-sphere bundle. Then $(W_q(\xi))^r = 0$.*

Here $W_q(\xi)$, as before, denotes the Euler class with integral coefficients when ξ is orientable, with mod 2 coefficients in any case. To deduce Corollary 19.18 from Proposition 19.17, we can assume $\dim B \le qr + 1$, without real loss of generality. Then polcat $\Sigma_B E \le r$ implies that secat $E \le r$ and so $(W_q(\xi))^r = 0$.

Our interest in polar category is due to

Proposition 19.19 *Let X be a normal fibrewise pointed space over B. Suppose that B admits a fibrewise pointed categorical neighbourhood in X. Then*

$$\text{cat}_B^B X \le 1 + \text{polcat } X \cdot \text{cat}_B(X - B).$$

To prove Proposition 19.19, let N be a fibrewise pointed categorical neighbourhood of B. Since X is normal and B is closed there exist neighbourhoods

N' and N'' of B such that $\bar{N}'' \subseteq N'$ and $\bar{N}' \subseteq N$. Rather than proceed at once to the general case let us suppose, to begin with, that X is polarized.

Let U be an open set of $X - \bar{N}'$ which is fibrewise contractible in $X - B$. So there exists a fibrewise homotopy $h_t : U \to X - B$ such that h_0 is the inclusion and $h_1 = \sigma p \mid U$ for some section σ of $X - B$. By assumption there exists a vertical homotopy $k_t : B \to X$ of σ into the standard section s. By performing first h_t and then $k_t p \mid U$ we obtain a fibrewise null-homotopy of the inclusion into $sp \mid U$. Since U and N'' are disjoint the open set $U \cup N''$ is fibrewise pointed categorical in X, where the fibrewise pointed null-homotopy on the union is given by the fibrewise null-homotopy on U and the fibrewise pointed contraction of N on N''.

In the general case X may not be polarized. However, we can use the above argument over each polar categorical open set of X and conclude that $X - \bar{N}'$ can be covered by $\mathrm{polcat}\, X \cdot \mathrm{cat}_B(X - B)$ fibrewise pointed categorical open sets of X. We supplement this covering of $X - \bar{N}'$ by the fibrewise categorical neighbourhood N of $\dot{B}$ and conclude that X can be covered by $1 + \mathrm{polcat}\, X \cdot \mathrm{cat}_B(X - B)$ fibrewise pointed categorical open sets, as stated in Proposition 19.19.

For example, take G to be the orthogonal group $O(q)$ acting on the sphere S^q in the usual way, with the poles as fixed points. The open northern hemisphere provides a pointed G-categorical neighbourhood of the north pole, while the complement of the north pole is obviously G-categorical. So when X is a fibrewise pointed sphere-bundle Proposition 19.19 reduces to

$$\mathrm{cat}_B^B X \leq 1 + \mathrm{polcat}\, X, \tag{19.20}$$

which since X is fibrant implies

$$\mathrm{cat}_B^B X \leq 1 + \mathrm{cat}\, B. \tag{19.21}$$

So far we have mainly been concerned with fibrewise spaces in general. For fibrant fibrewise spaces, however, more can be said and we have already had some examples of this. Fibre bundles form an important class of fibrewise spaces and it is usually unnecessary to assume the existence of a structural group.

Let X be a fibrewise pointed fibre bundle over B with fibre Y. The r-fold fibrewise product $\prod_B^r X$ is then a fibrewise pointed fibre bundle over B with fibre $\prod^r Y$. Also the fibrewise fat wedge $T_B^r(X, B)$ is a fibrewise pointed fibre bundle over B with fibre $T^r(Y, y_0)$. Some condition seems to be necessary before we can say that the fibrewise smash product $\bigwedge_B^r X$ is a fibrewise pointed fibre bundle over B with fibre $\bigwedge^r Y$. Certainly it is sufficient for Y to be compact. Note that when Y is $(q-1)$-connected then $(\prod^r Y, T^r(Y, y_0))$ and $\bigwedge^r Y$ are $(qr - 1)$-connected.

Various results about fibrewise pointed category in the new sense can be found in [93], including

Proposition 19.22 *Let B be a finite complex. Let X be a fibrewise pointed fibre bundle over B with $(q-1)$-connected fibre Y, also a complex, where $q \geq 1$. Then*

$$\mathrm{cat}_B^B X \leq [q^{-1} \dim(X-B)] + 1,$$

where $[x]$ means the integer part of a real number x.

For example, take B to be the real projective n-space P^n. Take X to be the fibrewise pointed circle bundle $S(\xi \oplus 1)$, where ξ is the Hopf line bundle. Then Proposition 19.22 shows that $\mathrm{cat}_B^B X \leq n+2$. On the other hand, the Thom space X/B of ξ is just P^{n+1} and so (19.10) shows that $\mathrm{wcat}_B^B X \geq n+2$. In this case, therefore, both fibrewise pointed category and weak fibrewise pointed category are equal to $n+2$. If we use multiples of the Hopf bundle rather than the Hopf bundle itself the cohomological lower bound no longer coincides with the upper bound given by Proposition 19.22.

Proposition 19.23 *Let B be a finite complex. Let X be a fibrewise pointed fibre bundle over B with $(q-1)$-connected fibre Y, also a complex, where $q \geq 1$. Suppose that Y is a finite complex and that (X, B) is a relative complex. Suppose that $\dim(X-B) \leq q(r+1) - 2$, for some r, and that $\mathrm{wcat}_B^B X \leq r$. Then $\mathrm{cat}_B^B X \leq r$.*

Since the conclusion is an immediate consequence of Proposition 19.22 when $q = 1$ we assume $q \geq 2$. The argument which follows is essentially a fibrewise version of that used by Berstein and Ganea to prove the corresponding result Theorem 3 of [12] in the ordinary theory.

Consider the diagram shown below, where E is the fibrewise mapping path-space $W_B(\lambda)$ of the natural projection $\lambda : \prod_B^r X \to \bigwedge_B^r X$. We recall that $W_B(\lambda)$ is defined as the pull-back of the fibrewise path-space $\mathcal{P}_B \bigwedge_B^r X$ over $\bigwedge_B^r X$.

$$
\begin{array}{ccccc}
T_B^r(X,B) & \xrightarrow{\ j\ } & \prod_B^r X & \xrightarrow{\ q\ } & \bigwedge_B^r X \\
\downarrow{\scriptstyle g} & & {\scriptstyle f'}\uparrow\downarrow{\scriptstyle f} & & \downarrow{\scriptstyle 1} \\
F & \xrightarrow{\ i\ } & E & \xrightarrow{\ p\ } & \bigwedge_B^r X
\end{array}
$$

Here f is given by the identity into the first factor of the pull-back, by the fibrewise constant path following λ into the second. Since the fibrewise pointed map $\mathcal{P}_B \bigwedge_B^r X \to \bigwedge_B^r X$ is a fibrewise pointed fibration, so is the composition $p : E \to \bigwedge_B^r X$ of the second projection with this fibrewise pointed map. Finally, f' is the projection of the pull-back into the first factor, $F = p^{-1}B$ is the fibrewise fibre (that is, the fibrewise mapping fibre $F_B(\lambda)$) and i, j are the inclusions.

Now consider the pair $(F_0, T^r(Y, y_0))$, where F_0 is the fibre of F over the basepoint of B and $T^r(Y, y_0)$, the corresponding fibre of $T_B^r(X, B)$, is embedded in F_0 through the map g. Berstein and Ganea show, in the proof of Theorem 3 of [12], that this pair is $(q(r+1)-2)$-connected. Since $\mathrm{wcat}_B^B X \leq r$, by hypothesis, we have that $\rho \circ \Delta$ is fibrewise pointed null-homotopic, and so $p \circ f \circ \Delta$ is fibrewise pointed null-homotopic. Since the lower row in the diagram is a fibrewise pointed fibration we can lift the fibrewise pointed null-homotopy of $p \circ f \circ \Delta$ to a fibrewise pointed homotopy of $f \circ \Delta$ into a fibrewise pointed map $\psi : X \to F$. Since $\dim(X - B) \leq q(r + 1) - 2$ a standard deformation argument, on the lines of the proof of (6.1) given in [93], shows that ψ is fibrewise pointed homotopic to $g \circ \phi$ for some fibrewise map $\phi : X \to T_B^r(X, B)$. Since

$$j \circ \phi = f' \circ f \circ j \circ \phi = f' \circ i \circ g \circ \phi$$
$$\simeq_B f' \circ i \circ \psi \simeq_B f' \circ f \circ \Delta = \Delta$$

we conclude that $\mathrm{cat}_B^B X \leq r$, as asserted.

Fibrewise coHopf structure

Let us now turn to the special case where the fibrewise pointed category does not exceed 2, in other words fibrewise pointed spaces which admit fibrewise coHopf structure. This case has already been discussed in [87] and, more thoroughly, in [130]. However, weak fibrewise coHopf structures have not previously been considered and so we need to include them in the discussion. Returning to the situation in Proposition 19.23, we have at once

Proposition 19.24 *Let B be a finite complex. Let X be a fibrewise pointed fibre bundle B with $(q - 1)$-connected fibre Y, also a complex, where $q \geq 1$. Suppose that*

$$\dim(X - B) \leq 3q - 2$$

and that X admits weak fibrewise coHopf structure. Then X admits fibrewise coHopf structure.

To make further progress let us assume that Y is a coHopf space as, for example, when Y is a sphere. More specifically, let us assume that we have a pointed homotopy

$$h_t : Y \to Y \times Y$$

of the diagonal into the inclusion of a coHopf structure $m : Y \to Y \vee Y$. Suppose that $(X, B \cup Y)$ is a relative complex, that B is $(n - 1)$-connected, that Y is $(q - 1)$-connected, and hence $(X, B \cup Y)$ is $(n + q - 1)$-connected.

The primary obstruction to extending m to a fibrewise map $X \to X \vee_B X$ is an element

$$\theta \in H^{n+q}(X, B \cup Y; \pi_{n+q-1}(Y \vee Y)).$$

The primary obstruction to extending h_t to a fibrewise pointed homotopy $X \to X \times_B X$ of the diagonal into the fibrewise wedge is an element

$$\phi \in H^{n+q}(X, B \cup Y; \pi_{n+q}(Y \times Y, Y \vee Y)).$$

The primary obstruction to extending the pointed null-homotopy h'_t of Δ' to a fibrewise pointed null-homotopy of $\Delta' : X \to X \wedge_B X$ is an element

$$\psi \in H^{n+q}(X, B \cup Y; \pi_{n+q}(Y \wedge Y)).$$

It is a simple exercise to show that

$$\theta = \delta_*(\phi), \tag{19.25}$$

where δ_* is the coefficient homomorphism given by the boundary operator

$$\delta : \pi_{n+q}(Y \times Y, Y \wedge Y) \to \pi_{n+q-1}(Y \vee Y),$$

and that

$$\psi = \lambda_*(\phi), \tag{19.26}$$

where λ^* is the coefficient homomorphism

$$\pi_{n+q}(Y \times Y, Y \vee Y) \to \pi_{n+q}(Y \wedge Y)$$

induced by the natural projection. Note that δ admits a left inverse, hence so does δ_*. So if we can calculate θ, in a particular case, then we can calculate ϕ and hence ψ. Examples will be given in the next section.

When Y is a sphere, say $Y = S^q$ with $q \geq 2$, the deformation h_t of the diagonal always exists and any two such deformations are, in a certain sense, equivalent. Thus ϕ is the primary obstruction to the fibrewise pointed deformation of Δ into $X \vee_B X$, θ is the primary obstruction to the existence of fibrewise coHopf structure on X, and ψ is the primary obstruction to the existence of a fibrewise pointed null-homotopy of $\Delta' : X \to X \wedge_B X$. The extension restrictions can be ignored, for the reasons stated.

20 The reduced product (James) construction

Reduced product spaces

Under certain conditions the reduced product space JX of a pointed space X has the same homotopy type as $\Omega \Sigma^* X$, the loop-space on the reduced suspension of X. Several proofs can be found in the literature. The original proof [77] made unnecessarily strong assumptions. Later, in the last chapter of [44], tom Dieck, Kamps and Puppe gave a proof under much weaker conditions and showed that they could not be further weakened.

Here, following [63], we give a simple proof of the original result, without striving for maximum generality, and show that the same method can be used to prove an equivariant version of the reduced product theorem and hence a fibrewise version.

To begin with we work in the category of pointed spaces. Unless otherwise stated maps and homotopies are basepoint-preserving. Suspension always means reduced suspension. Later we shall turn to the equivariant theory, but here again the action is to be basepoint-preserving.

Let X be a space with basepoint x_0, which we assume to be closed. As a set the reduced product space JX may be described as the free monoid on X, with x_0 acting as neutral element. Thus a point of JX may be represented by a finite sequence of points of X. Sequences with not more than n terms form a subset $J^n X \subseteq JX$ ($n = 0, 1, \ldots$). We topologize $J^n X$ as a quotient space of the n-fold topological product $\prod^n X$. Then we obtain a sequence of spaces

$$J^0 X \subseteq J^1 X \subseteq \ldots \subseteq J^n X \subseteq \ldots,$$

where each member of the sequence is contained in the next as a closed subspace. Finally, we topologize JX itself as the colimit of the sequence; this does not mean (see (17.10) of [44]) that the multiplication on JX is continuous. We prove

Theorem 20.1 *Let X be a well-pointed compact Hausdorff space. Suppose that X can be covered by open sets, each of which is contractible in X to the basepoint. Then JX has the same homotopy type as $\Omega \Sigma^* X$.*

The supposition is that the category of X is defined, in the pointed sense, and is therefore finite, by compactness. In the proof which follows we disregard basepoints. We shall show that JX and $\Omega \Sigma X$ have the same homotopy type in the non-pointed sense. For reasons given in (17.3) of [44] this will imply that they have the same homotopy type in the pointed sense.

First, observe that the natural projection from $\prod^n X$ to $J^n X$ ($n = 0, 1, \ldots$) is proper, since X is compact, and hence the function

$$X \times J^n X \to J^{n+1} X$$

defined by the multiplication on JX is continuous. Furthermore, $X \times JX$ is the colimit of the sequence

$$X \times J^0 X \subseteq X \times J^1 X \subseteq \ldots \subseteq X \times J^n X \subseteq \ldots,$$

and so the function

$$T : X \times JX \to JX$$

thus defined is continuous.

The first step in the proof of the theorem is to show that, for each n, the homotopy push-out of the cotriad

$$J^{n+1} X \longleftarrow X \times J^n X \xrightarrow{\pi_2} J^n X$$

is contractible, where the left-hand arrow is given by the restriction of T. This is trivial when $n = 0$; make the inductive hypothesis that it is true for some $n \geq 0$. Write

$$X \triangleright J^n X = x_0 \times J^n X \cup X \times J^{n-1} X \subseteq X \times J^n X$$

and consider the diagram shown below

$$
\begin{array}{ccc}
X \times J^n X & \xrightarrow{\pi_2} & J^n X \\
\downarrow & & \downarrow \\
X \triangleright J^{n+1} X \longrightarrow x_0 \times J^{n+1} X \longrightarrow & & J^{n+1} X \\
\downarrow & \downarrow & \downarrow \\
J^{n+1} X \longrightarrow & J^{n+2} X \longrightarrow & x_0
\end{array}
$$

The outer square is a homotopy push-out, by the inductive hypothesis. Also the top half of the diagram is a push-out in the topological sense, hence a homotopy push-out (since the inclusion $J^n X \to J^{n+1} X$ is a cofibration so is the inclusion $X \times J^n X \to X \triangleright J^{n+1} X$). Similarly, the bottom left-hand square is a homotopy push-out. Hence it follows from Theorem 10(ii) of Mather [102] that the bottom right-hand square is also a homotopy push-out, which proves the inductive step.

Now the colimit of this sequence of contractible homotopy push-outs is just the homotopy push-out of the cotriad

$$JX \xleftarrow{T} X \times JX \xrightarrow{\pi_2} JX.$$

We conclude, therefore, that this space is contractible. So far we have only used the assumptions that X is well-pointed and compact Hausdorff.

Now consider the diagram shown below, which depicts two adjacent faces of a cube

$$
\begin{array}{ccc}
X \times JX & \xrightarrow{\pi_2} & JX \\
& \searrow & \\
\downarrow & JX & \downarrow \\
X & \quad\longrightarrow\quad \downarrow \longrightarrow & x_0 \\
& \searrow & \\
& x_0 &
\end{array}
$$

Obviously, the back face in the diagram is a homotopy pull-back. We assert that the other face is also a homotopy pull-back. This amounts to showing that the 'shearing map'

$$\xi : X \times JX \to X \times JX,$$

with components (π_1, T), is a fibrewise homotopy equivalence, where $X \times JX$ is regarded as a fibre space over X through the first projection. However, X is path-connected and of finite category, and so this follows from Dold's theorem (Theorem 14.1).

Thus both faces are homotopy push-outs and so, completing the diagram of the cube as shown below, we can apply Theorem 11 of Mather [102] and conclude that the front face is a homotopy pull-back, from which Theorem 20.1 follows at once.

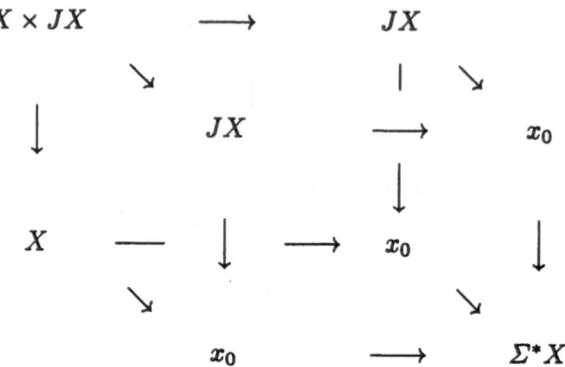

The equivariant version

We turn now to the equivariant version of the theorem, which does not seem to have been treated in the literature. Specifically, let X be a (pointed) G-space, where G is a compact group. Then JX and $\Omega\Sigma^*X$ are also G-spaces and we state

Theorem 20.2 *Let X be an equivariantly well-pointed compact Hausdorff G-space, where G is a compact Lie group. Suppose that X can be covered by invariant open sets each of which is G-contractible to the basepoint. Then JX has the same G-homotopy type as $\Omega\Sigma^*X$.*

The proof of Theorem 20.2 proceeds on the same lines as the proof of Theorem 20.1. An equivariant version of Mather's theory is needed, but this is completely routine. An equivariant version of Dold's theorem is also needed, but this too poses no problems. It seems unnecessary, therefore, to go through the details.

The supposition is equivalent to the assumption that the G-category of X is defined, in the pointed sense, and is therefore finite, by compactness. This is true, for example, if X is a finite G-complex such that the fixed point set X^H is connected for all closed subgroups H of G, as is the case when X is a double suspension. Thus take G to be the orthogonal group $O(n-1)$ acting on the sphere S^n in the usual way, where $n \geq 1$, so that the poles are left fixed and a point x_0 on the equator is also left fixed. Cover S^n by the enlarged hemispheres, which are $O(n-1)$ contractible to their respective poles, and then deform the poles into x_0, which we take as basepoint, along the line of longitude. We see that the condition is fulfilled in this case.

The fibrewise version

Finally, we turn to the fibrewise version of the reduced product theorem. We work in the category of fibrewise pointed spaces over a given base space B and require the standard section to be closed. Then the *fibrewise reduced product space* $J_B X$ is defined as in [88] for each fibrewise space X over B. The fibres of $J_B X$ are just the reduced product spaces of the corresponding fibres of X. We prove

Theorem 20.3 *Let X be a (fibrewise pointed) fibre bundle over the base space B with compact structure Lie group G and fibre the (pointed) G-space A. Suppose that A satisfies the conditions of Theorem 20.2. Then $J_B X$ has the same fibrewise homotopy type as $\Omega_B \Sigma_B^B X$.*

For let P be the principal G-bundle over B associated with X. The associated bundle functor $P_\#$ transforms the equivariant reduced product space JA into the fibrewise reduced product space $J_B X$, and similarly with the suspension and loop-space. Since JA has the same equivariant homotopy type as $\Omega \Sigma^* A$, by Theorem 20.2, it follows at once that $J_B X$ has the same fibrewise homotopy type as $\Omega_B \Sigma_B^B X$.

The original version of the fibrewise reduced product theorem is due to Eggar [59] but conditions are imposed which are rather inconvenient in practice. More recently, James [88] published a fibrewise version of the proof given by tom Dieck, Kamps and Puppe [44] in the ordinary case. This does not assume local triviality. However Theorem 20.3 seems adequate for most applications and its proof is a good deal simpler than the alternatives. For example, it applies in the case of an orthogonal sphere-bundle which admits a pair of mutually orthogonal sections. In Section 14 of Part II we shall resume the discussion of fibrewise reduced product spaces as part of a study of fibrewise configuration spaces and establish a fibrewise stable splitting theorem.

21 Fibrewise Hopf and coHopf structures

Fibrewise Hopf and coHopf spaces

Fibrewise Hopf and coHopf spaces have already been defined. In this section we establish some of their properties and give some examples. Other results may be found in Scheerer [120].

For fibrant fibrewise pointed spaces the following result is useful.

Proposition 21.1 *Let B be numerably categorical. Let X be a fibrant fibre-wise pointed space over B.*
(i) *If a Hopf structure on the fibre of X can be extended to a fibrewise multiplication on X then it can be extended to a fibrewise Hopf structure.*
(ii) *If a coHopf structure on the fibre of X can be extended to a fibrewise comultiplication on X then it can be extended to a fibrewise coHopf structure.*

Here we mean extend up to homotopy, rather than extend as a map. To prove (i) let $m : X \times_B X \to X$ be a fibrewise multiplication on X extending a Hopf structure on the fibre. Let $\sigma_j : X \to X \times_B X$ $(j = 1, 2)$ be the axial sections of the fibrewise product. Then the restriction of the fibrewise map $m \circ \sigma_j : X \to X$ to the fibre is homotopic to the identity. Hence $m \circ \sigma_j$ is a fibrewise pointed homotopy equivalence, by Dold's theorem. Let $\alpha_j : X \to X$ be a fibrewise pointed homotopy inverse of $m \circ \sigma_j$. Then

$$m \circ (\alpha_1 \times \alpha_2) : X \times_B X \to X$$

is a fibrewise Hopf structure on X which extends the given Hopf structure on the fibre. This proves (i) and the proof of (ii) is similar.

Proposition 21.2 *Let X be a numerably fibrewise pointed categorical and fibrewise well-pointed fibrewise Hopf space over B. Then X admits fibrewise homotopy inverses on each side.*

For let $m : X \times_B X \to X$ be the fibrewise Hopf structure. By using the fibrewise pointed homotopy extension property we may suppose, with no real loss of generality, that the section $s : B \to X$ is a strict neutral section for m, in the sense that

$$m \circ (c \times 1) \circ \Delta = 1 = m \circ (1 \times c) \circ \Delta,$$

where $c = s \circ p$ as before. Regard $X \times_B X$ as a fibrewise pointed space over X using the first projection and the section given by $(c \times 1) \circ \Delta$. Then the fibrewise shearing map

$$k : X \times_B X \to X \times_B X$$

is fibrewise over X, where the components of k are given by $\pi_1 \circ k = \pi_2$, $\pi_2 \circ k = m$. By Proposition 21.1 k is a fibrewise pointed homotopy equivalence. Hence the composition

$$X \xrightarrow{u} X \times_B X \xrightarrow{\ell} X \times_B X \xrightarrow{\pi_2} X$$

provides a right fibrewise homotopy inverse for m, where $u = (1 \times c) \circ \Delta$ and ℓ is a fibrewise pointed homotopy inverse of k. Similarly, m admits a left fibrewise homotopy inverse.

One would hope for a similar result to be true in the dual case. However, even for ordinary coHopf spaces there is a problem here, and the best way to proceed seems to be to prove

Proposition 21.3 *Let B be numerably categorical. Let X be a fibrant fibre-wise coHopf space over B. Suppose that the coHopf structure on the fibre admits a homotopy inverse on either side. Then the fibrewise coHopf structure on X admits a fibrewise homotopy inverse on the same side.*

For suppose that the fibrewise coHopf structure m, restricted to the fibre, admits a homotopy inverse on the right. Consider the fibrewise 'coshearing' map
$$k : X \vee_B X \to X \vee_B X$$
given by the identity on the first summand, by m on the second. By hypothesis k is a homotopy equivalence on the fibre and hence, by Proposition 10.2, a fibrewise pointed homotopy equivalence. Now the composition
$$X \xrightarrow{\sigma_2} X \vee_B X \xrightarrow{\ell} X \vee_B X \xrightarrow{u} X$$
provides a right fibrewise homotopy inverse for m, where $u = \nabla \circ (1 \vee c)$ and ℓ is a fibrewise homotopy inverse of k.

The existence of homotopy inverses on each side for ordinary coHopf structures has been discussed by Ganea [66] who shows (using coshearing maps) that it is sufficient for the fibres to be simply-connected.

Fibrewise retraction and coretraction

For each fibrewise pointed space X over B the right adjoint of the identity on $\Sigma_B^B X$ is a fibrewise pointed map $u : X \to \Omega_B \Sigma_B^B X$ while the left adjoint of the identity on $\Omega_B X$ is a fibrewise map $v : \Sigma_B^B \Omega_B X \to X$. Following the practice of Ganea [66] and others in the ordinary theory, let us describe a left inverse of u, up to fibrewise pointed homotopy, as a *fibrewise retraction*, and a right inverse of v, up to fibrewise pointed homotopy, as a *fibrewise co-retraction*. When X is fibrewise well-pointed the qualification 'up to fibrewise pointed homotopy' is unnecessary.

Clearly, if X is a fibrewise retract of $\Omega_B \Sigma_B^B X$ then X is a fibrewise Hopf space, while if X is a fibrewise coretract of $\Sigma_B^B \Omega_B X$ then X is a fibrewise coHopf space. Under certain conditions the converse implications hold.

Proposition 21.4 *Let X be a fibrewise Hopf space over B. Suppose that X satisfies the conditions of Theorem 20.3. Then X is a fibrewise retract of the fibrewise loop-space $\Omega_B \Sigma_B^B X$ on the reduced fibrewise suspension of X.*

For consider the fibrewise reduced product space $J_B X$ of X, i.e. the fibre-wise free monoid generated by X with topology as in Section 20. The fibre-wise Hopf structure on X determines a fibrewise retraction of $J_B X$ onto X

by extending multiplicatively. Now it is shown in Theorem 20.3, under the hypotheses of Proposition 21.4, that $J_B X$ has the same fibrewise pointed homotopy type as $\Omega_B \Sigma_B^B X$, hence the result.

Proposition 21.5 *Let X be a fibrewise well-pointed fibrewise coHopf space over B. Then X is a fibrewise coretract of the reduced fibrewise suspension $\Sigma_B^B \Omega_B X$ of the fibrewise loop-space on X.*

For each fibrewise pointed space Y we have a standard fibrewise coHopf structure

$$k : \Sigma_B^B Y \to \Sigma_B^B Y \vee_B \Sigma_B^B Y.$$

In particular, we have this when $Y = \Omega_B X$. Consider the diagram shown below where v is as before and j is the inclusion

$$
\begin{array}{ccc}
\Sigma_B^B \Omega_B X & \xrightarrow{\;v\;} & X \\
{\scriptstyle (v \vee v) \circ k} \big\downarrow & & \big\downarrow {\scriptstyle \Delta} \\
X \vee_B X & \xrightarrow[\;j\;]{} & X \times_B X
\end{array}
$$

It is not difficult to show (see (6.63) and (6.65) of [85]) that the fibrewise homotopy fibres of v and j are equivalent, in the sense of fibrewise pointed homotopy type. Moreover, the verticals in the diagram induce a fibrewise pointed homotopy equivalence of the fibrewise homotopy fibres. It follows that fibrewise coHopf structures $X \to X \vee_B X$ correspond precisely to fibrewise coretractions $X \to \Sigma_B^B \Omega_B X$.

In the ordinary case this result is due to Ganea [66] who works in the category of CW-spaces. Sunderland [130] uses Dold's theorem and Ganea's result instead.

It is hard to believe that there is not a proof of Proposition 21.4 on similar lines but we are not aware of one. Almost certainly the assumptions needed to make the proof we have given work are unnecessarily restrictive.

In the ordinary theory the existence of Hopf structure is equivalent to the vanishing of the Whitehead square, on the one hand, and to the existence of maps of Hopf invariant one on the other. Fibrewise versions of these results can be developed (see [59], for example), but these are somewhat routine. Again one can construct the fibrewise projective plane, from a fibrewise multiplication, and analyse its fibrewise cohomology. We shall not pursue these matters here, in full generality, although some special cases will be discussed later on.

Fibrewise pointed sphere-bundles

We turn now from the general theory to an important class of special cases, the fibrewise pointed sphere-bundles. As we shall see, there is plenty of interest in this range of examples. We may begin with a (real) n-plane bundle

ξ over B. The fibrewise compactification ξ_B^+ is a fibrewise pointed n-sphere bundle, the section being the complement of ξ, and every fibrewise pointed n-sphere bundle can be represented in this way. To look at the situation in another way, ξ_B^+ can be obtained from the sphere S^n, as a (pointed) $O(n)$-space, by applying the associated bundle functor $P_\#$ where P is a principal $O(n)$-bundle over B. Let us begin our discussion with some observations about $O(n)$-spaces.

Points of S^n are represented in the form

$$(\cos\alpha, x\sin\alpha) \quad (x \in S^{n-1}, \ 0 \le \alpha \le \pi)$$

with $e = (1,0)$ as basepoint. We regard S^n as an $O(n)$-space so that $g \in O(n)$ transforms $(\cos\alpha, x\sin\alpha)$ into $(\cos\alpha, g.x\sin\alpha)$. For each integer k we have an $O(n)$-map $\theta_k : S^n \to S^n$ of degree k, where

$$\theta_k(\cos\alpha, x\sin\alpha) = (\cos k\alpha, x\sin k\alpha).$$

Note that $\theta_{k_1 k_2} = \theta_{k_1}\theta_{k_2}$.

We regard $S^n \times S^n$ as an $O(n)$-space using the diagonal action. There is a well-known $O(n)$-map

$$\phi : S^n \times S^n \to S^n$$

of bidegree $(1 + (-1)^{n+1}, 1)$, which sends each pair (y, z) into the reflection of $\theta_{-1}z$ in the hyperplane orthogonal to y.

The coproduct $S^n \vee S^n$ may be identified with the invariant subspace

$$S^n \times \{e\} \cup \{e\} \times S^n \subseteq S^n \times S^n$$

Another $O(n)$-map

$$\psi : S^n \to S^n \vee S^n$$

of bidegree $(1 + (-1)^{n+1}, 1)$ is defined which sends $(\cos\alpha, x\sin\alpha)$ into

$$\begin{cases} ((\cos 3\alpha, x\sin 3\alpha), e) & (0 \le \alpha \le \tfrac{2}{3}\pi) \\ (e, (\cos 3\alpha, x\sin 3\alpha)) & (\tfrac{2}{3}\pi \le \alpha \le \pi). \end{cases}$$

We use these $O(n)$-maps, as described above, to construct fibrewise maps of the associated fibrewise pointed n-sphere bundles. Although we do not need to use Noakes' pioneering work in this area, a courtesy reference to [118] here is surely appropriate. Specifically let ξ, as before, be an n-plane bundle with fibrewise compactification ξ_B^+. Then for each integer k we have a fibrewise map $\Theta_k : \xi_B^+ \to \xi_B^+$ of degree k on the fibre. We also have the fibrewise multiplication

$$\Phi : \xi_B^+ \times_B \xi_B^+ \to \xi_B^+$$

of bidegree $(1 + (-1)^{n+1}, 1)$ on the fibres, and the fibrewise comultiplication

$$\Psi : \xi_B^+ \to \xi_B^+ \vee_B \xi_B^+$$

of the same bidegree on the fibres.

Proposition 21.6 *Let B be numerably categorical. Let ξ be an n-plane bundle over B, where n is odd. Suppose that the fibrewise pointed n-sphere bundle ξ_B^+ admits a fibrewise multiplication of bidegree $(2k+1, 2l+1)$ on the fibres, where k and l are integers. Then ξ_B^+ admits fibrewise Hopf structure.*

For let m be a fibrewise multiplication on ξ_B^+ of bidegree $(2k+1, 2l+1)$ on the fibres. Replace m by m', where

$$m'(y, z) = \Phi(\Theta_{-k}y, \Phi(\Theta_{-l}z, m(y, z)))$$

for y, z in the same fibre of ξ_B^+. Since Φ has bidegree $(2, 1)$ on each fibre we find that m' has bidegree $(1,1)$ on the fibres. By Proposition 21.1, therefore, ξ_B^+ admits fibrewise Hopf structure. Similarly, using Ψ instead of Φ we obtain

Proposition 21.7 *Let B be numerably categorical. Let ξ be an n-plane bundle over B, where n is odd. Suppose that the fibrewise pointed n-sphere bundle ξ_B^+ admits a fibrewise comultiplication of bidegree $(2k+1, 2l+1)$ on the fibres, where k, l are integers. Then ξ_B^+ admits fibrewise coHopf structure.*

As we shall see later, the conclusion no longer holds for even values of n (it is unnecessary to say this in the case of Proposition 21.5 since S^n cannot admit a multiplication of bidegree $(2k+1, 2l+1)$ when n is even). For odd values of n, however, these two results show that the existence of fibrewise Hopf structure and the existence of fibrewise coHopf structure on fibrewise pointed n-sphere bundles are 2-local problems. By using the machinery of fibrewise localization (see [105], for example) one can describe the situation with greater precision, but the message is that odd primes are irrelevant.

As we have already remarked, there is a relation between fibrewise multiplications

$$m : X \times_B X \to X$$

where $X = \xi_B^+$, and the corresponding fibrewise maps

$$c(m) : X *_B X \to \Sigma_B^B X,$$

obtained by the fibrewise version of the Hopf construction. Clearly, if m determines a Hopf structure on the fibre S^n then $c(m)$ determines a map $S^{2n+1} \to S^{n+1}$ of Hopf invariant one, and conversely. In fact one can go further and show that the existence of a fibrewise map

$$h : X *_B X \to \Sigma_B^B X$$

of Hopf invariant one on the fibres implies the existence of a fibrewise map

$$m : X \times_B X \to X$$

of bidegree $(1,1)$ on the fibres and hence the existence of fibrewise Hopf structure, by Proposition 21.1.

For which n-plane bundles ξ over B does ξ_B^+ admit fibrewise Hopf structure? The question only arises when $n = 0, 1, 3$ or 7, since otherwise the fibre does not admit Hopf structure. Ignoring the trivial case $n = 0$, we consider the other three cases in turn.

The classical Hopf structure on S^1, arising from complex multiplication, is $O(1)$-equivariant. Hence every line bundle ξ satisfies the condition and there is no more to be said.

The classical Hopf structure on S^3, arising from quaternionic multiplication, is $SO(3)$-equivariant. Hence every orientable 3-plane bundle ξ satisfies the condition.

The classical Hopf structure on S^7, arising from Cayley multiplication, is G_2-equivariant, where G_2 is the exceptional Lie group. Hence a 7-plane bundle ξ satisfies the condition if its structure group can be reduced to G_2.

What can be said in the other direction? It is only to be expected that the existence of fibrewise Hopf structure has implications for the characteristic classes. We find

Proposition 21.8 *Let B be a finite complex. A necessary condition on the n-plane bundle ξ for ξ_B^+ to admit fibrewise Hopf structure is that $w_i(\xi) = 0$ whenever $n + 1 - i$ is not a power of two.*

This was proved by Cook in his thesis through an analysis of the cohomology of the fibrewise projective plane determined by a fibrewise Hopf structure; another proof is given in [24].

The result shows, in particular that if $n = 3$ then $w_1(\xi) = 0$, so that ξ is orientable, while if $n = 7$ then $w_1(\xi) = 0$ and $w_2(\xi) = 0$ so that ξ admits spin structure. Thus we see that orientability is both necessary and sufficient when $n = 3$. Provided B is a finite complex and dim $B < 8$ we see that the existence of spin structure is both necessary and sufficient, since if the structural group is reducible to spin(7) it is also reducible to G_2.

Turning now to the dual problem, we recall that the Thom space B^ξ of the n-plane bundle ξ is equivalent to the space obtained from ξ_B^+ by collapsing the section. If ξ_B^+ admits fibrewise coHopf structure then B^ξ admits coHopf structure. This implies, in particular, that products are trivial in the cohomology ring of B^ξ. Since the Euler class $e(\xi)$ of ξ maps into the square of the Thom class under the Thom isomorphism

$$\tilde{H}^n(B) \to \tilde{H}^{2n}(B^\xi)$$

we conclude that $e(\xi) = 0$ when ξ_B^+ admits fibrewise coHopf structure.

This result is due to Sunderland [130] who established fibrewise versions of many of the results about coHopf spaces proved by Berstein, Ganea and

Hilton. For example, Sunderland obtained conditions for a fibrewise coHopf space to be a fibrewise suspension. In somewhat the same spirit we now prove two results which may illuminate the situation.

Proposition 21.9 *Suppose that X is a fibrewise pointed n-sphere bundle over a finite complex B. Let Y be a fibrewise pointed spherical fibration of dimension $n - 1$ over B. Suppose that (e) if n is even, there is a fibrewise pointed homotopy equivalence $X \to \Sigma_B^B Y$, (o) if n is odd, there is a fibrewise pointed map $X \to \Sigma_B^B Y$ of odd fibre degree. Then X admits fibrewise coHopf structure.*

Case (e) is clear. In case (o) it is elementary to check that there is a map $\Sigma_B^B Y \to X$ of odd degree in the opposite direction. By composition with these maps a coHopf structure on $\Sigma_B^B Y$ then determines a map of (odd, odd) bidegree on X. The assertion follows from Proposition 21.6.

It is possible that the existence of a spherical fibration Y as in the statement of the proposition is necessary for the existence of fibrewise coHopf structure. In a metastable range $\dim B < 2n - 2$ this is true by a result to be established in Part II, Proposition 4.16. We shall see shortly that it is true if the base B is a suspension.

Let A be a pointed finite complex and take B to be the reduced suspension ΣA. A clutching map $\alpha : A \to O(n)$ gives a fibrewise pointed n-sphere bundle X on B. (We assume $n > 1$, the case $n = 1$ being decided by the Euler class.) Let $\nu : S^n \to S^n \vee S^n$ be the comultiplication. Then X admits fibrewise coHopf structure if and only if $(J\alpha \vee J\alpha) \circ \nu = \nu \circ J\alpha$ in $[A \wedge S^n; S^n \vee S^n]$.

Lemma 21.10 *The obstruction $(J\alpha \vee J\alpha) \circ \nu - \nu \circ J\alpha$ is the image of α under the maps*

$$[A; O(n)] \to [A; S^{n-1}] \to [A \wedge S^n; S^{2n-1}] \to [A \wedge S^n; S^n \vee S^n]$$

(projection onto $S^{n-1} = O(n)/O(n-1)$, n-fold suspension, and the Whitehead product).

By the Hilton–Milnor theorem, the third map in Theorem 21.10 is a split injection. The composition of the first two maps is the Hopf invariant of $J\alpha$. Hence X has fibrewise coHopf structure if and only if $H(J\alpha) \in [A; \Omega^n S^{2n-1}]$ vanishes. From the EHP-sequence, localized at 2 if n is odd, we find that $J\alpha \in [A; \Omega^n S^n]$, or some odd multiple if n is odd, must desuspend to $[A; \Omega^{n-1} S^{n-1}]$. This establishes:

Proposition 21.11 *Suppose that B is a suspension. Then the existence of a spherical fibration Y as in Proposition 21.9 is a necessary and sufficient condition for the sphere-bundle X to admit fibrewise coHopf structure.*

(If X' is a fibrewise pointed n-sphere bundle over B classified by an odd multiple of α, then, if n is odd, there is a fibrewise pointed map $X \to X'$ of odd fibre degree.)

Sphere-bundles over spheres

Let us turn now to the case where the base space is a sphere. Specifically, consider the fibrewise pointed oriented n-sphere bundle X_α over S^m $(m > 1)$ corresponding, in the standard classification, to the element $\alpha \in \pi_{m-1}SO(n)$. For which α does X_α admit fibrewise Hopf structure or fibrewise coHopf structure? We take the second question first.

Proposition 21.12 *The fibrewise pointed space X_α admits fibrewise coHopf structure if and only if $\Sigma_*^n \rho_* \alpha = 0$.*

Here Σ_* denotes the suspension operator and ρ_* is induced by the evaluation map $\rho : SO(n) \to S^{n-1}$. We have $\Sigma_* \rho_* \alpha = 0$ if and only if X_α is polarized. In the metastable range, when $m < 2n - 1$, the iterated suspension Σ_*^{n-1} is an isomorphism and so the existence of a fibrewise coHopf structure implies that $\Sigma_* \rho_* \alpha = 0$ and hence that X_α is polarized. For an example where the converse of Proposition 19.15 breaks down, take $m = 9$ and $n = 5$. Since S^5 admits a 2-field the Whitehead square $w_5 \in \pi_9(S^5)$ can be desuspended to an element $\gamma \in \rho_* \pi_8 SO(5)$. Then $\Sigma_* \gamma \neq 0$ but $\Sigma_*^5 \gamma = 0$. Again, by taking α so that $\rho_* \alpha$ has order 3 in $\pi_6(S^3) = \rho_* \pi_6 SO(4)$ we see that Proposition 21.6 breaks down for even values of n.

To prove Proposition 21.12, we use the analysis of the homotopy theory of sphere-bundles over spheres developed in [94]. Writing $X_\alpha = X$, for simplicity, we have the direct sum decomposition

$$\pi_t(S^m) \oplus \pi_t(S^n) \approx \pi_t(X) \quad (t = 2, 3, \ldots),$$

where the isomorphism is given by the section $\iota_m \in \pi_m(X)$ on the first summand, by the inclusion $\iota_n \in \pi_n(X)$ of the fibre on the second. Recall from [94] that with appropriate sign conventions the relation

$$[\iota_m, \iota_n] = \iota_n \circ J\alpha$$

holds in $\pi_{m+n-1}(X)$, where square brackets denote the Whitehead product.

Similarly with the associated bundle $Z = X \vee_B X$ with fibre the coproduct $S_1^n \vee S_2^n$ of copies of S^n. We have the direct sum decomposition

$$\pi_t(S^m) \oplus \pi_t(S_1^n \vee S_2^n) \approx \pi_t(Z)$$

and the relations

$$[\iota_m, \iota_n^k] = \iota_n^k \circ J\alpha \quad (k = 1, 2)$$

where $\iota_n^k \in \pi_n(Z)$ is the class of the kth insertion of the coproduct.

Suppose that X admits a fibrewise coHopf structure $f : X \to Z$. Then we have a commutative diagram as shown below, where $g : S^n \to S_1^n \vee S_2^n$ is the induced coHopf structure on the fibre.

$$
\begin{array}{ccc}
\pi_t(S^m) \oplus \pi_t(S^n) & \approx & \pi_t(X) \\
{\scriptstyle 1} \Big\downarrow {\scriptstyle g_*} & & \Big\downarrow {\scriptstyle f_*} \\
\pi_t(S^m) \oplus \pi_t(S_1^n \vee S_2^n) & \approx & \pi_t(Z)
\end{array}
$$

From the first relation stated above we obtain that

$$[\iota_m, \iota_n^1 + \iota_n^2] = (\iota_n^1 + \iota_n^2) \circ J\alpha$$

in $\pi_{m+n-1}(Z)$. Using the second relation and the left distributive law this reduces to

$$[\iota_n^1, \iota_n^2] \circ HJ\alpha = 0,$$

where H denotes the generalized Hopf invariant. However, $[\iota_n^1, \iota_n^2] \in \pi_{2n-1}(Z)$ is the image of the basic Whitehead product of $\pi_{2n-1}(S_1^n \vee S_2^n)$ and so it follows that $HJ\alpha = 0$. Since $HJ\alpha = \pm \Sigma_*^n \rho_* \alpha$, this proves that the condition in Proposition 21.12 is necessary.

Conversely, suppose that $\Sigma_*^n \rho_* \alpha = 0$ and hence $HJ\alpha = 0$. Since the difference $[\iota_m, \iota_n] - \iota_n \circ J\alpha$ is the class of the attaching map of the $(m + n)$-cell $X \backslash (S^m \vee S^n)$ of X we find, by reversing the above argument, that there exists a fibrewise map $f : X \to Z$ extending the coHopf structure on S^n. Hence X admits fibrewise coHopf structure by Proposition 21.1.

Let us turn now to the problem of the existence of fibrewise Hopf structure. Instead of dealing with this directly we first consider a related problem: does there exist a pointed map

$$X *_B X \to \Sigma_B X$$

which is given, on each fibre, by a map of which the homotopy class $\beta \in \pi_{2n+1}(S^{n+1})$ has Hopf invariant one? Such questions are answered in [94] where it is shown that such a fibrewise map exists if and only if

$$\Sigma_* J\alpha \circ \Sigma_*^m \beta = (2\beta - w_{n+1}) \circ \Sigma_*^{n+1} J\alpha. \tag{21.13}$$

Here $w_{n+1} = [\iota_{n+1}, \iota_{n+1}] \in \pi_{2n+1}(S^{n+1})$ is the Whitehead square. We recall, incidentally, that $2\beta - w_{n+1} \in \Sigma_* \pi_{2n}(S^n)$. We see, therefore, that X admits fibrewise Hopf structure if and only if β can be chosen with Hopf invariant one so that Proposition 21.12 is satisfied.

In particular, take $n = 7$ and $m = 8$. Then $\pi_7 SO(7)$ is cyclic infinite and $\pi_{14}(S^7) = J\pi_7 SO(7)$ is cyclic of order 120 for $n = 7$. Let σ denote the Hopf class in $\pi_{15}(S^8)$. Then $w_8 = 2\sigma + \Sigma_* \sigma'$, where σ' generates $\pi_{14}(S^7)$. Now $\beta = \sigma + k\Sigma_* \sigma'$ and $J\alpha = l\sigma'$, for some integers k, l. The relation in (21.13) reduces to

$$(3 - 2k)l\Sigma_* \sigma' \circ \Sigma_*^8 \sigma = 0.$$

However, $\sigma' \circ \Sigma_*^7 \sigma$ has order 24 in $\pi_{21}(S^7)$ and so k can be found to satisfy the relation if and only if $l \equiv 0 \bmod 8$. Therefore the fibrewise pointed 7-sphere bundles over S^8 which admit fibrewise Hopf structure are precisely those which arise from elements of the subgroup $8\pi_7 SO(7)$ of $\pi_7 SO(7)$ by means of the clutching construction.

We now supplement this by showing that X admits weak fibrewise coHopf structure if and only if $\Sigma_*^{n+1} \rho_* \alpha = 0$.

To see this, let us regard X, in the usual way, as a CW complex consisting of the wedge $S^m \vee S^n$ of the section and the fibre with an $(n+m)$-cell attached by the element

$$[\iota_m, \iota_n] + \iota_n \circ J\alpha,$$

where $J\alpha \in \pi_{m+n-1}(S^n)$ is obtained from α by the Hopf construction. It is shown in [94] that the obstruction ϕ to the existence of fibrewise coHopf structure is just

$$[\iota_n^1, \iota_n^2] \circ HJ\alpha \in \pi_{n+m-1}(S^n \vee S^n),$$

where H is the generalized Hopf invariant. Hence it follows by (19.25) that the obstruction θ to the fibrewise pointed deformation of the diagonal is just

$$(\iota_n^1 \times \iota_n^2) \circ C_* HJ\alpha \in \pi_{m+n}(S^n \times S^n, S^n \vee S^n),$$

where C_* is the cone homomorphism. Finally, it follows by (19.26) that the obstruction ψ to the existence of weak fibrewise coHopf structure is just

$$(\iota_n^1 \wedge \iota_n^2) \circ \Sigma_* HJ\alpha \in \pi_{m+n}(S^{2n});$$

the term $(\iota_n^1 \wedge \iota_n^2) \in \pi_{2n}(S^{2n})$ is just the class of the identity and can be suppressed. To obtain the necessary and sufficient conditions in the form previously stated, use the relation $HJ\alpha = \pm \Sigma_*^n \rho_* \alpha$.

Finally, let us show that there exists a fibrewise pointed 8-sphere bundle over S^{22} which is weakly fibrewise coHopf but not fibrewise coHopf. Recall that S^{15} admits an 8-field and so the Whitehead square $w_{15} \in \pi_{29}(S^{15})$ can be desuspended 8 times to an element $\beta \in \pi_{21}(S^7)$. Since $SO(8)$ has a section over S^7 we have $\beta = \rho_* \alpha$ for some element $\alpha \in \pi_{21} SO(8)$. Then

$$\Sigma_*^8 \rho_* \alpha = \Sigma_*^8 \beta = w_{15} \neq 0,$$

since $\pi_{31}(S^{16})$ has no element of Hopf invariant one, while

$$\Sigma_*^9 \rho_* \alpha = \Sigma_*^9 \beta = \Sigma_* w_{15} = 0,$$

since Whitehead products are killed by suspension. Thus we see that the fibrewise pointed 8-sphere bundle over S^{22} obtained from α by the clutching construction is weakly fibrewise coHopf but not fibrewise coHopf.

22 Fibrewise manifolds

Basic notions

Manifolds over a base have been considered by Atiyah and Singer [5]; we prefer the term *fibrewise manifold*. In [5] a fibrewise manifold X over a base space B is defined to be a fibre bundle with fibre a compact manifold A and structural group the group $\mathrm{Diff}(A)$ of self-diffeomorphisms of A. Up to a point we could work with this definition but it is unsatisfactory to be restricted to compact fibres. For non-compact fibres the Atiyah–Singer definition is inappropriate although it provides a guide as to how to proceed.

Fibrewise manifolds form a category in which the morphisms are called fibrewise smooth maps. An example of a fibrewise manifold over B is the product $B \times A$ where A is a manifold in the ordinary sense. (In this section and the next, by a manifold we mean a finite-dimensional, smooth manifold without boundary, of constant dimension, which is Hausdorff and has a countable basis.) An example of a fibrewise smooth map is a fibrewise map

$$\theta : B \times A \to B \times A',$$

where A and A' are manifolds, such that the second projection

$$\pi_2\theta : B \times A \to A'$$

defines, for each point b of B, a smooth map $\tau_b : A \to A'$ for which derivatives of all orders exist and vary continuously with b. These special cases are required for the general definitions, as follows.

We say that a fibrewise space X over B is a *fibrewise manifold* if there is given a numerable open covering of B and for each member U of the covering a local trivialization

$$\phi_U : X_U \to U \times A_U,$$

where ϕ_U is fibrewise over U and A_U is a smooth manifold, such that the transition functions are fibrewise smooth. Specifically, if U, V are members of the covering then the map

$$(U \cap V) \times A_U \to (U \cap V) \times A_V$$

determined by $\phi_V \circ \phi_U^{-1}$ is fibrewise smooth. Note that the fibres are manifolds. If $\dim A_U = k$, independently of U, we say that X is k-dimensional.

For example, a fibre bundle over B with structural group a Lie group G and fibre a smooth G-manifold is a fibrewise manifold over B.

Returning to the general case, let $f : X \to X'$ be a fibrewise map, where X and X' are fibrewise manifolds over B. We describe f as *fibrewise smooth* if when

$$\phi_U : X_U \to U \times A_U, \quad \phi'_{U'} : X'_{U'} \to U' \times A'_{U'}$$

are trivializations, as above, the map

$$(U \cap U') \times A_U \to (U \cap U') \times A'_{U'}$$

determined by $\phi'_{U'} \circ f \circ \phi_U^{-1}$ is fibrewise smooth.

The definition of fibrewise manifold we have given does not ensure that open subsets of fibrewise manifolds are also fibrewise manifolds. However, the definition of fibrewise smooth map can still be used in the case of such open subsets.

Note that if X and X' are fibrewise manifolds over B then so is the fibrewise topological product $X \times_B X'$. To demonstrate this we combine the numerable coverings for X and X' by taking intersections in the usual way.

Fibrewise smooth fibre bundles

Let $\pi : X \to Y$ be a fibrewise smooth map of fibrewise manifolds over B. We say that X is a *fibrewise smooth fibre bundle* over Y if there exists an open covering of Y by subsets U over which there exists a smooth (over B) trivialization

$$\pi^{-1} U \to U \times A.$$

Here A is a smooth manifold and $U \times A$ is open in the fibrewise manifold $Y \times A$ over B. We prove

Proposition 22.1 *Let $\pi : X \to Y$ be a fibrewise smooth map, where X and Y are fibrewise manifolds over B. Suppose that there exists a numerable open covering of B and for each member U of the covering local trivializations*

$$\phi : X_U \to U \times A_U^X, \quad \psi : Y_U \to U \times A_U^Y$$

such that the map

$$\psi \circ \pi \circ \phi^{-1} : U \times A_U^X \to U \times A_U^Y$$

is of the form $1 \times \pi_U$, where A_U^X is a smooth fibre bundle over A_U^Y with projection π_U. Then π is a numerable fibrewise smooth fibre bundle.

To see this, use a partition of unity for A_U^Y for each U of the numerable family. This determines a partition of unity of $U \times A_U^Y$ and hence of Y_U for each U. In this way we obtain a partition of unity for Y itself from which it follows that X is a fibrewise smooth fibre bundle over Y, as asserted.

The term *fibrewise smooth vector bundle* is defined in a similar fashion. For our purposes the important example is the *fibrewise tangent bundle* $\tau_B X$ of a fibrewise manifold X, constructed as follows. As a fibrewise set

$$\tau_B X = \coprod_{b \in B} \tau X_b,$$

the disjoint union of the tangent bundles to the manifolds X_b. We topologize $\tau_B X$ using the smooth local trivializations of X. Specifically, if U is a member of the open covering of B, over which X is locally trivial, and

$$\phi : X_U \to U \times A$$

is the corresponding local trivialization, then $\tau_U X_U$ receives the topology induced by

$$\tau_\phi : \tau_U X_U \to U \times \tau A.$$

Then the open sets of $\tau_B X$ are the subsets which meet each of the $\tau_U X_U$ in an open set. The local trivializations make $\tau_B X$ a fibrewise manifold over B. Moreover, the projections

$$U \times \tau A \to U \times A$$

and the numerable local trivializations of τA over A combine to provide a numerable family of local trivializations of $\tau_B X$, as required to show that $\tau_B X$ is a numerable fibrewise smooth vector bundle over X.

Proposition 22.2 *Let B be a manifold and let X be a smooth fibre bundle over B. If X admits a smooth section s then X (with this section) is smoothly locally trivial as a fibrewise pointed space, so that $X - sB$ is a smooth fibre bundle over B.*

The proof is essentially the same as that of Proposition 9.1.

Proposition 22.3 *Let X be a fibrewise manifold over B. Then there exists a fibrewise smooth map*

$$e : \tau_B X \to X \times_B X$$

over X which sends the zero-section of $\tau_B X$ into the diagonal of $X \times_B X$ and is injective on each fibre over X.

Here we regard $X \times_B X$ as a fibrewise manifold over X using the second projection. When B is a point, so that X is just a manifold, the above result is standard. The map e is constructed as an appropriately scaled exponential map. An outline of the proof is as follows.

By a *fibrewise smooth metric* on X we mean, roughly, a family of smooth Riemannian metrics on the fibres X_b depending continuously on b (or, formally, a fibrewise smooth section of $\tau_B^* X \otimes \tau_B^* X$ over X defining a metric on each fibre). Fibrewise smooth metrics can be constructed by the following procedure. Let $\{U\}$ be the numerable open covering of B and for each member U let

$$\phi_U : X_U \to U \times A_U$$

be the corresponding local trivialization defining the fibrewise smooth structure of X. The smooth manifold A_U admits a smooth Riemannian metric g_U, say, and this determines a fibrewise smooth metric $\phi_U^* g_U$ on X_U. Choose a partition of unity $\{\alpha_U\}$ on B subordinated to $\{U\}$. By composing with the projection we obtain a partition of unity $\{\beta_U\}$ on X subordinated to $\{X_U\}$. Then

$$g = \sum_U \beta_U \phi_U^* g_U$$

is a fibrewise smooth metric on X.

Proposition 22.4 *Let X be a fibrewise manifold over B with fibrewise smooth metric. Then there exists a fibrewise smooth map $\delta : X \to (0, \infty)$ such that the exponential map exp is defined and injective on the open disc*

$$\{\|v\|_g < \delta(x)\}$$

in the tangent space $\tau_x X_b$ for each $x \in X_b \subseteq X$.

Here $\| \quad \|_g$ denotes the norm on $\tau_x X_b$ defined by the fibrewise metric g.

To prove Proposition 22.4 it is sufficient to find such a map δ_U on each $U \times A_U$. For then we can take

$$\delta = \sum_U \beta_U \phi_U^* \delta_U,$$

so that $\delta(x) \leq \max \delta_U(\phi_U(x))$ for all x. Without real loss of generality, therefore, we can assume X is of the form $B \times A$, for some smooth manifold A. The proof then proceeds much as in the classical theory when B is a point. The details are in [31].

The treatment we have just given is similar to that we gave previously in [31]. We shall be returning to the theory of fibrewise manifolds in Part II, Section 11.

23 Fibrewise configuration spaces

The original idea

Although configuration spaces arise in other branches of mathematics, for topologists they were first considered by Fadell and Neuwirth [62] in 1962.

Recall that the nth configuration space $\mathcal{F}^n(X)$ of a space X is defined as the subspace of the topological nth power $\prod^n(X)$ consisting of n-tuples of distinct points of X. We can also think of $\mathcal{F}^n(X)$ as the space $\mathrm{emb}(Q_n, X)$ of embeddings in X of the discrete space $Q_n = \{1, 2, \ldots, n\} \subseteq \mathbb{R}$. Thus $\mathcal{F}^1(X) = X$, while $\mathcal{F}^2(X)$ is the complement of the diagonal in $X \times X$.

The main results of Fadell and Neuwirth concern the case when X is a manifold. Then $\mathcal{F}^n(X)$ is also a manifold. Further, if X is connected then $\mathcal{F}^n(X)$ is a fibre bundle (without structural group) over $\mathcal{F}^r(X)$ for $r = 1, 2, \ldots, n \ldots - 1$. Some conditions are given for the existence of sections.

The special case when $X = \mathbb{R}^k$, the real k-plane, is of particular interest. Clearly $\mathcal{F}^2(\mathbb{R}^k)$ can be identified with $\mathbb{R}^k \times (\mathbb{R}^k - \{0\})$ through the transformation

$$(x_1, x_2) \mapsto (x_1 + x_2, \, x_1 - x_2).$$

We note, for later use, that this transformation is $\mathbb{Z}/2$-equivariant, where $\mathbb{Z}/2$ acts on $\mathcal{F}^2(\mathbb{R}^k)$ by switching factors, acts on $\mathbb{R}^k$ trivially, and acts on $\mathbb{R}^k - \{0\}$ by the antipodal transformation. Recently, Massey [100] has shown that $\mathcal{F}^3(\mathbb{R}^k)$ is a fibre bundle over $\mathcal{F}^2(\mathbb{R}^k)$ with structural group the orthogonal group $O(k-1)$. Moreover, the bundle is trivial if and only if $k = 1, 2, 4$ or 8. Another case of special interest is when $X = S^{k-1}$, the $(k-1)$-sphere. Then $\mathcal{F}^2(S^{k-1})$ can be identified with the tangent bundle $\tau(S^{k-1}) \subseteq S^{k-1} \times \mathbb{R}^k$ by projecting x_2 from $-x_1$ onto the tangent plane $\{x_1\} \times \mathbb{R}^k$ at x_1.

The term configuration space is also used for the space of unordered n-tuples, rather than the space of ordered n-tuples, but in this section we only use it in the latter sense.

Fibrewise configuration spaces

The first mention of *fibrewise* configuration spaces in the literature appears to be in a note by Duvall and Husch [52]. However, the idea had occurred to others about the same time. Our object, in this section, is to obtain fibrewise versions of some of the original results of Fadell and Neuwirth. Essentially the same theory appeared in [31].

By a *covering space*, in this section, we simply mean a fibre bundle with discrete fibre. Let B be a space and let E be a covering space of B, in this sense. If $\sigma : B \to E$ is a section of E the complement $E' = E - \sigma B$ is also a covering space of B. We shall mainly be concerned with finite coverings, such as $B \times Q_n$ $(n = 1, 2, \ldots)$.

Let X be a fibrewise space over B. For each finite covering space E of B the *fibrewise configuration space* $\mathcal{F}_B^E(X)$ is defined as the subspace $\mathrm{emb}_B(E, X)$ of the fibrewise mapping-space $\mathrm{map}_B(E, X)$ consisting of embeddings. Thus the fibre of $\mathcal{F}_B^E(X)$ at the point b of B is just the configuration space $\mathrm{emb}(E_b, X_b)$. When $E = B \times Q_n$ we may write $\mathcal{F}_B^n(X)$ instead of $\mathcal{F}_B^E(X)$.

Of course, the fibrewise orbit space of $\mathcal{F}_B^E(X)$ under the fibrewise action of the group of the covering E can also be considered.

As we shall soon see, the reduction formula

$$\mathcal{F}_B^n(X) = \mathcal{F}_{\mathcal{F}_B^m(X)}^{n-m}(\mathcal{F}_B^{m+1}(X)) \qquad (n \geq m \geq 1) \tag{23.1}$$

plays a useful rôle in the theory. In fact the fibrewise theory can be used to retrieve some of the original results in the ordinary theory.

We have $\mathcal{F}_B^1(X) = X$, of course, and $\mathcal{F}_B^2(X)$ is just the complement of the diagonal in $X \times_B X$. When X is a k-plane bundle over B we can identify $\mathcal{F}_B^2(X)$ with the fibrewise product $X \times_B (X - B)$, where B is embedded as the zero-section, using in each fibre the transformation mentioned earlier. This identification is $\mathbb{Z}/2$ equivariant, the group acting on $\mathcal{F}_B^2(X)$ by switching factors, on X by the identity, and on $X - B$ by the antipodal transformation.

In most cases of interest X is a fibre bundle over B. Note that $\mathcal{F}_B^E(X)$ is then also a fibre bundle over B. If X is a finite covering space of B so is $\mathcal{F}_B^E(X)$.

Suppose that X is a G-bundle over B with fibre A, where G is a topological group and A is a G-space. Then X may be identified with the mixed product $P \times_G A$, where P is the associated principal G-bundle. The configuration space $\mathcal{F}^n(A)$ is also a G-space, under the diagonal action, and the fibrewise configuration space $\mathcal{F}_B^n(X)$ may be identified in the same way with the mixed product $P \times_G \mathcal{F}^n(A)$.

Suppose now that A is a smooth manifold. To see that $\mathcal{F}^n(A)$ is a smooth fibre bundle over A we argue as follows. We have already noted that $\mathcal{F}^2(A)$ is the complement of the diagonal section in $A \times A$, regarded as a fibrewise space over A using the first projection. By Proposition 22.2, therefore, $\mathcal{F}^2(A)$ is a smooth fibre bundle over A. Now we can use the reduction formula

$$\mathcal{F}^n(A) = \mathcal{F}_A^{n-1}(\mathcal{F}^2(A))$$

to see at once that $\mathcal{F}^n(A)$ is a fibre bundle over A.

Similar arguments show more generally that $\mathcal{F}^n(A)$ is a smooth fibre bundle over $\mathcal{F}^r(A)$ for $1 \leq r < n$. Indeed $\mathcal{F}^{r+1}(A)$ as a fibrewise space over $\mathcal{F}^r(A)$ can be regarded as the complement of the union of the r canonical sections of the trivial bundle $\mathcal{F}^r(A) \times A$ over $\mathcal{F}^r(A)$. Removing the sections one by one we see from Proposition 22.2 that $\mathcal{F}^{n+1}(A)$ is a fibre bundle over $\mathcal{F}^r(A)$. Then

$$\mathcal{F}^n(A) = \mathcal{F}_{\mathcal{F}^r(A)}^{n-r}(\mathcal{F}^{r+1}(A))$$

is clearly a smooth fibre bundle over $\mathcal{F}^r(A)$. By applying Proposition 22.1 we can generalize this to

Proposition 23.2 *Let E_1 and E_2 be numerable finite covering spaces of B, and let X be a fibrewise manifold over B. Then $\mathcal{F}_B^{E_1 \sqcup E_2}(X)$ is a numerable fibrewise smooth fibre bundle over $\mathcal{F}_B^{E_1}(X)$.*

Under certain conditions the fibrewise fibrations considered above admit sections. In special cases *ad hoc* constructions can be used but for a general result we need to rely on the theory of fibrewise manifolds, as in

Proposition 23.3 *Let E be a numerable finite covering space of B with section σ and let X be a fibrewise manifold over B with projection p. Suppose that the fibrewise configuration space $\mathcal{F}_X^{p^*E}(\tau_B X)$ admits a section over X. Then $\mathcal{F}_B^E(X)$, regarded as a fibrewise space over X with projection σ^*, admits a section.*

For a section of $\mathcal{F}_X^{p^*E}(\tau_B X)$ over X determines a fibrewise embedding of p^*E in $\tau_B X$ over X. Without loss of generality we may assume that the pull-back of σ corresponds to the zero-section of $\tau_B X$. Composition with the map e in Proposition 22.3 gives the required section of $\mathcal{F}_B^E(X)$ over X. To be precise, for $x \in X_b$ let us denote by $i_x : E_b \to (\tau_B X)_x$ the embedding given by the section: thus $i_x(\sigma(b)) = 0$. Write $e_x : (\tau_B X)_x \to X_b$ for the restriction of e to fibres over x. Then $e_x \circ i_x : E_b \to X_b$ is the required embedding in $\mathcal{F}_B^E(X)_b$. When E is trivial this result simplifies to

Proposition 23.4 *Let X be a fibrewise manifold over B. Suppose that the fibrewise tangent bundle $\tau_B X$ admits a nowhere-zero section over X. Then the fibrewise configuration space $\mathcal{F}_B^n(X)$ admits a section over X for all $n \geq 1$.*

Indeed let $v : X \to \tau_B X$ be the nowhere-zero vector field. Then a section of $\mathcal{F}_X^n(\tau_B X)$ over X is defined by sending each point x into the n-tuple $(u_1, \ldots, u_n)$, where $u_i = (i-1)v(x) \in (\tau_B X)_x$ $(i = 1, \ldots, n)$. We deduce

Corollary 23.5 *Let X be a fibrewise manifold over B. Suppose that the pull-back $\tau_B X \times_B \mathcal{F}_B^r(X)$ of $\tau_B X$ to $\mathcal{F}_B^r(X)$ admits a nowhere-zero section for some $r \geq 1$. Then $\mathcal{F}_B^n(X)$ admits a section over $\mathcal{F}_B^m(X)$ for $n \geq m \geq r$.*

For the hypothesis implies that $\tau_{\mathcal{F}_B^m(X)}\mathcal{F}_B^{m+1}(X)$ admits a nowhere-zero section over $\mathcal{F}_B^m(X)$. Then Proposition 23.4, with B replaced by $\mathcal{F}_B^m(X)$ and X replaced by $\mathcal{F}_B^{m+1}(X)$, shows that

$$\mathcal{F}_{\mathcal{F}_B^m(X)}^{n-m}(\mathcal{F}_B^{m+1}(X)) = \mathcal{F}_B^n(X)$$

has a section over $\mathcal{F}_B^m(X)$, as asserted.

When B is a point we retrieve from Corollary 23.5 various results of Fadell and Neuwirth. For example, we see that sections exist when X is an open manifold, also when X is a compact connected manifold with Euler characteristic zero.

With general B the conclusion of Corollary 23.5 holds for affine bundles with $r = 2$, also for sphere-bundles with $r = 3$, using the identifications made earlier. Of course, direct geometric constructions can also be used.

For example, let X be a sphere-bundle over B, and let

$$\rho : \mathcal{F}_B^{n+1}(X) \to \mathcal{F}_B^n(X)$$

be defined by dropping the last point from each n-tuple, where $n \geq 3$. Given three distinct points p_1, p_2, p_3 in the same fibre X_b of X, consider the line segment L joining p_1 to p_2 in the associated affine bundle. If x is any point of L between p_1 and p_2 we can project x into X_b from p_3, thus obtaining a point x' of X_b. If x is chosen sufficiently close to p_1 the projection x' will be distinct from any previously given set $p_1, \ldots, p_n$ of distinct points of X_b. Thus a section of ρ is given by the transformation

$$(p_1, p_2, \ldots, p_n) \mapsto (p_1, p_2, \ldots, p_n, x').$$

Full details are given by Fadell [61] in the case where B reduces to a point.

Again we suppose that E is a finite covering space of B with section σ. We also suppose that X is a fibre bundle over B with section s. The projection

$$\sigma^* : \mathcal{F}_B^E(X) \to X$$

is defined as before, and the fibrewise fibre $\rho^{-1}(sB)$ is equivalent to the fibrewise configuration space $\mathcal{F}_B^{E'}(X')$, where $E' = E - \sigma B$ and $X' = X - sB$.

When X is a vector bundle over B and s is the zero section it is easy to see that $\mathcal{F}_B^E(X)$ is equivalent, as a fibrewise pointed space, to the fibrewise topological product $X \times_B \mathcal{F}_B^{E'}(X')$. In fact a trivialization

$$\xi : X \times_B \mathcal{F}_B^{E'}(X') \to \mathcal{F}_B^E(X)$$

is given in each fibre by the formula $\xi(x, u') = u$, where $x \in X_b$ $(b \in B)$, $u' : E_b' \to X_b'$ and $u : E_b \to X_b$ are related by

$$u(s(b)) = x - s(b), \quad u(e') = x - u'(e').$$

The same conclusion may be reached in other cases.

Hopf structures

By a (strict) *Hopf structure* on a pointed space A we mean a multiplication $m : A \times A \to A$, which coincides with the folding map on the wedge product $A \vee A$. We describe the Hopf structure as *special* if each of the left translations is homeomorphic. For example, the classical Hopf structures on S^q, for $q = 1, 3$ or 7, are special in this sense.

Now let A be a pointed G-space, where G is a topological group. Then $A \times A$ is a pointed G-space, with the diagonal action, and G-equivariant Hopf structures can be considered, as in [24].

For example, take the classical Hopf structure on S^q, for $q = 1, 3$ or 7. In the case of S^1 this is given by complex multiplication, which is $O(1)$ equivariant, in the case of S^3 by quaternionic multiplication, which is $SO(3)$-equivariant, and in the case of S^7 by Cayley multiplication, which is G_2-equivariant.

Equivariant Hopf structures lead to fibrewise Hopf structures, as follows. Let B be a space and let X be a fibrewise pointed G-bundle over B with fibre the pointed G-space A. Suppose that A admits a G-equivariant Hopf structure m. Then m defines a fibrewise Hopf structure $m' : X \times_B X \to X$ on X. Moreover, if m is special then m' is special in the sense that fibrewise left translation is homeomorphic. This implies that the fibrewise configuration space $\mathcal{F}_B^E(X)$ is trivial as a fibrewise fibre bundle over X with fibrewise fibre $\mathcal{F}_B^{E'}(X')$. Specifically, a trivialization

$$\xi : X \times_B \mathcal{F}_B^{E'}(X') \to \mathcal{F}_B^E(X)$$

is given in each fibre by the formula $\xi(x, u') = u$, where $x \in X_b$ ($b \in B$), $u' : E_b' \to X_b'$ and $u : E_b \to X_b$ are related by

$$u(s(b)) = m'(x, s(b)), \quad u(e') = m'(x, u'(e')).$$

The conclusion holds, in particular, for the fibrewise suspension of every (orthogonal) 0-sphere bundle, of every orientable 2-sphere bundle, and of every 6-sphere bundle admitting G_2-structure.

Sequences of fibrewise fibrations

As Fadell and Neuwirth show, the homotopy theory of ordinary configuration spaces can be investigated through a sequence of fibrations. Specifically, if A is a connected manifold the fibrations are those associated with the successive configuration spaces $\mathcal{F}^n(A)$, $\mathcal{F}^{n-1}(A - Q_1), \ldots, \mathcal{F}^1(A - Q_{n-1})$, where $q_1, \ldots, q_{n-1}$ are distinct points of A and $Q_r = \{q_1, \ldots, q_r\}$. By homogeneity, the spaces obtained by this procedure are independent of the choice of points to be deleted, up to diffeomorphism.

When we turn to the fibrewise theory this is no longer the case. For example, take $B = S^n$ and $X = S^n \times S^n$, regarded as a fibrewise space using the first projection. The complement of the diagonal section is the tangent bundle to S^n, while the complement of the second (axial) insertion is the trivial bundle. Except when S^n is parallelizable, the complements of these sections are not equivalent in the sense of fibrewise homeomorphism. This can be seen as follows. If the tangent bundle is fibrewise homeomorphic to the trivial bundle then its fibrewise one-point compactification is trivial. From such a trivialization we readily obtain a Hopf structure on S^n; hence $n = 1$, 3 or 7. It is therefore necessary to exercise caution.

So let X be a fibrewise manifold over B, with connected fibres. We let $s_1, \ldots, s_{n-1}$ be mutually non-intersecting sections of X and we write $Q_r B = s_1 B \cup \ldots \cup s_r B$, for $r = 1, \ldots, n - 1$. To investigate the fibrewise homotopy theory of $\mathcal{F}_B^n(X)$ we can proceed by inductive arguments through the sequence of fibrewise fibrations associated with the successive fibrewise configuration spaces $\mathcal{F}_B^n(X)$, $\mathcal{F}_B^{n-1}(X - Q_1 B), \ldots, \mathcal{F}_B^1(X - Q_{n-1} B)$. As we have seen, these fibrations may admit sections under certain conditions.

For example, let X be a Euclidean bundle of rank k, with associated sphere-bundle $S(X)$. We suppose that $S(X)$ admits a section s from which we construct the family of mutually non-intersecting sections $s, 2s, 3s, \ldots, ns$ of X. Then X is fibrewise contractible, $X - Q_1 B$ has the same fibrewise homotopy type as $S(X)$, $X - Q_2 B$ has the same fibrewise homotopy type as $S(X) \vee_B S(X)$, and so on.

Of course, it can also be useful to consider the sequence of fibrewise fibrations

$$\mathcal{F}_B^n(X) \to \mathcal{F}_B^{n-1}(X) \to \ldots \to \mathcal{F}_B^1(X) = X$$

where the successive fibres are $X - Q_1 B$, $X - Q_2 B, \ldots, X - Q_{n-1} B$.

We shall be returning to the theory of fibrewise configuration spaces in Section 14 of Part II.

Part II. An Introduction to Fibrewise Stable Homotopy Theory

Introduction

It is invariably difficult to trace the origins of a mathematical theory. But it is appealing to see the roots of Fibrewise Stable Homotopy Theory in Algebraic Geometry, especially as in Grothendieck's reformation of the subject, [69], both in the abstract framework and more specifically in Grothendieck's formulation of the Riemann–Roch Theorem. In the work of Atiyah and Singer [5] in the late 1960s on the Index Theorem for families of elliptic operators we already have a fully fledged fibrewise stable theory (K-theory) using many of the techniques (equivariant methods, transfer maps) which were to play a major rôle in Algebraic Topology in the next decade. Major applications within Algebraic Topology soon followed: the Kahn–Priddy theorem [96], the Becker–Gottlieb proof of the Adams conjecture [9], the fixed-point theory of Dold [47]. More recently, taking up the K-theory theme Kasparov has introduced a fibrewise non-commutative theory [97].

The account of Fibrewise Stable Homotopy Theory which we shall present here is intended as an introduction; its goals are modest, and there is much that is omitted. (For a taste of recent work we refer the reader to the preprint [54] of Dwyer, Weiss and Williams.) The geometric setting of classical Algebraic Topology is the topology of finite complexes (or even closed manifolds). In studying finite complexes there is much to be gained by enlarging the field of interest to include infinite CW-complexes (such as classifying spaces) and spectra, but these more complicated objects can be viewed as technical constructs (or machines) rather than as objects of intrinsic geometric interest in their own right, and this is the viewpoint which we take here. Stated bluntly, an infinite complex is the union of its finite subcomplexes.

When we do fibrewise topology over a space B, the base will normally be a compact ENR (thus a retract of a closed manifold), and the fibres will usually be finite complexes (at least up to homotopy). We thus sidestep the topological niceties which cannot be avoided if one wants to set up an abstract theory in full generality. The restriction that the base space be an ENR is, however, more than a technical convenience. Although many of the formal constructions in the fibrewise stable theory can be carried through with little or no modification for more general base spaces, it is far from clear that this produces the 'right' stable theory. (In most applications where the base is an infinite CW-complex one would want the fibrewise stable theory to be built up from the fibrewise stable maps over finite subcomplexes.)

More seriously, it is at present necessary to restrict the fibrewise spaces to be locally trivial, at least up to homotopy. Thus we exclude precisely the sort of singular fibres which are so important in Algebraic Geometry. Traditionally these fibrewise spaces are characterized by homotopy lifting properties and called homotopy-fibrations. (The abbreviation 'h-fibration' was used in [44], and the term 'weak fibration' in [45].) In order to emphasize the geometric aspect and the parallel with fibre bundles in Differential Geometry we shall use the non-standard term 'homotopy fibre bundle'. (Readers will no doubt substitute their own preferred name.) The word 'homotopy' will be used in this adjectival form in other contexts, too, some standard like 'homotopy-fibre' and 'homotopy-cofibre', others such as 'homotopy fixed-point set' (in our usage, see Definition 6.11) less familiar.

Fibrewise homotopy theory is often called *homotopy theory over a* (base) *space*. Partly for emphasis, partly because the base space may not be clear from the context, we shall sometimes retain this usage and refer to fibrewise spaces over a base B, fibrewise maps over B, and so on. In Part I, where the base space was normally fixed, this was unnecessary.

The exposition is set out as follows. Assuming that the reader, having at least skimmed the first two chapters of Part I, is familiar with the basic concepts of Fibrewise Homotopy Theory, we introduce in Section 1 the homotopy fibre bundles to which we have alluded above. These are the basic geometric objects which we shall study. Section 2 recapitulates the formal framework of fibrewise homotopy theory, especially the various exact sequences, and concludes with an application to homotopy-commutative fibrewise Hopf spaces, setting a pattern, of including applications of the theory at the first opportunity, which we shall follow throughout Part II. The basic definitions of fibrewise stable homotopy theory are presented in Section 3. In Section 4 we look at the stable cohomotopy Euler class of a real vector bundle, the prototypical characteristic class; it plays a fundamental rôle in the theory. We give applications to obstruction theory and fibrewise coHopf spaces.

Section 5 is a somewhat technical topological section on the definition and properties of fibrewise Euclidean and Absolute Neighbourhood Retracts. The results are used at once in Section 6 to describe Dold's version of fibrewise fixed-point theory for ENRs, and in Section 7 to extend the theory to ANRs. These sections are written with a view to geometric applications, as for example in the investigation of periodic orbits for semi-flows; see [6], [27] and [36]. But they contain, in particular, the construction of the transfer, which is of fundamental importance in fibrewise stable homotopy theory.

In Section 8 we extend our fibrewise stable homotopy category to include formal desuspensions, which we shall call 'fibrewise stable spaces'. (A more advanced account would introduce the general concept of a fibrewise spectrum. Our fibrewise stable spaces are fibrewise suspension spectra.) The construction is described alongside that of virtual vector bundles, which enter into Section 9 on the solution of the Adams conjecture using the transfer. The structure of the stable category is investigated further in Section 10 where we consider duality and its connection with the fixed-point theory.

Fibrewise differential topology was already an important tool in the work of Atiyah and Singer to which we have referred above. The fibrewise manifolds, which appeared there under the name of manifolds over a base, were discussed briefly in Part I. In Section 11 we continue the development of this theory. From the standpoint of homotopy theory, the Pontrjagin–Thom construction is central to the study of manifolds. In Section 12 we describe the fibrewise version of the construction and various ideas which flow from it: the definition of Gysin (or direct image) maps, the Poincaré–Hopf vector field index, and Poincaré–Atiyah duality for smooth manifolds.

Sections 13 and 14, although of considerable independent interest, are included primarily as applications of the methods of Section 12. In Section 13 we show how fibrewise techniques can be used to establish Miller's stable splitting [111] of the unitary group $U(n)$. In Section 14 we describe fibrewise generalizations of the stable splitting theorems for configuration spaces. The generalizations themselves are rather routine, but lead in a special case to a proof due to Bödigheimer and Madsen [15] of the stable splitting of the space of free loops on a suspension, a result of Carlsson and Cohen [21].

Finally, Section 15 deals with fibrewise homology theory. Earlier sections are mostly concerned with describing the structure of fibrewise stable homotopy theory, but with the introduction of homology theory we are able, at last, to make some concrete calculations.

Notation is a frequent barrier to understanding. We present below a key to the notational conventions which we have striven to maintain throughout Part II. There are, inevitably, a few exceptions to our general rules. The term 'map' is normally used for a continuous mapping. Occasionally, for emphasis, we add the adjective 'continuous'. As in Part I, the recurrent adjective 'fibrewise' is normally placed at the beginning of a sequence of adjectives and is understood to qualify all the following terms. However, we do sometimes, when confusion is unlikely, omit the word 'fibrewise' in the cause of readability. For the sake of consistency we have broken with tradition and use the term 'fibrewise homotopy equivalence' rather than the established 'fibre homotopy equivalence' and sometimes 'fibrewise product' instead of 'fibre product'.

We draw the reader's attention, in particular, to our conventions, emphasized in boldface in Section 1, that the generic base space B is normally an ENR and that all fibrewise pointed spaces (and all pointed spaces) are homotopy well-pointed.

A guide to notation

B, B'	base spaces, normally compact ENRs.
A	closed subspace of B, normally a sub-ENR.
$\alpha, \beta, \ldots$	Greek letters are normally used for maps between base spaces.
M, N	fibrewise spaces over B.
X, Y, Z	fibrewise pointed spaces over B.
$f, g, \ldots$	Roman letters are normally used for fibrewise maps.
U, V, W	open subspaces of topological spaces.
ξ, η, ζ	vector bundles, usually real and finite-dimensional.
E, F	vector spaces, usually real and finite-dimensional (sometimes complex, sometimes infinite-dimensional normed vector spaces).
F	We also use 'F' for fibre.
$+, +_B$	A subscript $+$ denotes adjunction of a disjoint basepoint, in fibres over B.
$+, \overset{+}{B}$	A superscript $+$ denotes one-point compactification (with basepoint at infinity), fibrewise over B.
$*$	A '$*$' is used as a generic symbol for a basepoint, sometimes in the fibrewise sense.
Γ	the space of sections of a bundle.
B	A boldface '**B**' is used for the classifying space so as to avoid a clash of notation with the base 'B'.
C, C_B	cone, fibrewise cone, defined for (fibrewise) pointed spaces, and also homotopy-cofibre.
P, P_B	path space, fibrewise path space, defined for (fibrewise) pointed spaces.
F, F_B	homotopy-fibre, fibrewise homotopy-fibre.
Σ, Σ_B	suspension, fibrewise suspension, defined for (fibrewise) pointed spaces.
Ω, Ω_B	loop space, fibrewise loop space, defined for (fibrewise) pointed spaces.
map_B	fibrewise mapping-space over B.
map_B^*	fibrewise pointed mapping-space over B.
$\mathcal{L}$	free loop space $\mathrm{map}(S^1, -)$.
$\mathcal{P}$	free path space $\mathrm{map}([0, 1], -)$.
τ, τ_B	tangent bundle, fibrewise tangent bundle of a smooth manifold, fibrewise manifold.
M^ξ, M_B^ξ	Thom space, fibrewise Thom space, of a finite-dimensional vector bundle ξ over a (fibrewise) space M.

Chapter 1. Foundations

1 Fibre bundles

We begin by reviewing some of the fundamental concepts of fibrewise homotopy theory to establish notation and set the scene for the introduction of new material.

It will be convenient to assume throughout Part II that all topological spaces considered are compactly generated and that products are formed in that category. (As in [91] we take the definition of the term 'compactly generated' to include the weak Hausdorff condition that the inclusion of the diagonal be closed.) We recall that metric spaces, in particular, are compactly generated.

Base spaces

As we have indicated in the Introduction, in order to develop a satisfactory elementary stable theory we have to restrict the base spaces considered to be Euclidean Neighbourhood Retracts (ENRs). This represents such a significant change of viewpoint from that taken in Part I that we restate for emphasis:

Unless the context clearly indicates otherwise (such as when we talk about bundles over a classifying space) all base spaces considered are assumed to be ENRs.

This restriction simplifies many aspects of the homotopy theory described in Part I and sometimes allows us to omit, in the statements of results, hypotheses which would otherwise be necessary assumptions. To avoid confusion we shall include frequent reminders that the base is understood to be an ENR.

At this point it may be useful to recall that a topological space B is an *ENR* if there exists an embedding $i : B \hookrightarrow U$ into some open subset U of a finite-dimensional Euclidean space E and a retraction $r : U \to B$ of U onto B: $r \circ i = 1_B$. Compact (smooth) manifolds are the basic examples. A compact manifold B can be embedded in a Euclidean space E and the normal bundle of the embedding included in E as a tubular neighbourhood

U; the retraction r is given by projecting the fibres of the normal bundle to B. Finite complexes are ENRs, and every compact ENR is a retract of a finite complex, indeed of a finite polyhedron.

A review of the basic properties of ENRs is included at the beginning of Section 5. We note at once that an ENR is locally compact Hausdorff and admits partitions of unity subordinate to any open covering. Another important property enjoyed by an ENR B is uniform local contractibility. This means that there is an open neighbourhood W of the diagonal $\Delta(B)$ in $B \times B$ and a homotopy $H_t : W \to B$ such that $H_0(a, b) = a$, $H_1(a, b) = b$ and $H_t(b, b) = b$, for all $(a, b) \in W$, $0 \le t \le 1$.

From now on, our generic base space B is supposed to be an ENR. It will usually be compact.

Fibrewise spaces

We shall use notation such as $M \to B$ (or, more carefully, $p : M \to B$) for a fibrewise space over B. The fibrewise space M should be thought of, informally, as a family of spaces M_b (the fibres $p^{-1}(b)$ of p) indexed by points $b \in B$. In the same way, one should think of a fibrewise map $f : M \to N$ between fibrewise spaces over B as a family of maps $f_b : M_b \to N_b$ between fibres.

Given a subspace A of B, we use the standard notation $M_A \to A$ for the restriction $p^{-1}A$ of the fibrewise space M and $f_A : M_A \to N_A$ for the family of maps $f_a : M_a \to N_a$ indexed by $a \in A$.

Fibre bundles: local triviality

The basic examples of fibrewise spaces are the fibre bundles. For any space F we can form the fibrewise space $B \times F \to B$ over B with fibre $\{b\} \times F$, often identified with F in the natural way, at b. A fibrewise space M is *trivial* if it is fibrewise homeomorphic to such a fibrewise space $B \times F \to B$ and *locally trivial* if there exists an open covering $\mathcal{U}$ of B such that M_U is trivial over U for each $U \in \mathcal{U}$. We use the term *fibre bundle* for any locally trivial fibrewise space. Note that we do not require all the fibres to be homeomorphic (although this will, clearly, be the case if B is connected).

Since the base spaces which we are considering admit partitions of unity, the fibre bundles are numerable. The standard theory then applies. Thus, if $\alpha : B' \to B$ is a continuous map from an ENR B' to B, the pull-back $\alpha^* M \to B'$ of a fibre bundle M over B is a fibre bundle over B'. Moreover, if α and β are homotopic maps $B' \to B$, then the induced bundles $\alpha^* M$ and $\beta^* M$ are fibrewise homeomorphic. In particular, any fibre bundle over a contractible base is trivial.

Fibrewise homotopy

Informally, two fibrewise maps $M \to N$ over B are fibrewise homotopic if one can be deformed continuously into the other through fibrewise maps. Precisely, two fibrewise maps f and g are fibrewise homotopic if there is a fibrewise map $H' : M \times [0,1] \to N$ over B such that $H'(x,0) = f(x)$, $H'(x,1) = g(x)$, for $x \in M$. Such a map H' corresponds to a fibrewise map $H : M \times [0,1] \to N \times [0,1]$ over $B \times [0,1]$ which restricts to f on $B \times \{0\}$ and to g on $B \times \{1\}$ (up to obvious identification): $H(x,t) = (H'(x,t),t)$. From the viewpoint of fibrewise topology it is perhaps more natural to think of H (rather than H') as a *fibrewise homotopy* between f and g. Homotopy is by its very definition a fibrewise concept.

Fibrewise homotopy equivalence

In fibrewise homotopy theory we study fibrewise spaces up to fibre homotopy equivalence (or fibrewise homotopy equivalence to be more consistent in our terminology). One of the first theorems of fibrewise homotopy theory is the result of Dold that a fibrewise map is a fibrewise homotopy equivalence if it is *locally* a fibrewise homotopy equivalence:

Theorem 1.1 *Let $f : M \to N$ be a fibrewise map of fibrewise spaces over B, and let $\mathcal{U}$ be an open covering of the base B (supposed to be an ENR). Then f is a fibrewise homotopy equivalence if and only if the restriction $f_U : M_U \to N_U$ over each set $U \in \mathcal{U}$ of the covering is a fibrewise homotopy equivalence.*

Note that this simplified statement is possible only because the base space B admits partitions of unity; in general, one needs to specify that $\mathcal{U}$ be a numerable covering.

Homotopy fibre bundles: local homotopy triviality

When we pass to homotopy theory it is natural to say that a fibrewise space $M \to B$ is *homotopy trivial* if it is fibrewise homotopy equivalent to a trivial fibrewise space $B \times F \to B$. It is *locally homotopy trivial* if there exists an open covering $\mathcal{U}$ of the base such that $M_U \to U$ is homotopy trivial for all $U \in \mathcal{U}$. We shall call a fibrewise space which is locally homotopy trivial a *homotopy fibre bundle* (although it should be noted that this is not standard terminology).

Fibrewise maps between homotopy fibre bundles are locally homotopic to products.

Proposition 1.2 *Let $f : M \to N$ be a map between homotopy fibre bundles over B. Then for each $b \in B$ there exists an open neighbourhood U*

of b and local fibrewise homotopy trivializations $\phi : M_U \to U \times M_b$ and $\psi : U \times N_b \to N_U$ such that f_U is fibrewise homotopic to $\psi \circ (1 \times f_b) \circ \phi$.

Since B is locally contractible we can reduce to the case in which the bundles are trivial:

$$f : B \times F \to B \times F',$$

and $b \in B$ is contained in an open neighbourhood U which is contractible in B to the point b. Write $f(a, x) = (a, f'(a, x))$. Let $H_t : U \to B$ be a homotopy with $H_0 = 1_B$ and H_1 the constant map at b. Define $F_t(a, x) = (a, f'(H_t(a), x))$. Then $F_0 = f$ and $F_1 = 1 \times f_b$. This completes the proof.

At the risk of excessive use of the word 'trivial', we refer to the property established in Proposition 1.2 as *local homotopy triviality* of the fibrewise map f.

Remark 1.3. Local homotopy triviality may originate in properties of the fibres rather than in properties of the base. Suppose that P is a compact Hausdorff topological space and Q is an ENR. Then the space $\mathrm{map}(P, Q)$ is an Absolute Neighbourhood Retract (ANR) and hence uniformly locally contractible. (See the discussion in Section 5.) Now let B be any topological space, not necessarily an ENR, and let f be a map $B \to \mathrm{map}(P, Q)$. Then it follows from uniform local contractibility that each point $b \in B$ has an open neighbourhood U such that $f \,|\, U$ is homotopic to a constant map.

The homotopy lifting property

Homotopy fibre bundles (over ENRs) are characterized by the homotopy lifting property up to homotopy. Let us begin with the straightforward homotopy lifting property.

Definition 1.4 Let $M \to B$ be a fibrewise space over B. We say that M has the *homotopy lifting property* if, for each base space B' and fibrewise space $M' \to B'$ over B', given a homotopy $\alpha_t : B' \to B$, $0 \leq t \leq 1$, and fibrewise map $f_0 : M' \to M$ over α_0 (that is, a fibrewise map $M' \to \alpha_0^* M$ over B'), there is a homotopy f_t over α_t extending f_0, as displayed in the diagrams:

$$
\begin{array}{ccc}
M' & \xrightarrow{f_t} & M \\
\downarrow & & \downarrow \\
B' & \xrightarrow[\alpha_t]{} & B
\end{array}
\qquad\qquad
\begin{array}{ccc}
M' \times [0,1] & \xrightarrow{f} & M \\
\downarrow & & \downarrow \\
B' \times [0,1] & \xrightarrow[\alpha]{} & B
\end{array}
$$

As it stands, the space B' seems to be redundant. We could take $M' = B'$ with the identity projection (at the expense of waiving our convention that base spaces should be ENRs). The property then reduces to the standard definition of a fibration over B. However, the change in emphasis is important conceptually and will be essential when we study fibrewise pointed spaces.

Now let us introduce, for any fibrewise space $M \to B$, an associated fibrewise space $M^{\sharp} \to B$ as follows. Let $\mathcal{P}B = \mathrm{map}([0,1], B)$ be the space of (free) paths in B. As a space

$$M^{\sharp} := \{(x, \omega) \in M \times \mathcal{P}B \mid p(x) = \omega(0)\},$$

with the projection $(x, \omega) \mapsto \omega(1)$. It is thus the topological fibre product $M \times_B \mathcal{P}B \to B$. There is a natural fibrewise map $M \to M^{\sharp}$, mapping $x \in M_b$ to $(x, b) \in M_b^{\sharp}$, where the second factor b denotes the constant path at b. The next lemma is an elementary exercise.

Lemma 1.5 *The fibrewise space $M^{\sharp} \to B$ has the homotopy lifting property (Definition 1.4).*

In the homotopy category it is desirable to weaken the homotopy lifting property to produce a property of fibrewise spaces which is invariant under (fibrewise) homotopy equivalence.

Definition 1.6 We say that a fibrewise space M over B has the *homotopy lifting property up to homotopy* if, for each base space B' and fibrewise space $M' \to B'$ over B', the following condition holds.

Given a homotopy $\alpha_t : B' \to B$, $0 \le t \le 1$, for which there exists an $\epsilon > 0$ such that $\alpha_t = \alpha_0$ for $0 \le t \le \epsilon$, and a fibrewise map $f_0 : M' \to M$ over α_0, there is a homotopy f_t over α_t extending f_0.

An equivalent condition is that, given any homotopy $\alpha_t : B' \to B$ and fibrewise map f_0 over α_0, there exists a homotopy g_t over α_t and a homotopy f_t over (constant) α_0 extending f_0 with $f_1 = g_0$. (To see that this follows from Definition 1.6, extend the range of definition of α_t to the interval $[-1, 1]$ by defining $\alpha_t = \alpha_0$ for $t < 0$. The converse is proved by reversing the process, with appropriate rescaling of the interval.) From this form of the condition it is clear that, if M satisfies the homotopy lifting property up to homotopy, then so does any fibrewise space which is fibrewise homotopy equivalent to M.

Remark 1.7. It is an elementary exercise, replacing B' by $B' \times [0, 1]$, to show that any two lifts f_t in the definition (Definition 1.4 or 1.6) of the homotopy lifting property are fibrewise homotopic. In particular, given $f_0 : M' \to M$ over α_0, we obtain a canonical map $f_1 : M' \to M$ over α_1 up to fibrewise homotopy.

It is routine to check that the homotopy lifting properties are inherited by pull-backs. These properties are *local* in the following sense.

Proposition 1.8 *Let $M \to B$ be a fibrewise space. Suppose that $\mathcal{U}$ is an open covering of the base B. Then M has the homotopy lifting property (or the homotopy lifting property up to homotopy) if and only if each $M_U \to U$, $U \in \mathcal{U}$, has that property.*

Again notice that this statement uses our standing hypothesis that the base space B is an ENR.

Since a trivial bundle clearly has the homotopy lifting property, we have:

Proposition 1.9 *Let $M \to B$ be a homotopy fibre bundle over an ENR B. Then M has the homotopy lifting property up to homotopy.*

More is true: a homotopy fibre bundle is fibrewise homotopy equivalent to a fibrewise space with the homotopy lifting property.

Proposition 1.10 *Let $M \to B$ be a homotopy fibre bundle. Then the natural map $M \to M^\sharp$, described above, is a fibrewise homotopy equivalence.*

This brief exposition of the theory of fibrations started from the concept of a homotopy fibre bundle. The standard development of the theory begins with the homotopy lifting properties and proves that every fibrewise space with the homotopy lifting property up to homotopy (over an ENR) is a homotopy fibre bundle.

Pull-backs

The pull-back $\alpha^* M \to B'$ of a homotopy fibre bundle $M \to B$ by a map $\alpha : B' \to B$ is transparently a homotopy fibre bundle.

Proposition 1.11 *Let α_0, $\alpha_1 : B' \to B$ be homotopic maps, and let $M \to B$ be a homotopy fibre bundle. Then the induced homotopy fibre bundles $\alpha_0^* M$ and $\alpha_1^* M$ are fibrewise homotopy equivalent. To be more precise, a homotopy α_t, $0 \le t \le 1$, determines a canonical fibrewise homotopy equivalence $\alpha_0^* M \to \alpha_1^* M$ up to homotopy.*

For consider the fibrewise map, f_0 say, $\alpha_0^* M \to M$ over $\alpha_0 : B' \to B$ which defines the pull-back. The homotopy lifting property (up to homotopy) gives maps $f_t : \alpha_0^* M \to M$ over α_t, which determine fibrewise maps $\alpha_0^* M \to \alpha_t^* M$ over B'. From Remark 1.7, the fibrewise homotopy class of f_1, and so of the map $\alpha_0^* M \to \alpha_1^* M$, is well-defined. One can check that this map is a fibrewise homotopy equivalence by constructing in the same manner a map in the opposite direction and using again the uniqueness of the lifts.

The particular case in which B' is a point has special significance. A path $\omega : [0,1] \to B$ from $a = \omega(0)$ to $b = \omega(1)$ determines a homotopy class: $M_a \to M_b$. Moreover, homotopic paths from a to b determine the same homotopy class. This defines a functor from the fundamental groupoid of B (with objects the points of B and morphisms from a to b the homotopy classes of paths from a to b) to the homotopy category (of compactly generated topological spaces).

Dold's theorem for homotopy fibre bundles

The next theorem of Dold is a key result in the development of fibrewise homotopy theory. It allows us to establish fibrewise generalizations of many classical results by merely formulating them correctly. For ease of reference we single it out from many theorems due to Dold as *Dold's theorem*.

Theorem 1.12 (Dold's theorem). *Let $f : M \to N$ be a fibrewise map between homotopy fibre bundles over B. Then f is a fibrewise homotopy equivalence if and only if $f_b : M_b \to N_b$ is a homotopy equivalence of fibres at each point $b \in B$ of the base.*

Notice that, if B is connected, then it suffices, by Proposition 1.2, to check that $f_b : M_b \to N_b$ is a homotopy equivalence at a single point $b \in B$ in order to conclude that f is a fibrewise homotopy equivalence.

This theorem plays the rôle in fibrewise homotopy theory that is played in equivariant homotopy theory by the theorem that recognizes an equivariant homotopy equivalence by its non-equivariant restriction to fixed subspaces. (To be precise, let G be a compact Lie group and let $f : M \to N$ be a G-map between compact G-ENRs. Then f is a G-equivariant homotopy equivalence if and only if, for each closed subgroup $H \leq G$, the restriction $f^H : M^H \to N^H$ to the subspaces fixed by H is a (non-equivariant) homotopy equivalence.)

The gluing construction

The well-understood and elementary procedure for gluing together fibre bundles defined on subspaces can be carried over to homotopy theory.

Proposition 1.13 *Suppose that the compact ENR B is a union of two closed sub-ENRs B_1 and B_2 with intersection an ENR $A = B_1 \cap B_2$. Let $M_1 \to B_1$ and $M_2 \to B_2$ be homotopy fibre bundles, and let $h : (M_1)_A \to (M_2)_A$ be a fibrewise homotopy equivalence. Suppose further that the fibres of M_1 and M_2 have the homotopy type of CW-complexes. Then there exists a homotopy fibre bundle $M \to B$ over B together with fibrewise homotopy equivalences $f_1 : M_1 \to M_{B_1}$ over B_1 and $f_2 : M_2 \to M_{B_2}$ over B_2 such that $(f_1)_A \simeq (f_2)_A \circ h$ over A.*

This result is significantly harder to establish that the corresponding one for topological, rather than homotopy, fibre bundles. It follows from the general theory of classifying spaces of such homotopy fibre bundles and the Mayer–Vietoris construction (2.12). See, for example, [104]. The force of the classification theorem is the following. We take a fixed CW-complex F as fibre. First of all, every homotopy fibre bundle, with fibre homotopy equivalent to F, over a compact ENR B is fibrewise homotopy equivalent to one of a set of bundles of this type, so that we may talk about the *set*, in *ad hoc* notation $\mathcal{X}(B)$ say, of fibrewise homotopy equivalence classes of homotopy fibre bundles over B with fibre F. Then we can introduce, for a closed sub-ENR $A \subseteq B$, the set $\mathcal{X}(B, A)$ of fibrewise homotopy equivalence classes of homotopy fibre bundles $M \to B$ with fibre F equipped with a fibrewise homotopy trivialization (up to homotopy) $M_A \to A \times F$ over A. The theorem asserts that this functor $\mathcal{X}$ is represented by a classifying space $\mathbf{X}$, which is a pointed CW-complex. To be precise, there is an equivalence of functors:

$$\mathcal{X}(B, A) \to [B/A; \mathbf{X}]$$

to the set of pointed homotopy classes of maps $B/A \to \mathbf{X}$.

Fibrewise cofibrations

We recall next some properties of fibrewise cofibrations, referring to Part I and [44] for details.

Definition 1.14 Let $i : M \to N$ be a fibrewise map over B. We say that i is a *fibrewise cofibration* if, for each fibrewise space P over B, each fibrewise map $g_0 : N \to P$ and homotopy $f_t : M \to P$ with $f_0 = g_0 \circ i$, there exists a homotopy $g_t : N \to P$ extending f, that is, $f_t = g_t \circ i$.

In homotopy theory it is often more natural to look at homotopy-cofibrations.

Definition 1.15 We say that i, as above, is a *fibrewise homotopy-cofibration* if it has the extension property described above for the data (P, g_0, f_t) whenever, for some ϵ, $0 < \epsilon < 1$, $f_t = f_0$ for $0 \le t \le \epsilon$.

The property of being a fibrewise cofibration or homotopy-cofibration is local:

Proposition 1.16 *Let $i : M \to N$ be a fibrewise map over an ENR B. Then i is a fibrewise (homotopy) cofibration if and only if each point $b \in B$ of the base has an open neighbourhood U such that $i_U : M_U \to N_U$ is a fibrewise (homotopy) cofibration over U.*

Let M now be a subspace of a fibrewise space N over B. There are two useful criteria for the inclusion $i : M \hookrightarrow N$ to be a fibrewise cofibration, the formulation of which involves a pair (ψ, h_t), where $\psi : N \to [0, 1]$ is a continuous function which is zero throughout M and $h_t : N \to N, 0 \le t \le 1$, is a fibrewise homotopy such that

$$\begin{aligned} &h_0(x) = x \text{ for all } x \in N, \\ &h_t(x) = x \text{ for all } x \in M, t \in [0, 1]. \end{aligned} \qquad (1.17)$$

Lemma 1.18 *Let M be a closed subspace of the fibrewise space N over B. Then the following conditions are equivalent.*

(i)   *The inclusion $i : M \hookrightarrow N$ is a fibrewise cofibration.*
(ii)   *There exists a pair (ψ, h_t) as in (1.17) such that (a) $\psi^{-1}(0) = M$, and (b) $h_t(x) \in M$ whenever $t > \psi(x)$.*
(iii) *There exists a pair (ψ, h_t) as in (1.17) such that (a) $\psi^{-1}(0) = M$, and (b) $h_1(x) \in M$ for all $x \in N$ such that $1 > \psi(x)$.*

A pair (ψ, h_t) satisfying (ii) is often called a fibrewise Strøm structure. (See Part I, Proposition 4.3). The condition (ii)(a) follows automatically from (ii)(b) and the fact that M is closed in N.) Condition (iii) is a neighbourhood deformation retraction property.

To state the corresponding characterization of a fibrewise homotopy-cofibration we need some notation. For any ϵ, with $0 < \epsilon < 1$, we write $\lambda_\epsilon : [0, 1] \to [0, 1]$ for the piecewise-linear function

$$\lambda_\epsilon(t) = \begin{cases} 0 & \text{for } 0 \le t \le \epsilon, \\ (t - \epsilon)/(1 - \epsilon) & \text{for } \epsilon < t \le 1. \end{cases}$$

Proposition 1.19 *Let $i : M \hookrightarrow N$ be the inclusion of a (not necessarily closed) subspace M of a fibrewise space N over B. Then the following conditions are equivalent.*

(i)   *The map i is a fibrewise homotopy-cofibration.*
(ii)   *There exists a pair (ψ, h_t) as in (1.17) such that $h_t(x) \in M$ whenever $\lambda_\epsilon(t) > \psi(x)$.*
(iii) *There exists a pair (ψ, h_t) as in (1.17) such that $h_1(x) \in M$ for all $x \in N$ such that $1 > \psi(x)$.*

The condition (ii) is independent of the choice of $\epsilon : 0 < \epsilon < 1$.

Fibrewise pointed spaces

Pointed spaces play an essential, if sometimes purely technical, rôle in homotopy theory. Geometric spaces are not usually equipped with a basepoint, and in order to do homotopy theory one has to choose or adjoin basepoints. The same is true in the fibrewise theory.

We shall try to maintain a distinction in notation between the unpointed and pointed theories, using letters such as M, N for geometric fibrewise spaces (such as fibrewise manifolds) and X, Y for fibrewise pointed spaces.

Informally, a fibrewise pointed space $X \to B$ is a family of pointed spaces X_b parametrized by $b \in B$. Precisely, it is given by a fibrewise space $X \to B$ and a map $s : B \to X$ such that $s(b) \in X_b$. The map s embeds B as a subspace of X. It is often convenient to write b for the basepoint $s(b)$ of the fibre X_b.

A fibrewise pointed map $f : X \to Y$ over B is a family of basepoint-preserving maps $f_b : X_b \to Y_b$.

The unique fibrewise pointed map $X \to Y$ which is null in each fibre is normally called the *fibrewise null map*; but when we pass to the stable theory it is natural to refer to it also as the *zero map*.

Adjoining a basepoint

Let $M \to B$ be a fibrewise space over B. We write $M_{+B} \to B$ for the fibrewise pointed space obtained by adjoining a disjoint basepoint to each fibre: as a space M_{+B} is the disjoint union $M \sqcup B$.

If $N \to B$ is another fibrewise space over B and $f : M \to N$ is a fibrewise map, then a fibrewise pointed map $f_+ : M_{+B} \to N_{+B}$ is defined in the obvious way. When M is connected (as a space), every fibrewise pointed map is of this form, with the exception of the null (or zero) map. In general, one can analyse pointed maps $M_{+B} \to N_{+B}$ in terms of the components of M.

Adjunction of a basepoint in this way allows us to absorb the geometric theory of fibrewise homotopy into the pointed theory.

Fibrewise one-point compactification

Another way in which basepoints arise geometrically is by one-point (or Alexandroff) compactification. (See Section 10 of Part I.) Let $M \to B$ be a locally compact Hausdorff fibrewise space over B. Then the fibrewise one-point compactification of M, obtained by adjoining a basepoint at infinity to each fibre, will be denoted by M_B^+.

An important example which will occur often is that of a finite-dimensional real vector bundle ξ over B. The fibrewise one-point compactification ξ_B^+ is a sphere-bundle. When ξ is equipped with a positive-definite inner product, ξ_B^+ can be identified, by stereographic projection, with the unit sphere-bundle $S(\mathbb{R} \oplus \xi)$ in $\mathbb{R} \oplus \xi$, the direct sum of the trivial bundle $B \times \mathbb{R} \to B$ and ξ, with $(1, 0)$ corresponding to the basepoint at infinity (in each fibre).

Pointed fibre bundles

We say that a fibrewise pointed space $X \to B$ over B is *trivial*, if it is fibrewise pointed homeomorphic to $B \times F \to B$ for some pointed space F. The fibrewise pointed space $X \to B$ is locally trivial, or a *pointed fibre bundle*, if there is an open covering $\mathcal{U}$ of B such that $X_U \to U$ is trivial, as a fibrewise pointed space, for all $U \in \mathcal{U}$. Thus a pointed fibre bundle is a *bundle of pointed spaces*. The fibrewise one-point compactification ξ_B^+ of a vector bundle ξ over B is a good example: it is a pointed sphere-bundle.

The complement $X - B$ of the basepoints in a pointed fibre bundle X is evidently a fibre bundle over B. We use this observation in the following example.

Example 1.20. Let $X := B \times B \to B : (a, b) \mapsto a$ be the fibrewise pointed space with basepoint $(b, b) \in X_b$. As a fibrewise space $X \to B$ is trivial. It is shown in Proposition 11.20 that $X \to B$ is a pointed fibre bundle when B is a (topological) manifold without boundary. On the other hand, if B is the closed interval $[0, 1]$, then X is not a pointed fibre bundle, for $X - B \to B$ is not locally trivial.

Homotopy well-pointed fibrewise spaces

To obtain a workable homotopy theory it is necessary to put some restrictions on basepoints. Let $X \to B$ be a fibrewise pointed space. It is reasonable to insist first of all that the basepoints form a closed subspace $B \subseteq X$, as is always the case when X is Hausdorff.

Let us define a fibrewise pointed space X^b to be the subspace

$$X^b := (X \times \{0\}) \cup (B \times [0, 1])$$

of $X \times [0, 1]$ with the basepoint $(b, 1)$ at $b \in B$. The projection defines a fibrewise pointed map

$$X^b \to X. \tag{1.21}$$

The fibrewise pointed space X is said to be (fibrewise) *well-pointed* if the inclusion $B \to X$ is a fibrewise cofibration or, in other words, if there exists a fibrewise retraction $r : X \times [0, 1] \to X^b$. (See Sections 4 and 16 of Part I.) This property of X is not invariant under fibrewise pointed homotopy equivalence. The corresponding invariant property is the following.

We say that X is (fibrewise) *homotopy well-pointed* (or *non-degenerate*) if the inclusion $B \to X$ is a fibrewise homotopy-cofibration.

It is elementary to check that for any X the fibrewise pointed space X^b is well-pointed and that the fibrewise pointed space X is homotopy well-pointed if and only if the map $X^b \to X$ is a fibrewise pointed homotopy equivalence.

The condition is usually satisfied in geometrically occurring examples. Clearly, for any fibrewise space $M \to B$, the fibrewise pointed space M_{+B} is well-pointed. A pointed fibre bundle is homotopy well-pointed if and only if each fibre is homotopy well-pointed, because the property is local, by Proposition 1.16.

Proposition 1.22 *Let $X \to B$ be a fibrewise pointed space and let $\mathcal{U}$ be an open covering of B. Then X is homotopy well-pointed if and only if X_U is homotopy well-pointed for all $U \in \mathcal{U}$.*

Example 1.23. Consider the path space $\mathcal{P}B = \mathrm{map}([0,1], B) \to B : \omega \mapsto \omega(0)$ as a fibrewise space over B. It is fibrewise pointed, with basepoint at b the constant path. Using the fact that B is an ENR, one can show that $\mathcal{P}B \to B$ is well-pointed. (In fact, the basepoint $B \to \mathcal{P}B$ includes B as a fibrewise sub-ENR of the fibrewise ANR $\mathcal{P}B$. See Lemma 5.4.)

Given a fibrewise pointed space X, the construction (Lemma 1.5) gives a fibrewise space $X^{\sharp}$. We can make it a fibrewise pointed space by taking as basepoint in the fibre at b the point (b, b), that is, the constant path at b.

Lemma 1.24 *Suppose that X is well-pointed. Then the fibrewise pointed space $X^{\sharp}$ constructed above is well-pointed.*

This follows rather easily from Example 1.23.

From now on all fibrewise pointed spaces (and all pointed spaces) are assumed to be homotopy well-pointed.

Pointed homotopy fibre bundles

Let $X \to B$ be a fibrewise pointed space. We say that X is a *pointed homotopy fibre bundle* if it is locally pointed homotopy trivial, that is, if each point of B has an open neighbourhood U such that X_U is pointed fibrewise homotopy equivalent to a trivial bundle of pointed spaces $U \times F$. Notice that our requirement that a pointed homotopy fibre bundle be homotopy well-pointed is equivalent to the requirement that each fibre be homotopy well-pointed, by Proposition 1.22.

Fibrewise pointed maps between pointed homotopy fibre bundles are locally trivial as fibrewise pointed maps. The proof of Proposition 1.2 carries through unchanged.

Proposition 1.25 *Let $f : X \to Y$ be a (fibrewise) map between pointed homotopy fibre bundles over B. Then for each $b \in B$ there exists an open neighbourhood U of b and local pointed fibrewise homotopy trivializations $\phi : X_U \to U \times X_b$ and $\psi : U \times Y_b \to Y_U$ such that f_U is fibrewise homotopic to $\psi \circ (1 \times f_b) \times \phi$.*

The pointed homotopy lifting property

The theory for pointed homotopy fibre bundles now proceeds along the same lines as the unpointed theory, although there are some subtleties in the detail.

Definition 1.26 Let $X \to B$ be a fibrewise pointed space over B. We say that X has the *pointed homotopy lifting property* if, for each base space B' and fibrewise pointed space $X' \to B'$ over B', given a homotopy $\alpha_t : B' \to B$, $0 \le t \le 1$, and fibrewise pointed map $f_0 : X' \to X$ over α_0, there is a pointed homotopy f_t over α_t extending f_0.

$$
\begin{array}{ccc}
X' & \xrightarrow{f_t} & X \\
\downarrow & & \downarrow \\
B' & \xrightarrow{\alpha_t} & B
\end{array}
\qquad\qquad
\begin{array}{ccc}
X' \times [0,1] & \xrightarrow{f} & X \\
\downarrow & & \downarrow \\
B' \times [0,1] & \xrightarrow{\alpha} & B
\end{array}
$$

A fibrewise pointed space X possessing this property is generally referred to in the literature [10, 104] as an *ex-* or *based fibration*.

There is a corresponding definition of the *pointed homotopy lifting property up to homotopy*, and this is invariant under fibrewise pointed homotopy equivalence.

One can again show that both properties are local (for ENR base spaces), and so deduce the first part of the following proposition.

Proposition 1.27 *A pointed homotopy fibre bundle possesses the pointed homotopy lifting property up to homotopy and is fibrewise pointed homotopy equivalent to a fibrewise pointed space with the pointed homotopy lifting property.*

Starting from a fibrewise pointed space $X \to B$ it is not so straightforward as in the unpointed case to construct an associated fibrewise pointed space with the pointed homotopy lifting property. The fibrewise pointed space $X^\sharp$, defined as the fibre product $X \times_B \mathcal{P}B$ with the basepoint (b, b) in $X_b^\sharp$, is not quite the right candidate.

Proposition 1.28 *Let $X \to B$ be a fibrewise pointed space. Then $(X^\flat)^\sharp$ is a well-pointed fibrewise space with the pointed homotopy lifting property.*

If X is a pointed homotopy fibre bundle, then this construction produces a well-pointed fibrewise space with the pointed homotopy lifting property which is canonically fibrewise pointed homotopy equivalent to X.

The discussion of pull-backs of homotopy fibre bundles carries over, with only notational changes, to the pointed theory.

Dold's theorem for pointed homotopy fibre bundles

The following characterization of fibrewise pointed homotopy equivalences between pointed homotopy fibre bundles will again be referred to as simply *Dold's theorem*.

Theorem 1.29 (Dold's theorem). *Let $f : X \to Y$ be a fibrewise map between pointed homotopy fibre bundles over B. Then f is a fibrewise pointed homotopy equivalence if and only if $f_b : X_b \to Y_b$ is a pointed homotopy equivalence of fibres at each point $b \in B$ of the base.*

Recall that the fibrewise pointed spaces X and Y are assumed to be homotopy well-pointed; so, too, are the fibres. This theorem therefore follows from the original theorem of Dold (Theorem 1.12) and Part I, Theorem 16.2. (The result in this form is due to Eggar [58].) If f is a fibrewise homotopy equivalence then it is a fibrewise pointed homotopy equivalence, and if f_b is a homotopy equivalence then it is a pointed homotopy equivalence.

The gluing construction for pointed homotopy fibre bundles

We shall need the pointed version of Proposition 1.11.

Proposition 1.30 *Suppose that the compact base B is a union of two closed sub-ENRs B_1 and B_2 with intersection an ENR $A = B_1 \cap B_2$. Let $X_1 \to B_1$ and $X_2 \to B_2$ be homotopy fibre bundles with fibres of the homotopy type of CW-complexes, and let $h : (X_1)_A \to (X_2)_A$ be a fibrewise pointed homotopy equivalence. Then there exists a pointed homotopy fibre bundle $X \to B$ over B together with fibrewise pointed homotopy equivalences $f_1 : X_1 \to X_{B_1}$ over B_1 and $f_2 : X_2 \to X_{B_2}$ over B_2 such that $(f_1)_A \simeq (f_2)_A \circ h$ over A.*

Moreover, the fibrewise space X together with the equivalences f_1 and f_2 is unique in the following sense. If X', f_1' and f_2' give a second solution to the gluing problem, then there is a fibrewise pointed homotopy equivalence $g : X \to X'$ such that $f_i \simeq f_i' \circ g_{B_i}$ for $i = 1, 2$. (This is a routine exercise on the Mayer–Vietoris sequence that we shall meet in the next section (Proposition 2.14).)

Local triviality of fibrewise Hopf spaces

Fibrewise Hopf spaces have already been discussed in some detail in Part I, Section 21. Let (X, m) and (X', m') be two fibrewise Hopf spaces over B. We say that a fibrewise pointed map $f : X \to X'$ is a *fibrewise H-map* if it is compatible with the multiplication: $m' \circ (f \times f) \simeq f \circ m$. The map f is a *fibrewise H-equivalence* if further it is a fibrewise pointed homotopy equivalence. It then follows that its inverse is a fibrewise H-map.

Proposition 1.31 *Let $X \to B$ be a pointed homotopy fibre bundle admitting a fibrewise Hopf structure m. Then (X, m) is locally trivial as a fibrewise Hopf space.*

We have to show that each point $b \in B$ is contained in an open neighbourhood U such that (X_U, m_U) is fibrewise H-equivalent to the trivial fibrewise Hopf space over U with fibre the Hopf space (X_b, m_b). This follows from the local contractibility of the base. There is no loss of generality in supposing that X is trivial: $X = B \times F$. We choose U contracting by a homotopy $H_t : U \to B$ to b. Thus, H_0 is the inclusion and H_1 is the constant map b. Then we get a homotopy m_{H_t} from m to the multiplication $(1 \times m_b) : U \times (F \times F) \to U \times F$.

Remark 1.32. Suppose that G is a pointed compact ENR admitting a Hopf structure. Then the space of Hopf structures on G is an ANR and so, in particular, locally contractible.

This can be seen as follows. We may assume that G is a subspace of a Euclidean space E with the basepoint at 0. Choose a retraction $r : U \to G$ of an open neighbourhood U onto G. Consider the set of Hopf structures on G as a subspace of the Banach space of continuous maps $G \times G \to E$ taking $(0,0)$ to 0. A neighbourhood retraction is given by mapping m to the Hopf structure $\bar{m}$ given by:

$$\bar{m}(x, y) = r(m(x, y) - m(x, 0) + x - m(0, y) + y).$$

(These ideas on local triviality of fibrewise Hopf spaces are taken from [38].)

2 Complements on homotopy theory

In this section we shall work entirely with fibrewise pointed spaces (assumed to be homotopy well-pointed).

Homotopy classes

Let X and Y be fibrewise pointed spaces over B. Suppose first of all that B is compact, as will usually be the case, and that $A \subseteq B$ is a closed sub-ENR. We write

$$[X; Y]_{(B,A)} \quad \text{or} \quad \pi^0_{(B,A)}[X; Y]$$

for the set of (fibrewise) homotopy classes of fibrewise pointed maps $f : X \to Y$ which are zero (that is, null) over A: $f_a = * : X_a \to Y_a$ for all $a \in A$. Fibrewise homotopies are understood also to be zero over A.

More generally, for $i \geq 0$, we set

$$\pi_{(B,A)}^{-i}[X;Y] := [\Sigma_B^i X; Y]_{(B,A)},$$

where Σ_B denotes the (reduced) fibrewise suspension:

$$\Sigma_B X = (X \times [0,1])/\sim,$$

where the equivalence relation $\sim$ identifies $X_b \times \{0,1\}$ and $\{b\} \times [0,1]$ to the basepoint b in the fibre at $b \in B$. (As we deal exclusively with fibrewise pointed spaces here, we omit the superscript 'B' used in Part I to indicate the reduced suspension. The same simplification in notation will be made elsewhere.) It will be useful later to think of the iterated fibrewise suspension Σ_B^i as the fibrewise smash product with the trivial sphere-bundle $(B \times \mathbb{R}^i)_B^+ = B \times S^i$.

For $i = 0$, $\pi_{(B,A)}^{-i}[X;Y]$ is simply a pointed set; for $i = 1$, it is a group, and, for $i > 1$, an Abelian group. The group structure on these *fibrewise homotopy groups* is defined just as in the classical theory.

The cohomological indexing is chosen with the stable theory, to be introduced in the next section, in mind. When the subspace A is empty, we abbreviate $\pi_{(B,\emptyset)}^*$ to π_B^*.

Compact supports

In some applications, when the base is not compact, it is useful to consider homotopy classes with compact supports. Suppose that B is not necessarily compact. We say that a fibrewise pointed map $f : X \to Y$ has *compact support* if it is zero outside a compact subset of the base, that is, if $f_b = *$ for b in the complement of some compact subset of B. In the same way, a (fibrewise) homotopy with compact support, regarded as a fibrewise map over $B \times [0,1]$, is required to have compact support in $B \times [0,1]$.

We write $_c\pi_B^{-i}[X;Y]$ for the set of (compactly supported) homotopy classes of compactly supported fibrewise pointed maps $\Sigma_B^i X \to Y$.

If the topology of B is sufficiently well behaved, the compactly supported theory over B will be the direct limit of the theories over pairs (B', A'), where B' is a compact sub-ENR of B and A' is a closed sub-ENR of B' such that $B' - A'$ is open in B. This will be the case if B is, for example, a smooth manifold; see Proposition 5.3.

The cofibre exact sequence

Let $f : X' \to X$ be a fibrewise pointed map over B.

Definition 2.1 The fibrewise *homotopy-cofibre* (or *mapping cone*) $C_B(f)$ of f is defined to be the fibrewise quotient of

$$(B \times \{0\}) \sqcup (X' \times [0,1]) \sqcup (X \times \{1\})$$

by the identification of $(x', 0)$ with $(p(x'), 0)$ and $(x', 1)$ with $(f(x'), 1)$ for $x' \in X'$, and (b, t) with $(b, 0)$ for all $b \in B$. In other words, we have a topological push-out diagram:

$$
\begin{array}{ccc}
X' & \xrightarrow{\;f\;} & X \\
\downarrow & & \downarrow \\
C_B X' & \longrightarrow & C_B(f)
\end{array}
$$

over B, where $C_B X'$ is the fibrewise cone on X'. (Consistent with the convention introduced above, the cone is understood to be reduced.)

The homotopy-cofibre $C_B(f)$ is equipped with maps $X \to C_B(f) \to \Sigma_B X'$. There is also an evident map from $C_B(f)$ to the *topological cofibre* $X/_B f(X')$, given by $(x', t) \mapsto [f(x')]$, for $x' \in X'$, $(x, 1) \mapsto [x]$ for $x \in X$. As in the ordinary theory, we have:

Lemma 2.2 *Let $f : X' \to X$ be a fibrewise homotopy-cofibration. Then the canonical map from the homotopy-cofibre to the topological cofibre:*

$$
C_B(f) \to X/_B f(X')
$$

is a fibrewise pointed homotopy equivalence.

(Indeed, the condition that f be a fibrewise homotopy-cofibration is necessary as well as sufficient.)

The construction of the homotopy-cofibre is compatible with homotopies. Consider a homotopy-commutative diagram:

$$
\begin{array}{ccc}
X' & \xrightarrow{\;f\;} & X \\
h' \downarrow & & \downarrow h \\
Y' & \xrightarrow[g]{} & Y
\end{array}
$$

with a given (fibrewise) homotopy H between hf and gh'. Then there is an associated map $H_* : C_B(f) \to C_B(g)$ induced by $(x, 1) \mapsto (h(x), 1)$,

$$
(x', t) \mapsto \begin{cases} (h'(x'), 2t) & \text{for } 0 \leq t \leq \tfrac{1}{2}, \\ (H(x', 2t - 1), 1) & \text{for } \tfrac{1}{2} \leq t \leq 1. \end{cases}
$$

As a first consequence of this homotopy invariance one can replace X and X' by the fibrewise pointed spaces $X^\flat$ and $(X')^\flat$, (1.21). It is then not too hard to check that $C_B(f)$, and $\Sigma_B X$, are homotopy well-pointed (using the technique described in Part I (Proposition 16.4)).

It is clear that the fibrewise suspension $\Sigma_B X$ of a pointed homotopy fibre bundle is again a pointed homotopy fibre bundle. The homotopy invariance noted above shows that the cofibre construction can also be made within the category of pointed homotopy fibre bundles. For a map between such bundles is locally homotopy trivial, by Proposition 1.25. We state the result for reference.

Lemma 2.3 *Let $f : X' \to X$ be a fibrewise pointed map between pointed homotopy fibre bundles. Then the homotopy-cofibre $C_B(f)$, and the fibrewise suspension $\Sigma_B X$, are pointed homotopy fibre bundles.*

The construction of the homotopy-cofibre is made precisely to achieve the property that: given a fibrewise pointed map $g : X \to Y$ and a null-homotopy H from the null map $*$ to gf, there is a canonical factorization of g through $C_B(f)$, given by $(x', t) \mapsto H(x', t)$ and $(x, 1) \mapsto g(x)$. Beginning with this observation, the usual arguments yield the *exact cofibre* (or *Puppe*) *sequence*. We assume in the statement that B is compact and that A is a closed sub-ENR (but only because we have limited our definition of the relevant terms to that case).

Proposition 2.4 *For any fibrewise pointed space Y over B, there is a long exact sequence:*

$$\cdots \to \pi^{-i}_{(B,A)}[C_B(f); Y] \to \pi^{-i}_{(B,A)}[X; Y] \to \pi^{-i}_{(B,A)}[X'; Y]$$
$$\to \pi^{-i+1}_{(B,A)}[C_B(f); Y] \to \cdots$$
$$\to \pi^0_{(B,A)}[C_B(f); Y] \to \pi^0_{(B,A)}[X; Y] \to \pi^0_{(B,A)}[X'; Y].$$

The sequence is exact, in the first place, as a sequence of pointed sets, and so of groups from the term $\pi^{-1}_{(B,A)}[X; Y]$. Just as when B is a point, a little more can be said at the next term. The group $\pi^{-1}_{(B,A)}[X'; Y]$ acts on the set $\pi^0_{(B,A)}[C_B(f); Y]$, and two elements lie in the same orbit if and only if they have the same image in $\pi^0_{(B,A)}[X; Y]$. (For further details see Section 12 of Part I.)

There is a corresponding exact sequence for homotopy with compact supports $_c\pi^*_B$ over a locally compact base B.

The fibre exact sequence

The treatment of the homotopy-fibre is formally similar. We write the space of based paths $\mathrm{map}^*_B(B \times [0,1], Y)$ as $P_B(Y)$ (taking 0 as the basepoint in $[0,1]$). Consider a fibrewise pointed map $g : Y \to Y''$.

Definition 2.5 The fibrewise *homotopy-fibre* (or *mapping fibre*) $F_B(g) \to B$ of g is the subspace of $Y \times_B P_B Y''$ consisting of those pairs (y, ω) where $y \in Y_b$ and $\omega : [0,1] \to Y''_b$ is a path from the basepoint $\omega(0) = b$ to $\omega(1) = g(y)$. The basepoint in the fibre at b is (b, b), where, as usual, the second component b denotes the constant path. We have a pull-back diagram

$$
\begin{array}{ccc}
F_B(g) & \longrightarrow & P_B(Y'') \\
\downarrow & & \downarrow \\
Y & \xrightarrow{\ g\ } & Y''
\end{array}
$$

There is a natural map from the fibrewise *topological fibre*, $\{y \in Y \mid g(y) = *\}$, to the homotopy-fibre, and this map is a fibrewise pointed homotopy equivalence if and only if g is a fibrewise homotopy-fibration. (See the discussion in Section 13 of Part I.)

The construction of the homotopy-fibre, like that of the homotopy-cofibre, is homotopy-theoretic. From our hypothesis that Y and Y'' are homotopy well-pointed it follows that $F_B(g)$ and the fibrewise loop space $\Omega_B Y$ are also homotopy well-pointed. The construction also preserves homotopy local triviality.

Lemma 2.6 *Let* $g : Y \to Y''$ *be a fibrewise pointed map of pointed homotopy fibre bundles over* B. *Then the homotopy-fibre* $F_B(g)$, *and the fibrewise loop space* $\Omega_B Y$, *are pointed homotopy fibre bundles.*

The derivation of the long exact fibre sequence (or Nomura sequence) follows the classical theory. For the statement we suppose that B is compact and that A is a closed sub-ENR. There is a corresponding sequence in the theory with compact supports.

Proposition 2.7 *For any fibrewise pointed space* X *over* B, *there is a long exact sequence:*

$$
\begin{aligned}
\cdots &\to \pi_{(B,A)}^{-i}[X; F_B(g)] \to \pi_{(B,A)}^{-i}[X; Y] \to \pi_{(B,A)}^{-i}[X; Y''] \\
&\to \pi_{(B,A)}^{-i+1}[X; F_B(g)] \to \cdots \\
&\to \pi_{(B,A)}^{0}[X; F_B(g)] \to \pi_{(B,A)}^{0}[X; Y] \to \pi_{(B,A)}^{0}[X; Y''].
\end{aligned}
$$

As in the case of the cofibre sequence, the group $\pi_{(B,A)}^{-1}[X; Y'']$ acts on the set $\pi_{(B,A)}^{0}[X; F_B(g)]$ and the sequence is exact at this point in the refined sense.

Pull-backs

Let $\alpha : (B', A') \to (B, A)$ be a map of compact ENR pairs. Given a fibrewise pointed map $f : X \to Y$ between fibrewise pointed spaces over B, we can form the pull-back $\alpha^* f : \alpha^* X \to \alpha^* Y$ over B'. If f is zero (null) over A, then $\alpha^* f$ will be zero over A'. This construction gives us pull-back maps

$$\alpha^* : \pi^{-i}_{(B,A)}[X;Y] \to \pi^{-i}_{(B',A')}[\alpha^* X; \alpha^* Y]$$

with evident functorial properties.

In the class of pointed homotopy fibre bundles the pull-back is homotopy invariant. To explain this statement, let us suppose that X and Y are both pointed homotopy fibre bundles and that $\alpha_t : (B', A') \to (B, A)$, $0 \le t \le 1$, is a homotopy. Then, as we have observed in Proposition 1.11, there are natural fibrewise pointed homotopy equivalences: $\alpha_0^* X \to \alpha_1^* X$ and $\alpha_0^* Y \to \alpha_1^* Y$.

Proposition 2.8 *In the situation described above,*

$$\alpha_0^* = \alpha_1^* : \pi^{-i}_{(B,A)}[X;Y] \to \pi^{-i}_{(B',A')}[\alpha_0^* X; \alpha_0^* Y] = \pi^{-i}_{(B',A')}[\alpha_1^* X; \alpha_1^* Y],$$

where the identifications are made using the canonical fibrewise pointed homotopy equivalences determined by the homotopy α_t.

This follows from the uniqueness, up to homotopy, of the homotopy lifting (Remark 1.7). For let $g_t : \alpha_0^* X \to \alpha_t^* X$ extend the identity $g_0 = 1$ on X and $h_t : \alpha_0^* Y \to \alpha_t^* Y$ extend the identity $h_0 = 1$ on Y. Consider a fibrewise map $f : X \to Y$. Then $\alpha_t^* f \circ g_t$ and $h_t \circ \alpha_0^* f : \alpha_0^* X \to \alpha_t^* Y$ both extend $\alpha_0^* f$ and so are homotopic.

There is a corresponding pull-back construction for fibrewise homotopy with compact supports over locally compact base spaces. The map $B' \to B$ will, of course, be required to be proper (so that the inverse image of a compact subspace is closed and, consequently, the pull-back of a compactly supported map will be compactly supported.)

The relative exact sequence

In this subsection we fix fibrewise pointed spaces X and Y over a compact base B and examine the relation between homotopy theory over B and homotopy theory over a closed sub-ENR A. There is an obvious *restriction map*

$$\pi^{-i}_B[X;Y] \to \pi^{-i}_A[X_A; Y_A].$$

We begin by extending the restriction map on π^0: $\pi^0_B[X;Y] \to \pi^0_A[X_A; Y_A]$ formally on the left to a long exact sequence. The next term in the sequence will classify maps $f : X \to Y$ over B together with a null-homotopy over A, that is, $F_t : X_A \to Y_A$, $0 \le t \le 1$, with $F_0 = f_A$ and $F_1 = *$. The pair (f, F_t) determines a class in

$$\pi^0_{(B \times \{0\} \cup A \times [0,1], A \times \{1\})}[X; Y].$$

Here we have written simply X and Y for the pull-backs of those fibrewise spaces over B to $B \times \{0\} \cup A \times [0, 1]$. We make similar abbreviations in what follows. For example, with this convention we have a natural identification

$$\pi^{-i}_{(B, A)}[X; Y] = \pi^0_{(B \times D^i, A \times D^i \cup B \times \partial D^i)}[X; Y],$$

which we shall use shortly.

The next term in the long exact sequence will be

$$\pi^0_{(A \times [0, 1], A \times \{0, 1\})}[X; Y] = \pi^0_A[\Sigma_A X_A; Y_A] = \pi^{-1}_A[X_A; Y_A].$$

We can also write this group as the set

$$\pi^0_{(B \times \{0\} \cup A \times [0, 1], B \times \{0\} \cup A \times \{1\})}[X; Y],$$

by excision, and this set maps by restriction to the set of homotopy classes over $(B \times \{0\} \cup A \times [0, 1], A \times \{1\})$. We need to check exactness of the segment

$$\pi^0_{(B \times \{0\} \cup A \times [0, 1], B \times \{0\} \cup A \times \{1\})}[X; Y] \to$$
$$\pi^0_{(B \times \{0\} \cup A \times [0, 1], A \times \{1\})}[X; Y] \to \pi^0_{B \times \{0\}}[X; Y].$$

But a map over $(B \times \{0\} \cup A \times [0, 1], A \times \{1\})$ and a null-homotopy (of its restriction) over $B \times \{0\}$ combine to give a map over the pair

$$(B \times [-1, 1] \cup A \times [0, 1], B \times \{-1\} \cup A \times \{1\}).$$

Now restrict this map to $(B \times \{-1\} \cup A \times [-1, 1], B \times \{-1\} \cup A \times \{1\})$ and change the second coordinate t to $(1 + t)/2$ to obtain the required lift.

In this way we construct, quite formally, a relative exact sequence:

Proposition 2.9 *Let X and Y be fibrewise pointed spaces over a compact ENR B and let A be a closed sub-ENR of B. Then there is a long exact sequence:*

$$\cdots \to \pi^{-i}_{(B \times \{0\} \cup A \times [0, 1], A \times \{1\})}[X; Y] \to \pi^{-i}_B[X; Y] \to \pi^{-i}_A[X_A; Y_A]$$
$$\to \cdots \to \pi^0_B[X; Y] \to \pi^0_A[X_A; Y_A].$$

Again this sequence is exact in the refined sense, with the homotopy group $\pi^{-1}_A[X_A; Y_A]$ acting on the term to its right in the sequence.

The result has more substance when X and Y are pointed homotopy fibre bundles. The following crucial lemma is due to Becker and Gottlieb [10].

Lemma 2.10 *Suppose that X and Y are pointed homotopy fibre bundles. Then the projection map $(B \times \{0\} \cup A \times [0, 1], A \times \{1\}) \to (B, A)$ induces an isomorphism (of sets, or groups when appropriate):*

$$\pi^{-i}_{(B, A)}[X; Y] \to \pi^{-i}_{(B \times \{0\} \cup A \times [0, 1], A \times \{1\})}[X; Y].$$

The proof uses both the fact that the inclusion $B \hookrightarrow A$ is a cofibration and the fact that X and Y have the homotopy lifting property (up to homotopy). From the first, we have a retraction $r : B \times [0,1] \to B \times \{0\} \cup A \times [0,1]$. This gives a map

$$(B, A) \to (B \times \{0\} \cup A \times [0,1], A \times \{1\}) : b \mapsto r(b,1),$$

which is an inverse homotopy equivalence, of pairs, to the projection. The assertion of the lemma now follows from Proposition 2.8.

We shall refer to the final result as the *relative exact sequence*:

Proposition 2.11 *Let X and Y be fibrewise pointed spaces over a compact ENR B, and let A be a closed sub-ENR of B. Suppose that X and Y are pointed homotopy fibre bundles. Then homotopy groups over B and A are related by a long exact sequence:*

$$\cdots \to \pi^{-i}_{(B,A)}[X; Y] \to \pi^{-i}_B[X; Y] \to \pi^{-i}_A[X_A; Y_A] \to \cdots$$
$$\to \pi^0_{(B,A)}[X; Y] \to \pi^0_B[X; Y] \to \pi^0_A[X_A; Y_A].$$

The Mayer–Vietoris sequence

We can also look at the problem of gluing together fibrewise maps defined over subspaces of the base and agreeing where both are defined. Let B_1 and B_2 be closed sub-ENRs of the compact ENR B, and suppose that their intersection $A = B_1 \cap B_2$ is also an ENR. Consider the pair of restriction maps

$$\pi^0_{B_1}[X_{B_1}; Y_{B_1}] \times \pi^0_{B_2}[X_{B_2}; Y_{B_2}] \rightrightarrows \pi^0_A[X_A; Y_A].$$

To extend this diagram to an exact sequence, we note that a pair of maps $f_i : X_{B_i} \to Y_{B_i}$, $i = 1, 2$, and a (fibrewise) homotopy $F_t : X_A \to Y_A$ between their restrictions to the intersection A define a map from $X \to Y$ over $B_1 \times \{0\} \cup A \times [0,1] \cup B_2 \times \{1\}$, thus:

$$\pi^0_{B_1 \times \{0\} \cup A \times [0,1] \cup B_2 \times \{1\}}[X; Y] \to \pi^0_{B_1}[X; Y] \times \pi^0_{B_2}[X; Y].$$

The next term is

$$\pi^{-1}_A[X; Y] = \pi^0_{(A \times [0,1], A \times \{0,1\})}[X; Y]$$
$$= \pi^0_{(B_1 \times \{0\} \cup A \times [0,1] \cup B_2 \times \{1\}, B_1 \times \{0\} \cup B_2 \times \{1\})}[X; Y],$$

and we can restrict from $(B_1 \times \{0\} \cup A \times [0,1] \cup B_2 \times \{1\}, B_1 \times \{0\} \cup B_2 \times \{1\})$ to $B_1 \times \{0\} \cup A \times [0,1] \cup B_2 \times \{1\}$.

Again, quite formally, we obtain a long exact sequence:

$$\cdots \to \pi_{B_1 \times 0 \cup A \times [0,1] \cup B_2 \times 1}^{-i}[X; Y] \to \pi_{B_1}^{-i}[X; Y] \oplus \pi_{B_2}^{-i}[X; Y] \to \pi_A^{-i}[X; Y]$$

$$\cdots \to \pi_{B_1}^{-1}[X; Y] \times \pi_{B_2}^{-1}[X; Y] \to \pi_A^{-1}[X; Y]$$

$$\to \pi_{B_1 \times 0 \cup A \times [0,1] \cup B_2 \times 1}^{0}[X; Y] \to \pi_{B_1}^{0}[X; Y] \times \pi_{B_2}^{0}[X; Y] \rightrightarrows \pi_A^{0}[X; Y].$$

The union $B_1 \cup B_2$ is necessarily a closed sub-ENR of B. (See, for example, [18].) To simplify notation we suppose that $B = B_1 \cup B_2$. Then the projection

$$B_1 \times \{0\} \cup A \times [0,1] \cup B_2 \times \{1\} \to B \qquad (2.12)$$

is a homotopy equivalence. An inverse map can be obtained by gluing together maps $B_1 \to B_1 \times \{0\} \cup A \times [0, \frac{1}{2}]$ and $B_2 \to A \times [\frac{1}{2}, 1] \cup B_2 \times \{1\}$ which coincide on A: $a \in A \mapsto (a, \frac{1}{2})$. The two maps are constructed from retractions $B_i \times [0,1] \to B_i \times \{0\} \cup A \times [0,1]$ as in the proof of Lemma 2.10. From Proposition 2.8 again, we obtain:

Lemma 2.13 *Let X and Y be pointed homotopy fibre bundles over $B = B_1 \cup B_2$. Then the projection map (2.12) induces a bijection*

$$\pi_B^{-i}[X; Y] \to \pi_{B_1 \times \{0\} \cup A \times [0,1] \cup B_2 \times \{1\}}^{-i}[X; Y].$$

This establishes the *Mayer–Vietoris sequence*:

Proposition 2.14 *Let B_1 and B_2 be closed sub-ENRs of B such that $B = B_1 \cup B_2$ and $A := B_1 \cap B_2$ is an ENR. Let X and Y be pointed homotopy fibre bundles over B. Then one has a long exact sequence:*

$$\cdots \to \pi_B^{-i}[X; Y] \to \pi_{B_1}^{-i}[X; Y] \oplus \pi_{B_2}^{-i}[X; Y] \to \pi_A^{-i}[X; Y] \to \cdots$$

$$\to \pi_B^{-1}[X; Y] \to \pi_{B_1}^{-1}[X; Y] \times \pi_{B_2}^{-1}[X; Y] \to \pi_A^{-1}[X; Y]$$

$$\to \pi_B^{0}[X; Y] \to \pi_{B_1}^{0}[X; Y] \times \pi_{B_2}^{0}[X; Y] \rightrightarrows \pi_A^{0}[X; Y].$$

Obstruction theory

The next result can be regarded as a refinement of Dold's theorem.

Proposition 2.15 *Let B be a finite complex of dimension $\leq m$ and let A be a subcomplex of B. Let $g : Y' \to Y$ be a fibrewise pointed map of pointed homotopy fibre bundles over B. Suppose that, for some integer n, for each $b \in B$ and any finite pointed complex P, composition with f_b induces a surjection $[P; Y_b'] \to [P; Y_b]$ if $\dim P \leq n$, a bijection if $\dim P < n$. Let X be a pointed homotopy fibre bundle over B with each fibre homotopy equivalent to a finite complex of dimension $\leq l$. Then the induced map*

$$g_* : \pi_{(B,A)}^{-i}[X; Y'] \to \pi_{(B,A)}^{-i}[X; Y]$$

is surjective for $i \leq n - (m + l)$, bijective for $i < n - (m + l)$.

There are several accounts of variants of this result in the literature [10, 79]. In the form stated, it is a routine application of classical obstruction theory.

Most of Proposition 2.15 can be established by cell-by-cell argument using the Mayer–Vietoris sequence, at least when A is empty. (The general case requires a relative version of the Mayer–Vietoris sequence.) The proof proceeds by induction over the cells of B, using the five-lemma. (To be precise, this will prove surjectivity for any i, but injectivity only for $i \geq 1$.) One writes $B = B_1 \cup B_2$, where B_2 is a disc and $B_1 \cap B_2$ is the bounding sphere. Since the bundles are all homotopy trivial over the disc, the assertion is true over B_2 and $B_1 \cap B_2$, by the hypothesis on the maps g_b on fibres. Given the result for the restriction of g to B_1, we can deduce the result for B.

Remark 2.16. We observe, for future reference, that it would suffice to assume that each fibre X_b is a *homotopy retract* of some pointed finite complex Q of dimension $\leq l$, that is, that there exist pointed maps $i : X_b \to Q$ and $r : Q \to X_b$ such that $r \circ i$ is homotopic to the identity.

The Serre exact sequence and the Blakers–Massey theorem

As a first application of the obstruction theory we give the fibrewise versions of the results of Serre and Blakers–Massey relating homotopy-fibres and homotopy-cofibres in a range of dimensions. Consider a fibrewise pointed map $g : Y' \to Y$ between pointed homotopy fibre bundles over B, and let $h : Y \to C_B(g)$ be the projection to the fibrewise homotopy-cofibre. Since the composition $h \circ g$ is canonically null-homotopic, we have a commutative diagram of fibre sequences:

$$
\begin{array}{ccccc}
F_B(g) & \longrightarrow & Y' & \xrightarrow{\;g\;} & Y \\
{\scriptstyle f}\downarrow & & {\scriptstyle p}\downarrow & & {\scriptstyle h}\downarrow \\
\Omega_B(C_B(g)) & \longrightarrow & B & \longrightarrow & C_B(g)
\end{array}
\qquad (2.17)
$$

Proposition 2.18 *Let B be a finite complex of dimension $\leq m$, let A be a subcomplex, and let $X \to B$ be a pointed homotopy fibre bundle with each fibre homotopy equivalent to a finite complex of dimension $\leq l$. Suppose that, in the situation described in diagram (2.17), the homotopy groups of the fibres of Y and $C_B(g)$ vanish in dimensions less than c and d, respectively. Then the fibrewise map f induces a map*

$$
f_* : \pi^{-i}_{(B,A)}[X; F_B(g)] \to \pi^{-i}_{(B,A)}[X; \Omega_B(C_B(g))] = \pi^{-i-1}_{(B,A)}[X; C_B(g)]
$$

which is surjective for $i \leq c+d-2-(m+l)$, bijective for $i < c+d-2-(m+l)$.

The assertion follows, by Proposition 2.15, from the classical Blakers–Massey theorem, deriving from the Serre spectral sequence. It allows us to substitute terms in the fibre exact sequence (Proposition 2.7) to obtain, in a range of dimensions: $i \leq N := c + d - 3 - (m + l)$, an exact sequence which we shall call the *Serre exact sequence*:

$$\pi_{(B,A)}^{-N}[X; Y'] \to \pi_{(B,A)}^{-N}[X; Y] \to \pi_{(B,A)}^{-N}[X; C_B(g)] \to \cdots \tag{2.19}$$

for the fibrewise cofibre sequence $Y' \to Y \to C_B(g)$. (Of course, the original Serre exact sequence relates the homology of the fibre, total space and base of a fibration.)

We have already met these methods in the proof of Proposition 19.23 in Part I.

Change of base

The main results of this section have been stated for pointed homotopy fibre bundles X and Y. This condition on X, whilst a natural one for the exposition, can often be relaxed by rephrasing the problem. Suppose that X, as space, is a compact ENR. The closed basepoint subspace B is then a sub-ENR of X. Now associated to a fibrewise pointed map $f : X \to Y$ over B there is a fibrewise pointed map $\tilde{f} : X \times S^0 \to X \times_B Y$ over X, namely, $\tilde{f}(x, -1) = (x, f(x))$, $\tilde{f}(x, 1) = (x, b)$, for $x \in X_b$. Moreover, $\tilde{f}$ is zero over the subspace $B \subseteq X$. This construction gives a bijection:

$$\pi_B^{-i}[X; Y] \to \pi_{(X,B)}^{-i}[X \times S^0; X \times_B Y]. \tag{2.20}$$

The pull-back $X \times_B Y$ of a pointed homotopy fibre bundle Y over B is a pointed homotopy fibre bundle over X.

There is a corresponding formal description of the relative groups:

$$\pi_{(B,A)}^{-i}[X; Y] \to \pi_{(X,B \cup X_A)}^{-i}[X \times S^0; X \times_B Y]. \tag{2.21}$$

For this to be useful (and to fit our conventions), $B \cup X_A$ should be a closed sub-ENR of X. This will often be the case in practice, but $B \cup X_A$ is not automatically an ENR. If the restriction X_A is an ENR, as, for example, when $X \to B$ is a fibrewise ENR (Corollary 5.9) then it does follow that $B \cup X_A$ is an ENR, but in that case X will be a homotopy fibre bundle, by Proposition 5.19.

Homotopy-commutativity of fibrewise Hopf spaces

As an application of some of the ideas introduced in this section we prove a theorem on fibrewise Hopf spaces over a compact ENR B.

We begin with a preliminary discussion of torus bundles. Let $L \to B$ be a bundle of free Abelian groups of (finite) rank l. Associated to L is the torus bundle $T = (L \otimes \mathbb{R})/L$ (the group-theoretic quotient of bundles of groups), which is the fibrewise classifying space $\mathbf{B}_B L$. Conversely, the torus bundle T determines L.

The fibrewise homotopy classes of maps between torus bundles can be described completely.

Proposition 2.22 *Let T and T' be torus bundles over B associated with bundles L and L' of free Abelian groups. Then*

$$\pi_B^0[T;\,T'] = \Gamma(\mathrm{Hom}(T,T')) = \Gamma(\mathrm{Hom}(L,L')).$$

In other words, every fibrewise map is homotopic to a unique fibrewise homomorphism.

Here, and throughout Part II, Γ is used for the space of sections of a bundle. The homomorphism bundles $\mathrm{Hom}(T,T')$ and $\mathrm{Hom}(L,L')$ have fibres at $b \in B$ the spaces of continuous homomorphisms $T_b \to T_b'$ and $L_b \to L_b'$ respectively. The proposition is easily proved by a Mayer–Vietoris argument. Since any compact ENR is a retract of a finite polyhedron, we may assume that B is a finite complex and argue inductively cell by cell.

Now let $X \to B$ be a pointed homotopy fibre bundle over B with each fibre of the pointed homotopy type of a connected finite complex. In each fibre of X we can form the fundamental group $\pi_1(X_b)$ and assemble these to form a locally trivial bundle of (finitely generated) discrete groups $\Pi_B(X) \to B$.

Lemma 2.23 *There is a unique fibrewise pointed map $X \to \mathbf{B}_B\Pi_B(X)$ (the fibrewise classifying space) which induces the identity map on the fundamental group of fibres.*

One first checks uniqueness if $X = B \times F \to B$ is trivial. Global existence and uniqueness are again established by a Mayer–Vietoris argument. The crucial point, which ensures uniqueness, is that $\pi_B^{-1}[X;\,T]$ is trivial for any torus bundle T.

Let us write $L_B(X)$ for the bundle of finitely generated free Abelian groups obtained by Abelianizing the fibres of $\Pi_B(X)$ and factoring out the torsion subgroup. Thus the fibre at $b \in B$ is $\mathrm{Hom}(\tilde{H}^1(X_b;\,\mathbb{Z}),\mathbb{Z})$. There is a map of fibrewise classifying spaces

$$\mathbf{B}_B\Pi_B(X) \to \mathbf{B}_B L_B(X) = (L_B(X) \otimes_{\mathbb{Z}} \mathbb{R})/L_B(X).$$

We shall call this torus bundle the *Albanese bundle* $\mathrm{Alb}_B(X)$ and write $\rho : X \to \mathrm{Alb}_B(X)$ for the induced map. It enjoys a universal property for maps from X to a torus bundle.

Proposition 2.24 *For any torus bundle $T \to B$ there is a natural equivalence:*

$$\pi_B^0[X;\,T] \to \pi_B^0[\mathrm{Alb}_B(X);\,T] = \Gamma(\mathrm{Hom}(\mathrm{Alb}_B(X),T)).$$

Remark 2.25. We can also interpret $\pi_B^0[X;T]$ as cohomology with twisted coefficients $H^1(X,B;p^*L)$.

Remark 2.26. Suppose that X, in language to be explained in Section 11, is a fibrewise smooth fibre bundle with fibre a (connected) closed manifold. Then we can give an explicit representation of ρ in differential geometric terms. Choose a fibrewise Riemannian metric on X. We define ρ in the fibre at $b \in B$. Let $x \in X_b$ and choose a smooth path γ from the basepoint of X_b to x. Any element of $H^1(X_b; \mathbb{R})$ has a unique harmonic representative ω. Mapping ω to $\int_\gamma \omega$ we obtain a linear map $H^1(X_b; \mathbb{R}) \to \mathbb{R}$. A different choice of path γ changes this linear map by a map which is integral (at least, up to a normalizing factor 2π) on the lattice $H^1(X_b; \mathbb{Z})$. We thus obtain a well-defined map to the torus $\mathrm{Hom}(H^1(X_b; \mathbb{R}), \mathbb{R})/\mathrm{Hom}(H^1(X_b; \mathbb{Z}), \mathbb{Z})$, which is the fibre of $\mathrm{Alb}_B(X)$ at b. The maps so defined on fibres fit together to give the fibrewise map ρ.

Assume now that X has a fibrewise Hopf structure, as in Proposition 1.31. Then ρ is a fibrewise H-map. This is now an easy calculation. For, by Proposition 2.22, $\pi_B^0[X \times_B X; \mathrm{Alb}_B(X)]$ is $\Gamma(\mathrm{Hom}(L_B(X) \oplus L_B(X), L_B(X)))$.

We can now describe completely the homotopy-commutative fibrewise Hopf spaces with fibre a finite complex, simply by applying Hubbuck's theorem [76] in each fibre and using Dold's theorem to recognize a fibrewise homotopy equivalence. Such fibrewise Hopf spaces are bundles of tori (up to equivalence).

Proposition 2.27 (Hubbuck's theorem). *Let X be a fibrewise Hopf space with each fibre homotopy commutative and of the homotopy type of a connected finite complex. Then the natural map $X \to \mathrm{Alb}_B(X)$ is an equivalence of fibrewise Hopf spaces.*

3 Stable homotopy theory

In this section we shall follow the naïve approach to stable homotopy theory, without introducing the machinery of spectra. Our account does not follow any particular source, but like the whole of Part II owes much to the seminal papers of Becker and Gottlieb [10] and Dold [47].

The base space B is supposed to be compact, unless explicit indication to the contrary is given. We work with fibrewise pointed spaces over B; these will normally be pointed homotopy fibre bundles.

Finite-dimensional real vector bundles over B will be denoted by Greek letters: ξ, η, ζ. Recall that ξ_B^+ is the fibrewise one-point compactification of ξ, the fibrewise pointed sphere-bundle obtained by adding a basepoint at infinity in each fibre. (Notation such as S_B^ξ, generalizing the classical S^n for the sphere, is widely used. We prefer to avoid the proliferation of superscripts which this would entail.) The fibrewise smash product $\xi_B^+ \wedge_B \eta_B^+$ is canonically identified with $(\xi \oplus \eta)_B^+$. (See Part I, Proposition 10.3.) A vector bundle isomorphism $a : \xi \to \xi'$ determines, on compactification, a fibrewise pointed topological equivalence $a_* : \xi_B^+ \to \xi_B'^+$.

Stable maps

Let us fix fibrewise pointed spaces X and Y over B. Although we shall give a formal definition of a stable fibrewise map from X to Y without imposing any restrictions on the spaces considered, the notion is unlikely to be useful unless X satisfies some finiteness condition. (The fibrewise space X will usually be a pointed homotopy fibre bundle with fibres of the pointed homotopy type of compact ENRs.)

A *fibrewise stable map* $X \to Y$ over B will be determined by a pair (f, ξ), where ξ is a finite-dimensional real vector bundle over B and f is a fibrewise pointed map

$$f : (\xi_B^+) \wedge_B X \to (\xi_B^+) \wedge_B Y.$$

We introduce the equivalence relation on such pairs generated by:

(i)   (Homotopy) $(f, \xi) \sim (f', \xi)$ if f and f' are homotopic as fibrewise pointed maps.

(ii) (Stability) If ζ is a vector bundle over B then $(f, \xi) \sim (1 \wedge f, \zeta \oplus \xi)$.

(iii) (Vector bundle isomorphism) If $a : \xi \to \xi'$ is a vector bundle isomorphism, then $(f, \xi) \sim (f', \xi')$ where f' is the composition

$$\xi_B'^+ \wedge_B X \xrightarrow{a_*^{-1} \wedge 1} \xi_B^+ \wedge_B X \xrightarrow{f} \xi_B^+ \wedge_B Y \xrightarrow{a_* \wedge 1} \xi_B'^+ \wedge_B Y.$$

A fibrewise stable map $X \to Y$ over B is defined to be an equivalence class of such representatives (f, ξ).

There are, of course, set-theoretic niceties which require some attention – but not too much. We should, perhaps, stipulate that each vector bundle ξ considered is a sub-bundle of one of the trivial bundles:

$$B \times \mathbb{R}^0 \subseteq \cdots \subseteq B \times \mathbb{R}^n \subseteq B \times \mathbb{R}^{n+1} \subseteq \cdots.$$

The definition which we have given fits well into the conceptual framework of fibrewise homotopy theory (and, as we explain briefly at the end of this section, has the advantage of carrying through virtually unchanged to the equivariant theory). In fact, there is no loss of generality in taking all the vector bundles in the definition to be trivial, indeed of the form $B \times \mathbb{R}^n$.

Lemma 3.1 *Every pair* (f, ξ) *is equivalent to a pair* $(g, B \times \mathbb{R}^n)$ *for some integer* n. *Moreover, two pairs* $(g, B \times \mathbb{R}^n)$ *and* $(g', B \times \mathbb{R}^{n'})$ *are equivalent if and only if*

$$1 \wedge g \simeq 1 \wedge g' : (B \times \mathbb{R}^N)_B^+ \wedge_B X \to (B \times \mathbb{R}^N)_B^+ \wedge_B Y,$$

for some $N \geq n, n'$.

The statement contains the implicit identification of $\mathbb{R}^{N-n} \oplus \mathbb{R}^n$ with $\mathbb{R}^N$. The first assertion is clear, because any finite-dimensional bundle over a compact base is isomorphic to a direct summand of a trivial bundle. The second depends upon the following elementary fact from K-theory.

Lemma 3.2 *Let* $a : \xi \to \xi$ *be a vector bundle isomorphism. Then*

$$a \oplus 1 \text{ and } 1 \oplus a : \xi \oplus \xi \to \xi \oplus \xi$$

are homotopic through vector bundle isomorphisms.

To see this, consider the homotopy

$$\theta_t = \begin{bmatrix} \cos(\pi t/2) & \sin(\pi t/2) \\ \sin(\pi t/2) & -\cos(\pi t/2) \end{bmatrix}, \quad 0 \leq t \leq 1$$

from

$$\theta_0 = \begin{bmatrix} 1 & 0 \\ 0 & -1 \end{bmatrix} \quad \text{to} \quad \theta_1 = \begin{bmatrix} 0 & 1 \\ 1 & 0 \end{bmatrix}$$

in $O(2)$. Tensor this with the identity on ξ to get a bundle isomorphism. Then $\theta_t(a \times 1)\theta_t^{-1}$ is a homotopy from $a \oplus 1$ to $1 \oplus a$.

Definition 3.3 We write $\omega_B^0\{X; Y\}$ for the set of *fibrewise stable maps* from X to Y over B.

According to the discussion above,

$$\omega_B^0\{X; Y\} = \varinjlim_{n \geq 0} \pi_B^0[\Sigma_B^n X; \Sigma_B^n Y]$$

is the direct limit of the iterated fibrewise suspension maps, just as in the classical theory. The set of stable maps thus has a natural Abelian group structure derived from the group structure on the π_B^{-i}, $i \geq 1$.

By construction, we have, for any vector bundle ζ over B, a *suspension isomorphism*

$$1 \wedge - : \omega_B^0\{X; Y\} \to \omega_B^0\{\zeta_B^+ \wedge_B X; \zeta_B^+ \wedge_B Y\},$$

given by the fibrewise smash product on the left with identity on the sphere-bundle ζ_B^+. We use this isomorphism to identify the two groups without comment. In particular, we can make the identification

$$\omega_B^0\{X; Y\} = \omega_B^0\{(B \times \mathbb{R}^N)^+ \wedge_B X; (B \times \mathbb{R}^N)^+ \wedge_B Y\}$$

for any integer $N \geq 0$.

We define $\omega_B^i\{X; Y\}$ for any $i \in \mathbb{Z}$ to be $\omega_B^0\{(\mathbb{R}^N)^+ \wedge_B X; (\mathbb{R}^{N+i})^+ \wedge_B Y\}$ for any $N \geq 0$ with $N + i \geq 0$. (Two such groups for different N are, as we have just observed, canonically identified by the suspension isomorphism.)

The extension of the definitions to the relative and locally compact theories presents no problem. Suppose that A is a closed sub-ENR of the compact base B. Then a stable map over (B, A) will be represented by a pair (f, ξ), where f is zero over A. Making the evident modifications, we arrive at the definition of the relative groups

$$\omega_{(B,A)}^i\{X; Y\}.$$

When defining stable maps with compact support over a locally compact base B, we need to restrict attention to pairs (f, ξ) where ξ is of *finite type*, that is, such that there exists a finite covering of B by open sets over which ξ is trivial. This ensures that ξ is a direct summand of a trivial bundle. (And such a bundle is necessarily of finite type, because it is the pull-back of the canonical bundle over a Grassmann manifold.) We denote the groups of stable maps with compact support by

$$_c\omega_B^i\{X; Y\}.$$

When the base B is a point we normally omit it from the notation. Thus, if E and F are pointed spaces, with E of the homotopy type of a compact ENR, $\omega^0\{E; F\}$ means the set of stable maps from E to F, traditionally written as $\{E; F\}$. We write

$$\tilde{\omega}^i(E) := \omega^i\{E; S^0\}$$
$$(= \omega^0\{E; S^i\} \quad \text{when } i \geq 0)$$

for the *reduced stable cohomotopy* of the pointed space E and

$$\tilde{\omega}_j(F) := \omega^{-j}\{S^0; F\}$$
$$(= \omega^0\{S^j; F\} \quad \text{when } j \geq 0)$$

for the *reduced stable homotopy* of F. The *unreduced* stable cohomotopy and homotopy groups of spaces P and Q (without basepoint) are defined by adjoining a basepoint:

$$\omega^i(P) := \tilde{\omega}^i(P_+), \qquad \omega_j(Q) := \tilde{\omega}_j(Q_+).$$

Stable maps to a trivial bundle

Suppose that Y is the trivial bundle $B \times F$, where F is a pointed space. A fibrewise map $X \to B \times F$ over B is prescribed by its second component, which is a map $X \to F$ taking the basepoint section B to the basepoint of F. Hence, we have a correspondence between fibrewise pointed maps $X \to B \times F$ and pointed maps $X/B \to F$. This evidently extends to stable maps.

Proposition 3.4 *There is a natural equivalence:*

$$\omega_B^*\{X; B \times F\} \to \omega^*\{X/B; F\}.$$

Note that, because X is well-pointed, the inclusion of B in X is a homotopy-cofibration and the pointed space X/B is well-pointed.

As a special case we have the identification:

$$\omega_B^i\{B \times S^0; B \times S^0\} = \omega^i(B).$$

Products

The stable homotopy groups have a composition/product structure exactly as in the classical theory. To describe *composition* of stable maps we consider another fibrewise pointed space $Z \to B$. Given a stable map $Y \to Z$ represented by (g, η) and a stable map $X \to Y$ represented by (f, ξ), we can form the composition $X \to Z$ represented by

$$(1 \wedge g) \circ (1 \wedge f) : (\xi \oplus \eta)_B^+ \wedge_B X \to (\xi \oplus \eta)_B^+ \wedge Y \to (\xi \oplus \eta)_B^+ \wedge_B Z.$$

(We use the canonical identification of $\xi \oplus \eta$ and $\eta \oplus \xi$.) The construction is compatible with the equivalence relation on representatives of stable maps and gives:

$$\circ : \omega_B^0\{Y; Z\} \otimes \omega_B^0\{X; Y\} \to \omega_B^0\{X; Z\}. \tag{3.5}$$

We obtain a category of fibrewise pointed spaces over B with morphisms from X to Y the fibrewise stable maps over B.

The composition extends to the graded groups:

$$\circ : \omega_B^i\{Y; Z\} \otimes \omega_B^j\{X; Y\} \to \omega_B^{i+j}\{X; Z\}. \tag{3.6}$$

One has to be a little careful about the order. This can be fixed by describing the composition of honest fibrewise maps $g : (B \times \mathbb{R}^i)_B^+ \wedge_B Y \to Z$ and $f : (B \times \mathbb{R}^j)_B^+ \wedge_B X \to Y$, where $i, j \geq 0$, as $g \circ (1 \wedge f)$:

$$(B \times \mathbb{R}^{i+j})_B^+ \wedge_B X = (B \times \mathbb{R}^i)_B^+ \wedge_B (B \times \mathbb{R}^j)_B^+ \wedge_B X \xrightarrow{1 \wedge f} (B \times \mathbb{R}^i)_B^+ \wedge_B Y \xrightarrow{g} Z.$$

There are two variants of the composition involving the relative groups, on the left and on the right:

$$\circ \, : \omega_B^i\{Y;\, Z\} \otimes \omega_{(B,A)}^j\{X;\, Y\} \to \omega_{(B,A)}^{i+j}\{X;\, Z\},$$

$$\circ \, : \omega_{(B,A)}^i\{Y;\, Z\} \otimes \omega_B^j\{X;\, Y\} \to \omega_{(B,A)}^{i+j}\{X;\, Z\},$$

(which can, of course, be combined with the restriction map $\omega_{(B,A)}^* \to \omega_B^*$).

Now we turn our attention to the closely related (smash) products. Let $X' \to B$ and $Y' \to B$ be fibrewise pointed spaces over B. The fibrewise smash product $X \wedge_B X'$ is homotopy well-pointed (Part I, Section 16). If X and X' are pointed homotopy fibre bundles, then $X \wedge_B X'$ is also a pointed homotopy fibre bundle.

The smash product in stable homotopy theory is in some respects analogous to the tensor product in the category of finite-dimensional vector spaces over a field. We shall pursue this analogy in the discussion of duality in Section 10. For the present we just note the canonical equivalence between $X \wedge_B X'$ and $X' \wedge_B X$: $[x, x'] \mapsto [x', x]$. It is an identification that we shall often make without further comment.

The *smash product*

$$\wedge \, : \omega_B^i\{X;\, Y\} \otimes \omega_B^{i'}\{X';\, Y'\} \to \omega_B^{i+i'}\{X \wedge_B X';\, Y \wedge_B Y'\} \qquad (3.7)$$

is constructed in the obvious way (preserving the order of the indices). In particular, if (f, ξ) and (f', ξ') are representatives of fibrewise stable maps $X \to Y$ and $X' \to Y'$, then $(f \wedge f', \xi \oplus \xi')$:

$$(\xi \oplus \xi')_B^+ \wedge (X \wedge_B X') = (\xi_B^+ \wedge_B X) \wedge_B (\xi_B'^+ \wedge_B X') \xrightarrow{f \wedge f'}$$

$$(\xi_B^+ \wedge_B Y) \wedge_B (\xi_B'^+ \wedge_B Y') = (\xi \oplus \xi')_B^+ \wedge (Y \wedge_B Y')$$

is a representative of the product.

Again there is a variant of the smash product with one factor a relative group over (B, A).

The smash product and composition are related as follows.

Lemma 3.8 *Let* $x \in \omega_B^i\{X;\, Y\}$, $x' \in \omega_B^{i'}\{X';\, Y'\}$. *Then the smash product* $x \wedge x' \in \omega_B^{i+i'}\{X \wedge_B X';\, Y \wedge_B Y'\}$ *is equal to the composition*

$$(x \wedge 1_{Y'}) \circ (1_X \wedge x') = (-1)^{ii'}(1_Y \wedge x') \circ (x \wedge 1_{X'}) \in \omega_B^{i+i'}\{X \wedge_B X';\, Y \wedge_B Y'\}.$$

The graded ring $\omega^*(B)$ is (super-) commutative. As a special case of the smash product we obtain, using the interpretation of $\omega_B^i\{B \times S^0;\, B \times S^0\}$ as the stable cohomotopy group $\omega^i(B)$ of the base, a graded (left) $\omega^*(B)$-module structure on $\omega_{(B,A)}^*\{X;\, Y\}$:

$$\omega^i(B) \otimes \omega_{(B,A)}^j\{X;\, Y\} \to \omega_{(B,A)}^{i+j}\{X;\, Y\}.$$

The corresponding right module structure is given by an appropriate change of sign:

$$x \cdot r = (-1)^{ij} r \cdot x \quad \text{for} \quad r \in \omega^i(B), \ x \in \omega^j_{(B,A)}\{X; Y\}.$$

The graded smash and composition products are $\omega^*(B)$-bilinear (in the super-commutative sense). In particular, the tensor products in (3.5) and (3.6) can be read as tensor products over $\omega^0(B)$, rather than over $\mathbb{Z}$.

We conclude this description of the formal structure of fibrewise stable homotopy theory by singling out the defining suspension isomorphism.

Proposition 3.9 (Suspension isomorphism). *For any finite-dimensional real vector bundle ζ over B, the smash product on the left with the identity $1 \in \omega^0_B\{\zeta^+_B; \zeta^+_B\}$ gives a suspension isomorphism:*

$$1\wedge \ : \omega^i_{(B,A)}\{X; Y\} \to \omega^i_{(B,A)}\{\zeta^+_B \wedge_B X; \ \zeta^+_B \wedge_B Y\}.$$

The cofibre exact sequences

We shall say that a sequence

$$X' \xrightarrow{f} X \xrightarrow{g} X''$$

of fibrewise pointed spaces over B is a *cofibre sequence* if we are given a fibrewise pointed homotopy equivalence $C_B(f) \to X''$ from the homotopy-cofibre of f such that the composition $X \to C_B(f) \to X''$ with the structure map of the homotopy-cofibre is homotopic to g. The standard map $C_B(f) \to \Sigma_B X'$ determines a stable map $\delta \in \omega^1_B\{X''; X'\}$.

The cofibre exact sequences on the left and right in classical stable homotopy theory generalize routinely to the fibrewise theory. The first sequence is the direct limit of exact sequences (Proposition 2.4) in unstable homotopy theory.

Proposition 3.10 *Let $X' \to X \to X''$ be a cofibre sequence over B. Then for any fibrewise pointed space Y over B we have a long exact cofibre sequence on the left:*

$$\cdots \to \omega^i_{(B,A)}\{X''; Y\} \to \omega^i_{(B,A)}\{X; Y\} \to \omega^i_{(B,A)}\{X'; Y\}$$
$$\xrightarrow{\delta} \omega^{i+1}_{(B,A)}\{X''; Y\} \to \cdots$$

of $\omega^*(B)$-homomorphisms.

The derivation of the second sequence is not so immediate.

Proposition 3.11 *Let $Y' \to Y \to Y''$ be a cofibre sequence over B. Then for any fibrewise pointed space X over B we have a long exact cofibre sequence on the right:*

$$\cdots \rightarrow \omega^i_{(B,A)}\{X; Y'\} \rightarrow \omega^i_{(B,A)}\{X; Y\} \rightarrow \omega^i_{(B,A)}\{X; Y''\}$$
$$\xrightarrow{\delta} \omega^{i+1}_{(B,A)}\{X; Y'\} \rightarrow \cdots$$

of $\omega^*(B)$-module homomorphisms.

We shall indicate only the key ingredient in the proof of exactness. (The ideas have already appeared in our discussion of the Serre exact sequence (2.19).) Let the maps in the cofibre sequence be denoted by $f : Y' \rightarrow Y$ and $g : Y \rightarrow Y''$. Since the composition is null-homotopic, the composition

$$\pi^0_{(B,A)}[X; Y'] \xrightarrow{f_*} \pi^0_{(B,A)}[X; Y] \xrightarrow{g_*} \pi^0_{(B,A)}[X; Y'']$$

is zero. Working in the opposite direction, consider a map $h : X \rightarrow Y$ such that $g \circ h$ is null-homotopic (over (B,A)). We show that the fibrewise suspension of h: $\Sigma_B X \rightarrow \Sigma_B Y$ lifts to a map $\Sigma_B X \rightarrow \Sigma_B Y'$. The map h lifts to the homotopy-fibre: $X \rightarrow F_B(g)$, by the construction (Definition 2.5) of the homotopy-fibre. Now we have a (homotopy) commutative diagram:

$$
\begin{array}{ccccc}
F_B(g) & \longrightarrow & Y & \xrightarrow{g} & Y'' \\
\downarrow & & \downarrow & & \downarrow{\delta} \\
\Omega_B \Sigma_B Y' & \longrightarrow & B & \longrightarrow & \Sigma_B Y'
\end{array}
$$

given by the naturality of the construction. For $\delta \circ g$ is null-homotopic and $\Omega_B \Sigma_B Y'$ is the homotopy-fibre of the inclusion of the basepoint: $B \rightarrow \Sigma_B Y'$. Composing the lift of h with the map $F_B(g) \rightarrow \Omega_B \Sigma_B Y'$ so constructed, we obtain a map $X \rightarrow \Omega_B \Sigma_B Y'$. Its adjoint $\Sigma_B X \rightarrow \Sigma_B Y'$ lifts the suspension of h.

By passage to the direct limit, this argument establishes exactness at Y. Since $Y \rightarrow Y'' \rightarrow \Sigma_B Y'$ and $Y'' \rightarrow \Sigma_B Y' \rightarrow \Sigma_B Y$ are also cofibre sequences (Part I, Section 12), exactness at Y'' and at Y' follows.

The relative exact sequence

The relative exact sequence and Mayer–Vietoris sequence for pointed homotopy fibre bundles arise as direct limits of the corresponding unstable sequences (Propositions 2.11 and 2.14).

Proposition 3.12 Let X and Y be pointed homotopy fibre bundles over a compact ENR B, and let A be a closed sub-ENR of B. Then there is a long exact sequence of graded $\omega^*(B)$-modules

$$\cdots \rightarrow \omega^i_{(B,A)}\{X; Y\} \rightarrow \omega^i_B\{X; Y\} \rightarrow \omega^i_A\{X_A; Y_A\}$$
$$\xrightarrow{\delta} \omega^{i+1}_{(B,A)}\{X; Y\} \rightarrow \cdots$$

As one would expect, there is generalization to a triple $B \supseteq A \supseteq A'$, where A' is a closed sub-ENR of A. The groups over B and A are replaced by relative groups over (B, A') and (A, A'). (See Proposition 15.6.)

The Mayer–Vietoris sequence

Proposition 3.13 *Let B_1 and B_2 be closed sub-ENRs of the compact ENR B, such that $B_1 \cup B_2 = B$ and $A := B_1 \cap B_2$ is a sub-ENR. Then, for any pointed homotopy fibre bundles X and Y over B, there is a long exact Mayer–Vietoris sequence*

$$\cdots \to \omega_B^i\{X; Y\} \to \omega_{B_1}^i\{X_{B_1}; Y_{B_1}\} \oplus \omega_{B_2}^i\{X_{B_2}; Y_{B_2}\} \to \omega_A^i\{X_A; Y_A\}$$
$$\xrightarrow{\delta} \omega_B^{i+1}\{X; Y\} \to \cdots$$

of graded $\omega^*(B)$-modules. The first homomorphism is the direct sum of the two restriction maps, and the second is the difference of the restriction maps on the two factors.

Again there is a generalization involving relative groups over (B, A'), (B_1, A'), (B_2, A') and (A, A'), where A' is a closed sub-ENR of A.

Nilpotence

The technology developed for the relative and Mayer–Vietoris exact sequences can be used, as in the classical theory, to establish various nilpotence results. We suppose in this subsection that $X_0, \ldots, X_n$ are pointed homotopy fibre bundles over B.

Lemma 3.14 *Let $B_1, \ldots, B_n$ be closed sub-ENRs of the compact ENR B with union equal to B. Let $x_k \in \omega_B^{i_k}\{X_k; X_{k-1}\}$, for $1 \leq k \leq n$, be a class which restricts to zero on B_k. Then the (composition) product*

$$x_1 \cdot x_2 \cdot \ldots \cdot x_n \in \omega_B^{i_1 + \cdots + i_n}\{X_n; X_0\}$$

vanishes.

By Proposition 3.12, x_k lifts to (B, B_k). The product of the classes lifts to the pair $(B, B_1 \cup \ldots \cup B_n)$ and is, thus, clearly zero.

From this lemma and Proposition 2.8 we obtain:

Proposition 3.15 *Suppose that B is the union of n closed sub-ENRs B_k, such that each inclusion $B_k \to B$ is homotopic to a locally constant map (that is, a map which is constant on each component). Let $x_k \in \omega_B^{i_k}\{X_k; X_{k-1}\}$, for $1 \leq k \leq n$, be a class which restricts to zero in each fibre. Then*

$$x_1 \cdot x_2 \cdot \ldots \cdot x_n = 0.$$

Note that if B is connected, then a class $x \in \omega_B^i\{X; Y\}$ which restricts to zero in the fibre $\omega^i\{X_b; Y_b\}$ at one point $b \in B$ restricts to zero at every point of the base.

It is possible to write any finite complex B as a union of closed sub-ENRs as in Proposition 3.15 for some natural number n. Recall that B has *category* $\leq n$ if it can be written as the union of n open subsets U_k such that each inclusion $U_k \to B$ is homotopic to a constant map. In that case, one can find closed sub-ENRs $B_k \subseteq U_k$ with union equal to B, by Proposition 5.3.

A relative version of this nilpotence result can be established in a similar fashion.

Proposition 3.16 *Let A be a closed sub-ENR of a compact ENR B. Suppose that B is a union of n closed sub-ENRs $B_1, \ldots, B_n$, such that each inclusion $B_k \to B$ is homotopic to a map into A. Further, suppose that, for some k, $A \subseteq B_k$ and the inclusion of pairs $(B_k, A) \to (B, A)$ is homotopic to a map into (A, A). Then the product $x_1 \cdot \ldots \cdot x_n$ of classes $x_k \in \omega_{(B,A)}^{i_k}\{X_k; X_{k-1}\}$ is zero.*

The suspension theorem

For the remainder of this section we assume that both X and Y are pointed homotopy fibre bundles over B and that the fibres of X have the pointed homotopy type of finite complexes. (To develop a satisfactory stable theory without some such finiteness condition on the codomain of a stable map, one cannot avoid a systematic treatment of fibrewise spectra.)

Freudenthal's suspension theorem tells us that there is a stable range: that is, that stabilization

$$\pi_{(B,A)}^0[\xi_B^+ \wedge_B X; \xi_B^+ \wedge_B Y] \to \omega_{(B,A)}^0\{X; Y\} \qquad (3.17)$$

is an isomorphism if the dimension of the vector bundle ξ is sufficiently large.

Proposition 3.18 *Let B be a finite complex of dimension $\leq m$ and let A be a subcomplex of B. Suppose that each fibre of X is pointed homotopy equivalent to a finite complex of dimension $\leq l$ and that the connectivity of each fibre of Y is $\geq c$. Then, for any real vector bundle ξ over B, the smash product with the identity on ξ_B^+:*

$$\pi_{(B,A)}^0[X; Y] \to \pi_{(B,A)}^0[\xi_B^+ \wedge_B X; \xi_B^+ \wedge_B Y]$$

is surjective if $m \leq 2c - l + 1$ and bijective if $m < 2c - l + 1$.

This follows by applying the obstruction theory, Proposition 2.15, to the fibrewise suspension map:

$$f : Y \to \mathrm{map}_B^*(\xi_B^+, \xi_B^+ \wedge_B Y).$$

Here map_B^* denotes the pointed homotopy fibre bundle of pointed maps, with fibre at $b \in B$ the space $\mathrm{map}^*(\xi_b^+, \xi_b^+ \wedge Y_b)$ of pointed maps of fibres. (For more details see Section 15 of Part I, where the notation map_B^B is used.) By the classical suspension theorem, f_b is at least $(2c + 1)$-connected, that is, satisfies the hypothesis of Proposition 2.15 with $n = 2c + 1$: for a pointed finite complex P,

$$[P; Y_b] \to [(\xi_b)^+; (\xi_b)^+ \wedge Y_b]$$

is surjective if $\dim P \leq 2c + 1$, bijective if $\dim P < 2c + 1$.

By replacing X and Y by their fibrewise smash products with ξ_B^+ we deduce:

Corollary 3.19 (Stability). *Let X and Y be as in the statement of Proposition 3.18. Then the stabilization map*

$$\pi_{(B,A)}^0[\xi_B^+ \wedge_B X; \xi_B^+ \wedge_B Y] \to \omega_{(B,A)}^0\{X; Y\}$$

is surjective if $\dim \xi \geq l + m - (2c + 1)$, bijective if $\dim \xi > l + m - (2c + 1)$.

These results have been obtained under the assumption that the base B is a finite complex, but any compact ENR is a retract of a finite complex. Given a compact ENR pair (B, A), one can find a finite complex B' and subcomplex A' together with maps $\iota : (B, A) \to (B', A')$ and $\rho : (B', A') \to (B, A)$ such that $\rho \circ \iota = 1 : (B, A) \to (B, A)$. The reader is again referred to Section 5. From these considerations, we see, using Proposition 2.8, that the statement that the stabilization map in Corollary 3.19 is an isomorphism for ξ of sufficiently large dimension holds in full generality.

Serre's theorem

One of the foundations of stable homotopy theory is Serre's theorem that the homotopy groups of spheres are finitely generated. By the suspension theorem, the stable homotopy groups of spheres are also finitely generated. From the relative exact sequence and induction over cells of the base, when the base is a finite complex, we obtain:

Proposition 3.20 *Let X and Y be pointed homotopy fibre bundles over a compact ENR B. Suppose that the fibres of X and Y are pointed homotopy equivalent to finite complexes. Then the stable homotopy groups*

$$\omega_{(B,A)}^i\{X; Y\}$$

are finitely generated Abelian groups.

The equivariant theory

We close this section with a few remarks about the equivariant fibrewise theory, which has proved to have important geometric applications (notably in the theory of differential equations). Let G be a compact Lie group. We have already observed that the formal side of fibrewise stable homotopy theory readily extends to a G-equivariant theory. The base B should be a G-ENR. This means that B is an equivariant retract of an open G-subspace U of some finite-dimensional real G-module E. Thus the inclusion $i : B \to U$ and retraction $r : U \to B$ are required to be G-maps.

Suppose that B is a compact G-ENR. To specify a G-equivariant fibrewise stable map $X \to Y$ between fibrewise pointed G-spaces over B we need a pair (f, ξ), where ξ is a finite-dimensional real G-vector bundle over B and f is a fibrewise pointed G-map. Fibrewise G-equivariant stable homotopy groups

$$_G\omega^i_B\{X; Y\}$$

indexed by integers $i \in \mathbb{Z}$ are defined just as in the non-equivariant theory. These groups are stable in the equivariant sense that, for any finite-dimensional real G-vector bundle ζ over B, the smash product with the identity gives an isomorphism:

$$1\wedge : {}_G\omega^i_B\{X; Y\} \to {}_G\omega^i_B\{\zeta^+_B \wedge_B X; \zeta^+_B \wedge_B Y\}.$$

Most of the constructions to be described in later sections can be pushed through, with little difficulty, in the equivariant setting.

4 The Euler class

Throughout this section ξ will be a finite-dimensional real vector bundle over a compact ENR B. There is no loss of generality in assuming that ξ is equipped with a positive-definite inner product. (For two such inner products g_0 and $g_1 : \xi \otimes \xi \to B \times \mathbb{R}$ are homotopic: $g_t = (1 - t)g_0 + tg_1$. Moreover, one has vector bundle automorphisms $a_t : \xi \to \xi$ such that $g_t(u, v) = g_0(a_t u, a_t v)$, for $u, v \in \xi_b$.) We write $S(\xi)$ and $D(\xi)$ for the unit sphere and disc-bundles in ξ. The one-point compactification of the zero vector space 0 is identified with S^0 (with basepoint 1), and the fibrewise one-point compactification of the zero vector bundle $B \times 0$ is likewise identified with $B \times S^0$.

The Gysin sequence

There is a fibrewise pointed homeomorphism:

$$c : D(\xi)/_B S(\xi) \to \xi^+_B$$

from the fibrewise quotient to the fibrewise one-point compactification given by the map $tv \mapsto \phi(t)v$, for $v \in S(\xi)$, $0 \le t < 1$, where $\phi : [0, 1) \to [0, \infty)$ is any homeomorphism with $\phi(0) = 0$. The fibrewise pointed homotopy class of c is independent of the choice of ϕ and gives a natural fibrewise pointed homotopy equivalence between the two bundles.

One can verify from first principles that the inclusion $S(\xi) \to D(\xi)$ is a fibrewise cofibration. (Alternatively, use the fact that the bundles are fibrewise ENRs (Proposition 5.11).) Hence, making the identification of $D(\xi)_{+B}/_B S(\xi)_{+B}$ with $D(\xi)/_B S(\xi)$ and, up to homotopy, with ξ_B^+, we have a cofibre sequence:

$$S(\xi)_{+B} \to D(\xi)_{+B} \to D(\xi)/_B S(\xi) = \xi_B^+$$

over B. Now the zero-section $z : B \times 0 \to D(\xi)$ is a fibrewise homotopy equivalence, and the associated pointed map $z_+ : B \times S^0 \to D(\xi)_{+B}$ is a fibrewise pointed homotopy equivalence. So we can rewrite the sequence above, in the homotopy category, as:

$$S(\xi)_{+B} \to B \times S^0 \to \xi_B^+, \tag{4.1}$$

where the first map is induced by the projection $S(\xi) \to B$ and the second by the inclusion of the zero-section $B \times 0 \to \xi$.

The long exact sequences arising from the cofibre sequence (4.1) are often referred to as *Gysin sequences*:

$$\cdots \to \omega_B^*\{X; S(\xi)_{+B}\} \to \omega_B^*\{X; B \times S^0\} \xrightarrow{z_*} \omega_B^*\{X; \xi_B^+\} \to \cdots$$
$$\cdots \to \omega_B^*\{\xi_B^+; Y\} \xrightarrow{z^*} \omega_B^*\{B \times S^0; Y\} \to \omega_B^*\{S(\xi)_{+B}; Y\} \to \cdots \tag{4.2}$$

(for fibrewise pointed spaces X and Y over B).

The Euler class

The classical cohomology Euler class of a vector bundle, the archetypal characteristic class, is the Hurewicz image of a stable cohomotopy class. (To be precise, the classical Euler class is defined for an oriented vector bundle and the identification of the Hurewicz image involves the Thom isomorphism.) The definition which follows emphasizes the fibrewise nature of the Euler class.

Definition 4.3 The zero-section of ξ determines, as above, a fibrewise pointed map $B \times S^0 \to \xi_B^+$. Its stable class

$$\gamma(\xi) \in \omega_B^0\{B \times S^0; \xi_B^+\}$$

is called the *stable cohomotopy Euler class*, or simply *Euler class*, of the vector bundle ξ.

The maps z_* and z^* in the Gysin sequences (4.2) are thus given by multiplication by the Euler class $\gamma(\xi)$.

The basic properties of γ are elementary consequences of the definition.

Proposition 4.4 (Properties of the Euler class). *Let ξ be a vector bundle over B.*

(i)   *Let $\alpha : B' \to B$ be a map from a compact ENR B'. Then*

$$\gamma(\alpha^*\xi) = \alpha^*\gamma(\xi).$$

(ii)   *Let ξ' be a second vector bundle over B. Then*

$$\gamma(\xi \oplus \xi') = \gamma(\xi) \cdot \gamma(\xi').$$

(iii)   *Let $a : \xi_B^+ \to \xi_B^+$ be a fibrewise pointed map preserving the zero-section, that is, with $a(0) = 0$ in each fibre. Then*

$$[a] \cdot \gamma(\xi) = \gamma(\xi),$$

where $[a] \in \omega^0(B)$ is the stable class defined by a.

In (iii) we have used the suspension isomorphism (Proposition 3.9) between $\omega_B^0\{\xi_B^+; \xi_B^+\}$ and $\omega^0(B) = \omega_B^0\{B \times S^0; B \times S^0\}$.

The third property has some interesting consequences. The restriction $a_b : \xi_b^+ \to \xi_b^+$ of the fibrewise map a to the fibre at $b \in B$ is a self-map of a sphere and characterized by its degree $d \in \mathbf{Z}$. This *fibre degree* is constant on components of B; if it is constant on B we say that a has fibre degree d. (In general, the fibre degree will be an element of $H^0(B; \mathbf{Z})$.) Suppose that this is the case. Then, by Proposition 3.15 (applied to a finite complex of which B is a retract), $x := [a] - d \in \omega^0(B)$ is nilpotent. Suppose that $x^N = 0$. Then, since $(d-1)\gamma(x) = -x \cdot \gamma(\xi)$, we deduce that $(d-1)^N\gamma(\xi) = 0$.

Multiplication by $(-1) : \xi \to \xi$ determines a map $(-1)^+ : \xi_B^+ \to \xi_B^+$ with fibre degree (-1) if ξ has odd dimension ($+1$ if the dimension is even). So we obtain:

Proposition 4.5 *Suppose that $\dim \xi$ is odd (or, more generally, that ξ contains a sub-bundle of odd dimension). Then $\gamma(\xi)$ is a 2-primary torsion class.*

Here is another example.

Example 4.6. Let λ be a complex line bundle over B with multiplicative order a power of a prime p. Then the Euler class $\gamma(\lambda)$ is p-primary torsion.

Suppose that the p^Mth tensor power of λ is trivial. Write $d = p^M + 1$. Then we have an isomorphism

$$\lambda^{\otimes d} \to \lambda.$$

Composing this with the tensor power map:

$$\lambda \to \lambda^{\otimes d} : z \mapsto z^d,$$

we obtain a fibrewise map $\lambda \to \lambda$ which compactifies to give a map $a : \lambda_B^+ \to \lambda_B^+$ with fibre degree d. The result follows from the discussion above.

The relative Euler class

The Euler class is an obstruction to the existence of a nowhere-zero cross-section.

Proposition 4.7 *Suppose that ξ admits a nowhere-zero cross-section s. Then $\gamma(\xi) = 0$.*

Without loss of generality we may take s to be a section of the unit sphere-bundle $S(\xi)$. It is homotopic, by the linear homotopy ts, $0 \le t \le 1$, through sections of $D(\xi)$ to the zero-section. But s determines the zero map $B \times S^0 \to D(\xi)/_B S(\xi) = \xi_B^+$.

To investigate the obstruction to existence of a nowhere-zero section more carefully, we introduce the relative Euler class.

Definition 4.8 Let s be a nowhere-zero section of $\xi \,|\, A$ defined over the closed sub-ENR A of B. We first replace s by a homotopic section of $S(\xi)$ over A. Then s extends to a section $\tilde{s}$ of $D(\xi)$ over B. We now follow the construction of the Euler class from the zero-section. The map $\tilde{s} : B \to D(\xi)$ determines a pointed map

$$B_{+B} = B \times S^0 \to D(\xi)_{+B} \to D(\xi)/_B S(\xi) = \xi_B^+$$

which is zero over A. Its stable class will be called *the relative Euler class*

$$\gamma(\xi; s) \in \omega^0_{(B,A)}\{B \times S^0; \xi_B^+\}$$

of the nowhere-zero section s over A. This is well-defined, since any two extensions $\tilde{s}_0$ and $\tilde{s}_1$ of s are homotopic (as sections extending s) by a linear homotopy $\tilde{s}_t = (1-t)\tilde{s}_0 + t\tilde{s}_1$.

From the construction it is evident that $\gamma(\xi; s) \mapsto \gamma(\xi)$ under the restriction homomorphism

$$\omega^0_{(B,A)}\{B \times S^0; \xi_B^+\} \to \omega^0_B\{B \times S^0; \xi_B^+\}.$$

The relative Euler class $\gamma(\xi; s)$ depends only on the homotopy class of the nowhere-zero section s over A and vanishes, clearly, if s extends to a nowhere-zero section over B. In a stable range, $\gamma(\xi; s)$ is the precise obstruction to extension.

Proposition 4.9 *Let A be a subcomplex of a finite complex B, and let ξ be a real vector bundle of dimension n over B. Suppose that $\dim B < 2(n-1)$. Then a nowhere-zero section s of ξ defined over A extends to B if and only if the relative Euler class*

$$\gamma(\xi; s) \in \omega^0_{(B,A)}\{B \times S^0; \xi_B^+\}$$

vanishes.

The difference class

To explain the proof of Proposition 4.9, we introduce the difference class of two nowhere-zero sections.

Definition 4.10 Let s_0 and s_1 be nowhere-zero sections of ξ over B coinciding on the sub-ENR A. Consider the pull-back of ξ to $B \times [0,1]$ by the projection π to B. A section s of $\pi^*\xi$ is defined over $B \times \{0,1\} \cup A \times [0,1]$ by s_0 on $B \times \{0\}$, s_1 on $B \times \{1\}$, and their common value on $A \times \{t\}$. The relative Euler class $\gamma(\pi^*\xi; s)$ lies in the group

$$\omega^0_{(B,A)\times([0,1],\{0,1\})}\{(B \times [0,1]) \times S^0; \pi^*\xi_{B\times[0,1]}^+\},$$

which we can identify with

$$\omega^{-1}_{(B,A)}\{B \times S^0; \xi_B^+\}.$$

The element corresponding to $\gamma(\pi^*\xi; s)$ will be called the *difference class*

$$\delta(s_0, s_1) \in \omega^{-1}_{(B,A)}\{B \times S^0; \xi_B^+\}.$$

Remark 4.11. The definition can be extended to define the difference class of sections s_0 and s_1 with a (given) homotopy between their restrictions to A.

It is immediate from the definition that, if s_0, s_1 and s_2 are three sections agreeing on A, then

$$\delta(s_0, s_2) = \delta(s_0, s_1) + \delta(s_1, s_2).$$

The difference class is related to the Euler classes in the following way.

Lemma 4.12 *Let s_0 and s_1 be nowhere-zero sections of $\xi \mid A$ over A. Then $\delta(s_0, s_1)$ maps to $\gamma(\xi, s_0) - \gamma(\xi, s_1)$ under the coboundary map*

$$\omega^{-1}_A\{A \times S^0; (\xi_B^+)_A\} \xrightarrow{\delta} \omega^0_{(B,A)}\{B \times S^0; \xi_B^+\}$$

of the relative exact sequence.

Proposition 4.13 *Let A be a subcomplex of a finite complex B, and let ξ be a real vector bundle of dimension n over B admitting a nowhere-zero section s_0. Then the map*

$$s_1 \mapsto \delta(s_0, s_1)$$

from the set of homotopy classes of nowhere-zero sections of ξ coinciding with s_0 on A to $\omega^{-1}_{(B,A)}\{B \times S^0; \xi_B^+\}$ is surjective if $\dim B < 2(n-1)$, bijective if $\dim B + 1 < 2(n-1)$.

The basic obstruction-theoretic results, Propositions 4.9 and 4.13, are first established when B is a disc D^m and A is its bounding sphere S^{m-1}. Both then reduce to the suspension theorem. Proposition 4.9 can then be established by induction over the cells of $B - A$. Suppose that $A' \subseteq B$ is obtained by adding a disc D to A: $A' = A \cup D$, $A' \cap A = \partial D$. The relative Euler class $\gamma(\xi \,|\, A'; s)$, assumed to be zero, corresponds by excision to the obstruction $\gamma(\xi \,|\, D; s \,|\, \partial D)$ to extending s over D. Let s_0 be one such extension, and consider the exact sequence of the triple (B, A', A):

$$\omega^{-1}_{(A',A)}\{A' \times S^0; (\xi_B^+)_{A'}\} \xrightarrow{\;\cong\;} \omega^{-1}_{(D,\partial D)}\{D \times S^0; (\xi_B^+)_D\}$$

$$\delta \downarrow$$

$$\omega^0_{(B,A')}\{B \times S^0; \xi_B^+\}$$

$$\downarrow$$

$$\omega^0_{(B,A)}\{B \times S^0; \xi_B^+\}$$

The relative Euler class $\gamma(\xi, s_0)$ is equal to $\delta(x)$ for some x. From Proposition 4.13 for the disc, there is a section s_1, agreeing with s_0 on A, such that $\delta(s_0, s_1) = x$. By Lemma 4.12, $\gamma(\xi; s_1) = 0$ and the induction can proceed.

The proof of Proposition 4.13 is similar. It is instructive to relate this obstruction theory to the stabilization map, σ say,

$$\pi^0_B[B \times S^0; S(\xi)_{+B}] \to \omega^0_B\{B \times S^0; S(\xi)_{+B}\}.$$

A section s of $S(\xi) \to B$ gives a splitting of the Gysin sequence

$$\omega^{-1}_B\{B \times S^0; \xi_B^+\} \to \omega^0_B\{B \times S^0; S(\xi)_{+B}\} \underset{s_*}{\overset{\longrightarrow}{\longleftarrow}} \omega^0_B\{B \times S^0; B \times S^0\} = \omega^0(B).$$

Classes in the image of σ map to $1 \in \omega^0(B)$; and if s_0 and s_1 are two sections the difference $(s_0)_*(1) - (s_1)_*(1)$ is (the image of) $\delta(s_0, s_1)$.

An example

As a generalization of the basic obstruction theory, taking up ideas explored in Section 19 of Part I, we establish the following result.

Proposition 4.14 *Let* $\xi_1, \ldots, \xi_r$ *be finite-dimensional real vector bundles over a finite complex* B *and let* $\xi = \xi_1 \oplus \cdots \oplus \xi_r$ *denote their direct sum. Suppose that* $\dim B < 2(n-1)$, *where* $n = \dim \xi$. *Then the following conditions are equivalent.*

(i)   *There is a covering of* B *by closed sub-ENRs* $B_1, \ldots, B_r$ *such that the stable cohomotopy Euler class of the restriction of* ξ_i *to* B_i *vanishes:* $\gamma(\xi_i \,|\, B_i) = 0$, *for* $1 \leq i \leq r$.

(ii)   *The stable cohomotopy Euler class of* ξ *is zero:* $\gamma(\xi) = 0$.

(iii)   *The vector bundle* ξ *admits a nowhere-zero section.*

(iv)   *There is a covering of* B *by closed sub-ENRs* $B_1, \ldots, B_r$ *such that* $\xi_i \,|\, B_i$ *has a nowhere-zero section for* $1 \leq i \leq r$.

The implication (i) $\Rightarrow$ (ii) follows from Lemma 3.14 and the multiplicativity of the Euler class (Proposition 4.4(ii)). Condition (ii) follows from (iii), and (i) from (iv), since the vanishing of the Euler class is necessary for the existence of a nowhere-zero section.

The conditions (iii) and (iv) are easily seen to be equivalent. If $s = (s_1, \ldots, s_r)$ is a nowhere-zero section of ξ, then the open sets $U_i \subseteq B$ where $s_i(b) \neq 0$ cover B and, as the base is assumed to be a finite complex, we can find a covering by closed sub-ENRs B_i with $B_i \subseteq U_i$.

Finally, since we are in the stable range, (ii) implies (iii).

Fibrewise coHopf structures on sphere-bundles

In this paragraph, which is based on the work of Sunderland [130], we sketch an application of the methods of this section to the existence and classification of fibrewise coHopf structures on sphere-bundles. Recall that a *fibrewise coHopf structure* on the sphere-bundle ξ_B^+ is a lifting μ of the diagonal Δ in the diagram:

$$\xi_B^+ \vee_B \xi_B^+ \ \xrightarrow{\ \subseteq\ }\ \xi_B^+ \times_B \xi_B^+ \ \xrightarrow{\ \pi\ } \xi_B^+ \wedge_B \xi_B^+ \ = (\xi \oplus \xi)_B^+$$

with μ the diagonal lifting from ξ_B^+ via Δ.

The horizontal line is a cofibre sequence over B. The composition map $\pi \circ \Delta : \xi_B^+ \to (\xi \oplus \xi)_B^+$ is the fibrewise one-point compactification of the diagonal vector bundle inclusion $\xi \to \xi \oplus \xi$ and is clearly fibrewise homotopic to the fibrewise one-point compactification of the inclusion of the first factor. So the stable class

$$[\pi \circ \Delta] \in \omega_B^0 \{\xi_B^+; \, \xi_B^+ \wedge_B \xi_B^+\}$$

is the fibrewise suspension $1 \wedge \gamma(\xi)$ of the Euler class.

We have shown:

Proposition 4.15 *The vanishing of the stable cohomotopy Euler class $\gamma(\xi)$ is a necessary condition for ξ_B^+ to admit a fibrewise coHopf structure*

In a stable range we have the following more precise result.

Proposition 4.16 *Let ξ be an n-dimensional real vector bundle over a finite complex B.*
(a) *Suppose that* $\dim B < 2n - 2$. *Then the following are equivalent.*
    (i)   *The stable cohomotopy Euler class $\gamma(\xi)$ is zero.*
    (ii)   *The sphere-bundle $S(\xi) \to B$ admits a section.*
    (iii) *The pointed sphere-bundle ξ_B^+ admits a fibrewise coHopf structure.*
(b) *Suppose that* $\dim B < 2n - 4$ *and that $\gamma(\xi) = 0$. Then there is a 1-1 correspondence between the set of homotopy classes of sections of $S(\xi)$ and that of fibrewise coHopf structures on ξ_B^+. The group $\omega_B^{-1}\{B \times S^0; \xi_B^+\}$ acts transitively and freely on both sets.*

For (a), note that if $S(\xi)$ admits a section, then ξ_B^+ is a fibrewise suspension $\Sigma_B \eta_B^+$, where η is the sub-bundle of ξ orthogonal to the section, and so has a fibrewise coHopf structure. Proposition 4.9 completes the proof.

For (b), we use the Serre exact sequence (2.19). Since the fibre of $\xi_B^+ \vee_B \xi_B^+$ is $(n-1)$-connected and that of $(\xi \oplus \xi)_B^+$ is $(2n-1)$-connected, in the range $\dim B < 2n - 4$, we get an exact sequence of groups:

$$\pi_B^{-1}[\xi_B^+; \xi_B^+ \vee_B \xi_B^+] \to \pi_B^{-1}[\xi_B^+; \xi_B^+ \times_B \xi_B^+] \to \pi_B^{-1}[\xi_B^+; (\xi \oplus \xi)_B^+]$$
$$\to \pi_B^0[\xi_B^+; \xi_B^+ \vee_B \xi_B^+] \to \pi_B^0[\xi_B^+; \xi_B^+ \times_B \xi_B^+] \to \pi_B^0[\xi_B^+; (\xi \oplus \xi)_B^+].$$

Now the first homomorphism (of Abelian groups) is a split epimorphism, for the inclusion $\Omega_B(\xi_B^+ \vee_B \xi_B^+) \to \Omega_B(\xi_B^+ \times_B \xi_B^+)$ is a fibrewise homotopy retraction. It follows that the group $\pi_B^{-1}[\xi_B^+; (\xi \oplus \xi)_B^+]$ acts freely and transitively on the set of elements lifting $\Delta \in \pi_B^0[\xi_B^+; \xi_B^+ \times_B \xi_B^+]$. The rest of the argument is straightforward.

Chapter 2. Fixed-point Methods

5 Fibrewise Euclidean and Absolute Neighbourhood Retracts

In this somewhat technical section we look at the theory of fibrewise ENRs and ANRs. The results are mostly due to Dold [47]. Our restriction to base spaces which are ENRs allows us to simplify the exposition at several points. We begin with a discussion of some of the properties of ENRs and ANRs which we have already used in earlier sections.

Properties of ENRs

The results which follow are formulated in the notation which we have reserved for base spaces, but are, of course, not restricted to that context.

We look first at uniform local contractibility.

Lemma 5.1 *Let B be an ENR. Then there is an open neighbourhood W of the diagonal $\Delta(B)$ in $B \times B$ and a homotopy $H_t : W \to B$ such that $H_0(a, b) = a$, $H_1(a, b) = b$ and $H_t(b, b) = b$, for all $(a, b) \in W$, $0 \leq t \leq 1$.*

Indeed, if we embed B as a retract of an open subset U of a Euclidean space E: $i : B \hookrightarrow U$, $r : U \to B$, $r \circ i = 1_B$, then we can define H by $H_t(a, b) = r((1 - t)i(a) + ti(b))$ on the subset W consisting of those points (a, b) such that the (compact) line segment joining $i(a)$ to $i(b)$ lies inside U.

One can show by an induction on cells that every finite complex is a compact ENR. In the opposite direction, it follows easily from the definition that a compact ENR is a retract of a finite polyhedron.

Lemma 5.2 *Let (B, A) be a compact ENR pair. Then there is a finite polyhedral pair (B', A') together with maps $\iota : (B, A) \to (B', A')$ and $\rho : (B', A') \to (B, A)$ such that $\rho \circ \iota = 1 : (B, A) \to (B, A)$.*

Indeed, one can embed B as a retract of an open subspace U of some Euclidean space $\mathbb{R}^{k+1}$ in such a way that A is the intersection of B with $\mathbb{R}^k$. Choose $\epsilon > 0$ so small that no cube of side ϵ in $\mathbb{R}^{k+1}$ intersects both

B and the complement of U. Take B' to be the union of all cubes of side ϵ with vertices in $(\mathbb{Z}\epsilon)^{k+1}$ which are contained in U, and take A' to be its intersection with $\mathbb{R}^k$.

It is a theorem of West [133] that every compact ENR B (and, more generally, any compact ANR) is homotopy equivalent to a finite complex. More precisely, if the homology of B vanishes above dimension m, then B is homotopy equivalent to a finite complex of dimension less than or equal to $\max\{m, 3\}$. (See, for example, the exposition [116].) The proof when B is simply connected is relatively elementary; hence, so too is the result that a pointed compact ENR is equivalent in the stable homotopy category to a (desuspension of a) finite complex.

While the theory to be explored in this chapter fits most naturally into the ENR setting, these remarks indicate that for most purposes we can work with finite complexes rather than ENRs. The topology of finite complexes (including finite polyhedra and compact smooth manifolds) has an important feature which is not enjoyed by general ENRs. (See Chapter VI, Section 4, of [18].)

Proposition 5.3 *Let B be a finite complex. Suppose that K is a compact subspace of an open subset U in B. Then there is a compact sub-ENR A in B such that $K \subseteq A \subseteq U$.*

For a finite polyhedron B, one can find such a sub-ENR A by subdivision. A similar argument works for a general complex. For a manifold B, one can construct A as a submanifold $\psi^{-1}[c, \infty)$ of codimension zero, where c is a regular value, $0 < c < 1$, of a smooth function $\psi : B \to \mathbb{R}$ which is 1 on K and 0 outside U.

ANRs

We shall say that a space M is an *absolute neighbourhood retract* (ANR) if it can be embedded as a retract of an open subspace U of a normed (real) vector space E: $i : M \to U$, $r : U \to M$, with $r \circ i = 1_M$. (The absolute neighbourhood retracts which we are considering are, technically, the metrisable ANRs.)

The proof above for ENRs extends immediately to show that an ANR is uniformly locally contractible.

Function spaces

Lemma 5.4 *Let P be a compact Hausdorff space and Q be an ANR. Then the space $\mathrm{map}(P, Q)$ is an ANR.*

Choose an embedding $Q \subseteq U$ of Q as a retract, $r : U \to Q$, of an open subset of a normed vector space E. The compact-open topology on map(P,Q) coincides with its topology as a subspace of the normed vector space map(P, E), with the supremum norm. The subspace map(P, U) is open, and it retracts onto map(P, Q) by: $f \mapsto r \circ f$.

Fibrewise ENRs and ANRs

Definition 5.5 A fibrewise space M over B is said to be a *fibrewise ANR* (Absolute Neighbourhood Retract) if there is an open covering $\mathcal{U}$ of B and, for each $U \in \mathcal{U}$, a normed (real) vector space E_U, an open subspace W_U of $U \times E_U$, and fibrewise maps $i_U : M_U \to W_U, r_U : W_U \to M_U$, with $r_U \circ i_U = 1$. If each normed vector space E_U may be taken to be finite-dimensional we say that M is a *fibrewise ENR* (Euclidean Neighbourhood Retract).

As a basic example of a fibrewise ENR we have a finite-dimensional real vector bundle over B. (Our definition is slightly more general than that in [47]. The vector bundle over $B = \mathbf{N}$ with fibre $\mathbb{R}^n$ at n is not an ENR$_B$ in Dold's sense.) A bundle of normed vector spaces is a fibrewise ANR. More generally, a fibre bundle whose fibres are ANRs (or ENRs) is a fibrewise ANR (or ENR).

It is immediate from the definition that, if $\alpha : B' \to B$ is a map of ENRs and $M \to B$ is a fibrewise ANR (or ENR), then the pull-back $\alpha^* M \to B'$ is a fibrewise ANR (or ENR, respectively). In particular, each fibre M_b of a fibrewise ANR (or ENR) is an ANR (or ENR).

Several further properties are readily checked. An open subspace $W \subseteq M$ of a fibrewise ANR (or ENR) over B is itself a fibrewise ANR (or ENR). The product $M \times_B N \to B$ of two fibrewise ANRs (or ENRs) M and N over B is a fibrewise ANR (or ENR). A fibrewise retract of a fibrewise ANR (or ENR) is a fibrewise ANR (or ENR). (We say that M is a *fibrewise retract* of N if there exist fibrewise maps $i : M \to N$ and $r : N \to M$ such that $r \circ i = 1 : M \to M$.)

Lemma 5.6 *Let M be a fibrewise ANR over B. Then there is a fibrewise embedding $M \to B \times E$ as a closed subspace of a trivial vector bundle with fibre some normed vector space E. If M is a fibrewise ENR and B is compact, then E may be taken to be finite-dimensional.*

We use the notation from Definition 5.5. When B is compact we can take the open covering $\mathcal{U}$ to be finite. (It can always be taken to be countable.) Let (ϕ_U) be a partition of unity subordinate to the covering.

We begin by embedding W_U as a closed subspace of $U \times (\mathbb{R} \oplus E_U)$. Choose a metric d on $U \times E_U$ and define $\rho : W_U \to [0, \infty)$ to be the reciprocal of the distance to the complement of W_U: $\rho(x) = \sup\{1/d(x,y) \mid y \notin W_U\}$ (zero if the complement is empty). Then $W_U \to U \times (\mathbb{R} \oplus E_U)$: $x \mapsto (b, (\rho(x), x))$ in the fibre at $b \in B$ is the required closed embedding. Composing this map with i_U, we obtain a fibrewise closed embedding

$$i'_U : M_U \to U \times E'_U,$$

where $E'_U = \mathbb{R} \oplus E_U$.

The maps i'_U are assembled using the partition of unity to construct the required fibrewise closed embedding $i : M \to B \times E$, where

$$E := \bigoplus_{U \in \mathcal{U}} E'_U$$

(with the L^1-norm: $\|(v_U)\| = \sum_{U \in \mathcal{U}} \|v_U\|$). In the fibre at b, i is the (effectively finite) sum $\sum_{U \in \mathcal{U}} \phi_U(b) i'_U$.

It follows that the total space M of a fibrewise ANR is metrisable.

At the root of the theory of ANRs is the extension of Tietze's theorem to functions with values in a normed vector space. See [67].

Proposition 5.7 *Let P be a closed subspace of a metric space Q, and let E be a normed vector space. Then any map $f : P \to E$ extends to a map $g : Q \to E$: $g(x) = f(x)$ for $x \in P$.*

We use this to show:

Lemma 5.8 *Let $M \to B$ be a fibrewise ANR embedded as a closed subspace $M \subseteq B \times E$, where E is a normed vector space. Then there is a fibrewise retraction $r : W \to M$ of an open neighbourhood W of M onto M.*

Again we use the notation of Definition 5.5, and assume that $\mathcal{U}$ is finite or countable. Without loss of generality we can assume that $\mathcal{U}$ is locally finite, that is, that any point is contained in a neighbourhood which meets only a finite number of sets of the covering, and that there exist functions

$$\psi_U : B \to [0,1] \quad \text{such that} \quad \overline{\psi_U^{-1}(0,1]} \subseteq U$$

with the property that, for each $b \in B$, there is a $U \in \mathcal{U}$ with $\psi_U(b) = 1$.

Consider the closed subspace M_U of the metric space $U \times E$. By Tietze's theorem, there is a map $f_U : U \times E \to E_U$ such that $i_U(x) = (b, f_U(x))$ for all $x \in M_b$, $b \in U$:

$$
\begin{array}{ccc}
M_U & \xrightarrow{\ i_U\ } & W_U \subseteq U \times E_U \\
\cap \downarrow & & \downarrow \\
U \times E & \xrightarrow[\ f_U\ ]{} & E_U
\end{array}
$$

Define $W'_U \subseteq B \times E$ to be the set of points $x = (b,v)$ such that $(b, f_U(b,v)) \in W_U$ where $\psi_U(b) > 0$. It is an open subset of $B \times E$ containing M. We define a map $q_U : W'_U \to B \times E$ by

$$q_U(b,v) := \begin{cases} (b,v) & \text{if } \psi_U(b) = 0, \\ \psi(b) r_U(b, f_U(b,v)) + (1 - \psi(b))(b,v) & \text{if } \psi_U(b) > 0. \end{cases}$$

Notice that q_U is the identity on M and maps into M in those fibres where $\psi_U(b) = 1$.

Now we enumerate the elements of $\mathcal{U}$ as $U_1, U_2, \ldots$ and make the abbreviations $W_i' = W_{U_i}'$, $q_i = q_{U_i}$. The subset

$$W := \{x \in B \times E \mid x \in W_1', \, q_1(x) \in W_2', \, q_2(q_1(x)) \in W_3', \ldots\}$$

is open in $B \times E$, by the local finiteness condition, and $r := \ldots \circ s_2 \circ s_1 : W \to M$ is a retraction onto M.

Corollary 5.9 *Let $M \to B$ be a fibrewise ANR. Then the total space M is an ANR. If $M \to B$ is a fibrewise ENR over a compact base, then M is an ENR.*

For if M is a retract of an open subset of $B \times E$ and B is a retract of an open subset of a finite-dimensional vector space E', then M is a retract of an open subset of $E' \oplus E$.

As an immediate generalization of Lemma 5.8 we obtain:

Proposition 5.10 *Let $M \to B$ be a closed fibrewise sub-ANR of a fibrewise ANR $N \to B$. Then M is a fibrewise retract of an open subset of N, that is, there exist an open neighbourhood W of M in N and a fibrewise map $r : W \to M$ such that $r(x) = x$ for all $x \in M$.*

We can assume that N is a closed subspace of $B \times E$ for some normed vector space E, by Lemma 5.6. This embeds M as a closed subspace of $B \times E$. Apply Lemma 5.8, and then intersect with N.

In fact we can do better: we can find an open neighbourhood W and a homotopy $r_t : W \to N$, $0 \le t \le 1$, with $r_0(x) = x$ for $x \in W$, $r_t(x) = x$ for $x \in M$, and $r_1(W) = M$. For let W' be an open neighbourhood of M in N and $r : W' \to M$ a retraction, and let W'' be an open neighbourhood of N in $B \times E$ and $s : W'' \to N$ a retraction. Then we can define

$$W := \{x \in W' \mid (1-t)x + tr(x) \in W'' \text{ for all } t \in [0,1]\}$$

and $r_t(x) := s((1-t)x + tr(x))$.

Choose a function $\psi : N \to [0,1]$ with zero-set $\psi^{-1}(0)$ equal to M and taking the value 1 on a neighbourhood of the complement $N - W$. Then we can give a retraction $q : N \times [0,1] \to N \times \{0\} \cup M \times [0,1]$ by

$$q(x,t) := \begin{cases} (x,0) & \text{for } x \notin W, \\ (r_{(2\psi(x)-t)/(2-t)}(x), 0) & \text{for } x \in W, \, t < 2\psi(x), \\ (r_0(x), t - 2\psi(x)) & \text{for } x \in W, \, t \ge 2\psi(x). \end{cases}$$

We have proved:

Proposition 5.11 *Let $M \to B$ be a closed fibrewise sub-ANR of a fibrewise ANR $N \to B$. Then the inclusion $M \hookrightarrow N$ is a fibrewise cofibration over B.*

Remark 5.12. The converse of this proposition is also true. Let N be a fibrewise ANR and let M be a closed subspace such that the inclusion $M \hookrightarrow N$ is a fibrewise cofibration. Then M is a fibrewise ANR.

For let $q : N \times [0,1] \to N \times \{0\} \cup M \times [0,1]$ be a fibrewise retraction. Set $W = \{x \in N \mid q(x,1) \in M \times (0,1]\}$ and define $r : W \to M$ by taking $r(x)$ to be the first factor of $q(x,1)$. Thus, M is a fibrewise retract of an open subspace of a fibrewise ANR, and hence a fibrewise ANR.

It follows that a fibrewise pointed space $X \to B$ which is a fibrewise ANR is automatically well-pointed, provided that the inclusion $B \to X$ of the basepoints is closed. For the identity map $B \to B$ is clearly a fibrewise ENR.

Local characterization

The arguments used in Lemma 5.6 to embed a fibrewise ANR M over B as a closed subspace of a trivial vector bundle $B \times E$ and in Lemma 5.8 to show that M is then a fibrewise retract of an open neighbourhood require only minor modification to establish:

Proposition 5.13 *Let $M \to B$ be a fibrewise space, where the topology of M is metrisable. Suppose that each point of M has an open neighbourhood which is a fibrewise ANR over B. Then M is a fibrewise ANR.*

In other words, subject to the metrisability condition which guarantees the existence of partitions of unity, M is a fibrewise ANR if each point of M is contained in an open subspace which is a fibrewise retract of an open subspace of some trivial vector bundle $B \times E$. When we consider fibrewise manifolds in Section 11 we shall take a local property of this type as our starting point.

Extension properties

Fibrewise ANRs inherit the following extension property from normed vector spaces.

Proposition 5.14 *Let M and N be fibrewise spaces over B, with the total space M metrisable and N a fibrewise ANR. Let $f : C \to N$ be a fibrewise map defined on a closed subspace C of M. Then f extends to a fibrewise map on an open neighbourhood of C in M. If N is a vector bundle (of normed vector spaces), then f extends to the whole of M.*

For let $i : N \to W$ be a fibrewise embedding of N in an open subset W of a trivial vector bundle $B \times E$ and $r : W \to N$ be a retraction. By Tietze's theorem $i \circ f$, defined on C, extends to a fibrewise map $g : M \to B \times E$. On $g^{-1}(W)$, the map $r \circ g$ extends f.

When N is a vector bundle extensions of f defined over the open sets of a covering $\mathcal{U}$ of B, such that N is trivial over each $U \in \mathcal{U}$, can be pieced together using a partition of unity.

The germ of the extension of f in Proposition 5.14 is unique up to homotopy.

Proposition 5.15 *Let M and N be fibrewise spaces over B with the total space M metrisable and N a fibrewise ANR. Let V_0 and V_1 be open neighbourhoods of a closed set C in M, and let $f_0 : V_0 \to N$ and $f_1 : V_1 \to N$ be fibrewise maps which coincide on C. Then there is an open neighbourhood U of C in $V_0 \cap V_1$ such that $f_0 | U$ and $f_1 | U$ are homotopic through fibrewise maps agreeing on C.*

This is a consequence of Proposition 5.14 applied to the closed subspace $D \times \{0, 1\} \cup C \times [0, 1] \subseteq M \times [0, 1]$ for some closed neighbourhood D of C in $V_0 \cap V_1$. But it seems more illuminating to give a direct proof. For this, choose i and r as in the proof of Proposition 5.14, put $g_t = (1 - t)if_0 + tif_1$, $0 \le t \le 1$, on $V_0 \cap V_1$, and set

$$U := \{x \in V_0 \cap V_1 \mid g_t(x) \in W \text{ for all } t \in [0, 1]\}.$$

Then $f_t := rg_t$ is the homotopy sought.

Uniform local contractibility

We have already made use of the uniform local contractibility of ENR base spaces. Fibrewise ANRs are fibrewise uniformly local contractible in the following sense.

Definition 5.16 A fibrewise space M over B is said to be *fibrewise uniformly locally contractible* if there is an open neighbourhood W of the diagonal in $M \times_B M$ and a fibrewise homotopy $H_t : W \to M$, $0 \le t \le 1$, such that $H_0(x, y) = x$, $H_1(x, y) = y$ and $H_t(x, x) = x$, for $x, y \in M_b$.

Proposition 5.17 *Let $M \to B$ be a fibrewise ANR. Then M is fibrewise uniformly locally contractible.*

The proof that we sketched for Lemma 5.1 easily adapts. Choose a fibrewise embedding $i : M \to V$ and retraction $r : V \to M$, where V is an open subset of $B \times E$ for some normed vector space E. Let

$$W := \{(x, y) \in M \times_B M \mid (1 - t)i(x) + ti(y) \in V \text{ for all } t \in [0, 1]\},$$

and define $H_t(x, y)$ to be $r((1 - t)i(x) + ti(y))$.

Fibrewise compactness

Fibrewise compactness has homotopy-theoretic consequences.

Lemma 5.18 *Let $M \to B$ be a fibrewise compact ANR over B. Then, for each point $b \in B$ there is an open neighbourhood U of b, a normed real vector space E, an open subspace V of E, and fibrewise maps*

$$M_U \xrightarrow{\ i\ } U \times V \xrightarrow{\ r\ } M_U$$

with $r \circ i = 1$.

We may assume that M is a fibrewise retract of an open subset W of $B \times E$. Let $b \in B$. By compactness, there is an open neighbourhood U of b and open subset V of E such that $M_U \subseteq U \times V \subseteq W$. (See Part I, Proposition 1.14.)

The lemma says that M_U is a fibrewise retract of a trivial bundle. It is, therefore, a fibration. But any fibrewise retract of a fibration is a fibration. Since the property of being a fibration is local, we have proved:

Proposition 5.19 *Let $M \to B$ be a fibrewise compact ANR over B. Then $M \to B$ has the homotopy lifting property and is, therefore, a homotopy fibre bundle.*

In the opposite direction we have:

Proposition 5.20 *Let $M \to B$ be a fibrewise space over B having the homotopy lifting property. Suppose that the space M is an ANR (or ENR). Then $M \to B$ is a fibrewise ANR (or ENR).*

We have an embedding $i : M \to V$ and retraction $r : V \to M$, where V is an open subspace of a normed vector space E, finite-dimensional if M is an ENR. The uniform local contractibility of the base gives us a homotopy $H_t : U \to B$, defined on an open neighbourhood U of the diagonal in $B \times B$, with $H_0(a, b) = b$, $H_1(a, b) = a$ and $H_t(b, b) = b$, for $a, b \in B$, $0 \leq t \leq 1$.

Choose a function $\psi : U \to [0, 1]$ with zero-set the diagonal $\Delta(B)$. (So ψ could be a distance function, truncated at 1.) Now define a homotopy $\alpha_t : U \to B$ by

$$\alpha_t(a, b) := \begin{cases} b & \text{if } \psi(a, b) = 0, \\ H_{t/\psi(a,b)}(a, b) & \text{for } t \leq \psi(a, b), \text{ if } \psi(a, b) > 0, \\ H_1(a, b) = b & \text{for } t \geq \psi(a, b), \text{ if } \psi(a, b) > 0. \end{cases}$$

Thus, $\alpha_0(a, b) = a$ and $\alpha_t(a, b) = b$ for $t \geq \psi(a, b)$.

Consider the diagram:

$$M \xrightarrow{(p,i)} B \times V \supseteq W \xrightarrow{f_t} M$$

$$p \downarrow \qquad \downarrow \qquad \downarrow \qquad \downarrow p$$

$$B \xrightarrow[\Delta]{} B \times B \supseteq U \xrightarrow[\alpha_t]{} B$$

in which the projection $B \times V \to B \times B$ is the map $(b, v) \mapsto (b, p(r(v)))$, W is the inverse image of U, and the map f_t is yet to be defined. The retraction r gives a lift f_0 of α_0: $(b, v) \mapsto r(v) \in M$. Now the homotopy lifting property provides the homotopy f_t.

The map $q : B \times V \to M$ given by f_t at the value t determined by ψ:

$$q(b, v) = f_{\psi(b, pr(v))}(b, v),$$

is a fibrewise retraction of the inclusion of M in $B \times V$ over B.

Fibrewise mapping-spaces

The mapping-space construction that we have already noted, in Lemma 5.4, extends with only minor modification to the fibrewise set-up.

Lemma 5.21 Let $M \to B$ be a fibre bundle with compact Hausdorff fibre and let $N \to B$ be a fibrewise ANR. Then the fibrewise mapping-space $\mathrm{map}_B(M, N) \to B$ is a fibrewise ANR over B.

For if we embed N as a fibrewise retract of an open subspace W of $B \times E$ for some normed vector space E, then $\mathrm{map}_B(M, N)$ will be a retract of the open subspace $\mathrm{map}_B(M, W)$ of the vector bundle $\mathrm{map}_B(M, B \times E)$.

Corollary 5.22 Suppose that $M \to B$ is a fibrewise compact ENR and $N \to B$ is a fibrewise ANR. Then $\mathrm{map}_B(M, N) \to B$ is a fibrewise ANR.

To deduce the corollary, we see from Lemma 5.18, using the fact that E there can be chosen to be finite-dimensional for a fibrewise ENR, that M is locally a fibrewise retract of a fibre bundle with compact Hausdorff fibre. Hence $\mathrm{map}_B(M, N)$ is a fibrewise retract of a fibrewise ANR, and so itself a fibrewise ANR.

A homotopy construction

The construction which we describe next will be needed in the discussion (following Dold and Puppe [49]) of duality theory for fibrewise ENRs.

Definition 5.23 Let M be a fibrewise sub-ENR of a fibrewise ENR N over B. Suppose that M is fibrewise compact, and hence closed in N. Then we define $C_B(N, N - M)$ to be the fibrewise mapping cone of the inclusion $(N - M)_{+B} \to N_{+B}$.

Because $N - M$ is an open subspace of N, the fibrewise topological space $C_B(N, N - M)$ is not particularly pleasant; we are interested more in its fibrewise homotopy type.

Lemma 5.24 *Let $\rho : N \to [0, 1]$ be a continuous function which takes the value 1 on a neighbourhood of M and has fibrewise compact support (that is, vanishes outside a fibrewise compact subspace). Then $C_B(N, N - M)$ is fibrewise pointed homotopy equivalent to the quotient of*

$$(B \times \{0\}) \sqcup (\{(x, t) \in (N - M) \times [0, 1] \mid t \le \rho(x)\} \cup (N \times \{1\}))$$

by the identification of $(x, 0)$ with the basepoint $(p(x), 0)$ in the fibre.

Indeed, the space described is clearly a fibrewise deformation retract of $C_B(N, N - M)$: retract (x, t) to $(x, \min\{t, \rho(x)\})$.

Remark 5.25. The map $N \to C_B(N, N - M)$ taking $x \in N$ to $(x, \rho(x))$ is trivial outside a fibrewise compact subspace and so extends to the fibrewise one-point compactification and determines a homotopy class

$$N_B^+ \to C_B(N, N - M),$$

which is easily seen to be independent of the choice of ρ.

It follows from this description that the fibrewise homotopy type of $C_B(N, N - M)$ is determined in a neighbourhood of M.

Proposition 5.26 *In the notation of Definition 5.23, let W be an open neighbourhood of M in N. Then the map*

$$C_B(W, W - M) \to C_B(N, N - M)$$

induced by the inclusions is a fibrewise pointed homotopy equivalence.

This follows immediately from Lemma 5.24, by choosing a function ρ vanishing outside a compact subset of W.

In practice one can often find an open neighbourhood W such that the closed subspace $N - W$ is a deformation retract of the open subspace $N - M$. (When M and N are fibrewise manifolds, one can take W to be a tubular neighbourhood; see Section 11.)

Proposition 5.27 *With the notation of Definition 5.23, suppose that the closure of W in N is fibrewise compact and that there is a fibrewise deformation retraction of $N - M$ onto $N - W$, that is, a homotopy $r_t : N - M \to N - M$,* with $r_0 = 1$, $r_t(x) = x$ for $x \in N - W$, $r_1(N - M) = N - W$. Then $C_B(N, N - M)$ is fibrewise pointed homotopy equivalent to the fibrewise one-point compactification W_B^+.

The retraction r_1 gives a fibrewise pointed homotopy equivalence from $C_B(N, N - M)$ to the mapping cone of the inclusion $(N - W)_{+B} \to N_{+B}$. But $N - W$ is a closed fibrewise sub-ENR of N. Hence the inclusion is a fibrewise cofibration and the fibrewise mapping cone is fibrewise homotopy equivalent to the topological quotient $N/_B(N - W)$. By Proposition 10.2 in Part I, the quotient $\overline{W}/_B(\overline{W} - W)$ is homeomorphic to the fibrewise one-point compactification of W.

Remark 5.28. Composition with the map $N_B^+ \to C_B(N, N - M)$ of Remark 5.25:

$$N_B^+ \to C_B(N, N - M) \xrightarrow{\sim} C_B(W, W - M) \xrightarrow{\sim} W_B^+$$

is the Pontrjagin–Thom construction which collapses the complement of the open subspace W of N to the basepoint at infinity (in each fibre).

Example 5.29. Let ξ be a finite-dimensional real vector bundle over B with a Euclidean metric. We take N to be ξ, M to be the zero-section $B \times 0$ and W to be the interior of the closed unit ball $D(\xi)$. Then we have a fibrewise pointed homotopy equivalence:

$$C_B(\xi, \xi - (B \times 0)) \to \xi_B^+.$$

We note, finally, that the construction behaves well under smash products.

Lemma 5.30 *Let $M \subseteq N$ and $M' \subseteq N'$ be inclusions of fibrewise ENRs over B, with M and M' fibrewise compact. Then there is a natural fibrewise pointed homotopy equivalence:*

$$C_B(N, N - M) \wedge_B C_B(N', N' - M') \to C_B(N \times_B N', (N \times_B N') - (M \times_B M')).$$

To see why this is true, we introduce the abbreviations W and W' for the open subspaces $(N - M) \times_B N'$ and $N \times_B (N' - M')$ of $N \times_B N'$. It is not difficult to see that the projection map

$$(W \times \{0\}) \cup ((W \cap W') \times [0, 1]) \cup (W' \times \{1\}) \to W \cup W'$$

is a fibrewise homotopy equivalence. Indeed, since the spaces are metrisable we can choose continuous functions $\psi : N \to [0, 1]$ and $\psi' : N' \to [0, 1]$ with zero-sets M and M' respectively. Let $\phi : W \cup W' \to [0, 1]$ be the function:

$$\phi(x, x') = \psi'(x')/(\psi(x) + \psi'(x')),$$

which takes the value 0 on $(N-M) \times_B M'$, and 1 on $M \times_B (N'-M')$. Then the map $(x, x') \mapsto (x, x', \phi(x, x'))$ is an inverse fibrewise homotopy equivalence.

Now, by definition $C_B(N \times_B N', (N \times_B N') - (M \times_B M'))$ is the fibrewise mapping cone of the inclusion $(W \cup W')_{+B} \to (N \times_B N')_{+B}$. It is, therefore, fibrewise pointed homotopy equivalent to the fibrewise mapping cone of the induced map

$$((W \times \{0\}) \cup ((W \cap W') \times [0, 1]) \cup (W' \times \{1\}))_{+B} \to (N \times_B N')_{+B}.$$

This fibrewise space is actually homeomorphic to the fibrewise smash product $C_B(N, N - M) \wedge_B C_B(N', N' - M')$.

Change of base: the direct image

We shall sometimes want to use the fibrewise ENR condition for a map $\alpha : B' \to B$ between ENRs B' and B, both playing the rôle of base spaces.

Definition 5.31 Given a fibrewise space $p : M \to B'$ over B' we shall write $\alpha_* M \to B$ for the fibrewise space $\alpha \circ p : M \to B$ over B. So as a space $\alpha_* M$ is just M. We shall refer to $\alpha_* M$ as the *direct image* of M.

It is straightforward to check:

Lemma 5.32 *Let* $\alpha : B' \to B$ *be a fibrewise ENR, where* B *and* B' *are both ENRs, and let* $M \to B'$ *be a fibrewise ANR (or ENR) over* B'. *Then* $\alpha_* M \to B$ *is a fibrewise ANR (or ENR) over* B.

As a special case, from a different point of view, we have:

Lemma 5.33 *Let* $\alpha : B' \to B$ *and* $\beta : B'' \to B'$ *be fibrewise ENRs. Then the composition* $\alpha \circ \beta : B'' \to B$ *is a fibrewise ENR.*

The direct image construction α_* is evidently functorial. Given a fibrewise map $f : M \to N$ over B', $\alpha_* f : \alpha_* M \to \alpha_* N$ is just the map f considered as a fibrewise map over B. The terminology 'direct image' (which is not well established) for this construct is to some extent justified by its adjoint relationship with the pull-back, which is sometimes called the 'inverse image'.

Lemma 5.34 *The functor* α_* *is left adjoint to the pull-back* α^*. *To be precise, let* $M \to B'$ *and* $N \to B$ *be fibrewise spaces. Then there is a natural equivalence between fibrewise maps* $\alpha_* M \to N$ *over* B *and fibrewise maps* $M \to \alpha^* N$ *over* B'.

We shall normally be concerned with the situation in which $\alpha : B' \to B$ is fibrewise compact. This means that the map α is proper and also, by Proposition 5.19, a fibration, so locally homotopy trivial.

Lemma 5.35 *Let $\alpha : B' \to B$ be a fibrewise compact ENR. Suppose that $M \to B'$ is a homotopy fibre bundle over B'. Then $\alpha_* M \to B$ is a homotopy fibre bundle over B.*

This follows from the fact that a composition of fibrations is a fibration, but it can also be seen in terms of local homotopy trivializations and this is the approach that is needed in the pointed case which we look at next.

Definition 5.36 Let $\alpha : B' \to B$ be a fibrewise compact ENR. Given a fibrewise pointed space $X \to B'$ over B', we define the *direct image* $\alpha_* X \to B$ to be the fibrewise pointed space

$$\alpha_* X := X/_B B'$$

over B constructed by collapsing the basepoints $B' \subseteq X$ to B. Thus $\alpha_* X$ is the quotient of $X \sqcup B$ by the equivalence relation which identifies $b' \in B' \subseteq X$ with $\alpha(b') \in B$. The construction is functorial. To a fibrewise pointed map $f : X \to Y$ over B' we associate a fibrewise pointed map $\alpha_* f : \alpha_* X \to \alpha_* Y$ over B' by passage to quotients.

Notice that, since X in the definition is fibrewise homotopy well-pointed, the inclusion $B' \hookrightarrow X$ over B is a fibrewise homotopy-cofibration. (See Sections 4 and 16 of Part I.) So $\alpha_* X$ is, up to homotopy, the fibrewise (unreduced) mapping cone, that is, the union of X and the fibrewise cone on B'.

Again in the pointed category the direct image is left adjoint to the pullback. We formulate the property in the homotopy category as

Proposition 5.37 *Let $X \to B'$ and $Y \to B$ be fibrewise pointed spaces over B' and B respectively. Then there is a natural equivalence:*

$$\pi^0_{B'}[X;\, \alpha^* Y] \overset{\simeq}{\longrightarrow} \pi^0_B[\alpha_* X;\, Y].$$

There is a corresponding stable result; we shall make use of it on several occasions. A special case has appeared already, in Proposition 3.4.

Remark 5.38. It follows formally from the adjoint property that, if $X \to B'$ is a fibrewise coHopf space over B', then $\alpha_* X \to B$ is a fibrewise coHopf space over B.

As we have already intimated, the construction remains within the category of pointed homotopy fibre bundles.

Lemma 5.39 *Let $\alpha : B' \to B$ be a fibrewise compact ENR. Suppose that $X \to B'$ is a pointed homotopy fibre bundle over B'. Then $\alpha_* X \to B$ is a pointed homotopy fibre bundle over B.*

Remark 5.40. The pull-back and direct image correspond algebraically to extension and restriction of scalars. Suppose briefly that B' and B are compact and that X and Y are compact Hausdorff fibrewise pointed spaces over B' and B, respectively. Let $\mathcal{C}(P)$ for a compact Hausdorff space P denote the commutative C^*-algebra of complex-valued continuous functions on P, and let $\mathcal{C}(P, Q)$ for a closed subspace Q be the ideal of functions which are zero on Q. A map $\alpha : B' \to B$ determines a C^*-algebra homomorphism $\alpha^* : \mathcal{C}(B) \to \mathcal{C}(B')$. The projection $Y \to B$ makes $\mathcal{C}(Y, B)$ an algebra (without identity) over $\mathcal{C}(B)$.

Now the C^*-algebra of the fibrewise pointed space $\alpha^* Y$ is the (appropriately completed) tensor product

$$\mathcal{C}(\alpha^* Y, B') = \mathcal{C}(B') \,\hat{\otimes}_{\mathcal{C}(B)}\, \mathcal{C}(Y, B)$$

obtained by extension of scalars, and the C^*-algebra of $\alpha_* X$ is just

$$\mathcal{C}(\alpha_* X, B) = \mathcal{C}(X, B')$$

considered as a $\mathcal{C}(B)$-algebra by restriction of scalars.

The left adjoint of the pull-back arises naturally in the stable theory. We shall not need the homotopy-theoretic right adjoint, which is constructed as a fibrewise mapping-space, but include a brief description in the interests of completeness.

The right adjoint of the pull-back

Let $\alpha : B' \to B$ again be a fibrewise compact ENR, and let $Y \to B'$ be a fibrewise pointed space over B'. We want to construct a fibrewise pointed space $_*\alpha Y \to B$ over B such that is a natural equivalence

$$\pi_{B'}^0 [\alpha^* X; Y] \xrightarrow{\;\cong\;} \pi_B^0 [X; \,_*\alpha Y]$$

for any fibrewise pointed space $X \to B$ over B.

The fibre of $_*\alpha Y$ at $b \in B$ is defined to be the space of sections $\Gamma(\alpha^{-1}(b); Y \mid \alpha^{-1}(b))$ of Y restricted to the fibre of B' at b. The space is topologized as a subspace of the fibrewise mapping-space $\mathrm{map}_B(B', Y)$. Indeed, $_*\alpha Y$ is the inverse image of $B \times \{1\}$ in the sequence:

$$_*\alpha Y \subseteq \mathrm{map}_B(B', Y) \to \mathrm{map}_B(B', B'). \tag{5.41}$$

(We are using 1 here for the identity map on a fibre of B' over B.) This sequence is a fibrewise fibration, and $_*\alpha Y$ is a fibrewise fibre; see Sections 5 and 8 of Part I.

As a dual of Remark 5.38 we note that the right adjoint $_*\alpha$ sends fibre-wise Hopf spaces to fibrewise Hopf spaces. This generalizes the elementary observation that the space of sections of a fibrewise Hopf space is a Hopf space.

6 Lefschetz fixed-point theory for fibrewise ENRs

In this section we present Dold's version of fixed-point theory for fibrewise ENRs and in the next we shall look more generally at the fixed-point theory for fibrewise ANRs. The primary sources are the two papers [47] and [48] of Dold. See also Ulrich's account [132] of the equivariant fibrewise theory. The definitive account of the classical theory on which the fibrewise generalization is based is in [46].

The Lefschetz–Hopf fixed-point index

We are concerned with the following situation. Let U be an open subset of a fibrewise ENR M over B and let $f : U \to M$ be a fibrewise map. The *fixed-point set* of f is the closed subset

$$\mathrm{Fix}(f) := \{x \in U \mid f(x) = x\}$$

of U. We say that the map f is *compactly fixed* if its fixed-point set is compact. (Note the stipulation that $\mathrm{Fix}(f)$ be compact as a space, which is stronger, when B is not compact, than requiring that $\mathrm{Fix}(f) \to B$ be fibrewise compact.)

The primary Lefschetz–Hopf fixed-point index of such a compactly fixed map f is a fibrewise stable map $B \times S^0 \to U_{+B}$, with compact supports.

In defining the index we begin with the case in which M is a trivial vector bundle $B \times E$, where E is a finite-dimensional real normed vector space. For the moment we assume that B is compact. We choose, first, an open neighbourhood V of $\mathrm{Fix}(f)$ in $B \times E$ such that $\overline{V}$ is compact and contained in U, and then a real number $\epsilon > 0$ so small that $\|f(x) - x\| \geq \epsilon$ for all $x \in \overline{V} - V$. (To be precise, the norm and the difference are interpreted on the second component.)

Now let

$$q : (B \times E)^+_B \to (B \times E)^+_B \wedge_B U_{+B}$$

be the fibrewise map defined on the fibrewise one-point compactification of $B \times E$ by

$$q(x) := \begin{cases} * \text{ (the basepoint)} & \text{if } x \notin V, \\ [c(x - f(x)), x] & \text{if } x \in \overline{V}, \end{cases}$$

where $c : E \to E^+$ is given by

$$c(v) = \begin{cases} (\epsilon^2 - \|v\|^2)^{-1/2}\, v & \text{if } \|v\| < \epsilon, \\ * & \text{otherwise.} \end{cases} \qquad (6.1)$$

(Thus c maps the open ball of radius ϵ, centre 0, in E homeomorphically onto E and maps the complement to infinity.) The homotopy class of c (or, more precisely, of its extension to a pointed map $E^+ \to E^+$), and so of q, is clearly independent of the choice of ϵ. It is not difficult to see that the fibrewise homotopy class of q does not depend on the choice of V. For let $V' \subseteq V$ be another open neighbourhood of $\mathrm{Fix}(f)$. Choose ϵ such that $\|f(x) - x\| \geq \epsilon$ for all $x \in \overline{V} - V'$. Then the constructions using V and V' produce the same result.

The stable homotopy class of q is the *Lefschetz–Hopf fixed-point index*

$$\tilde{L}_B(f, U) \in \omega_B^0\{B \times S^0; U_{+B}\}.$$

We shall also need a relative version. Suppose that A is a closed sub-ENR of B and that f has no fixed-points over A: $\mathrm{Fix}(f_A) = \emptyset$. Then the open neighbourhood V used in the construction of the index can be chosen so that $\overline{V}$ lies entirely over the complement of A. The map q is zero over A and defines the *relative index*

$$\tilde{L}_{(B,A)}(f, U) \in \omega_{(B,A)}^0\{B \times S^0; U_{+B}\}.$$

Similar considerations apply when B is possibly non-compact. The compact set $\overline{V}$ projects onto a compact subset of B. The map q is zero outside this compact set and defines the *index with compact supports*

$$_c\tilde{L}_B(f, U) \in {}_c\omega_B^0\{B \times S^0; U_{+B}\}.$$

We turn now to the general case, assuming that B is compact and that A is a closed sub-ENR. The modifications required for a locally compact base and compact supports are straightforward. Let U be an open subset of a fibrewise ENR M over B, and let $f : U \to M$ be a compactly fixed fibrewise map with no fixed-points over A.

Definition 6.2 Choose a fibrewise embedding $i : M \to W$ as a fibrewise retract $r : W \to M$ of an open subspace W of $B \times E$ for some Euclidean space E. Then the *Lefschetz–Hopf fixed-point index*

$$\tilde{L}_{(B,A)}(f, U) \in \omega_{(B,A)}^0\{B \times S^0; U_{+B}\}$$

is defined as follows. Consider the map $i \circ f \circ r : r^{-1}U \to B \times E$. The maps i and r restrict to inverse fibrewise homeomorphisms

$$\mathrm{Fix}(f) \xrightarrow{\; i \;} \mathrm{Fix}(i \circ f \circ r) \xrightarrow{\; r \;} \mathrm{Fix}(f).$$

So the map $i \circ f \circ r$ is compactly fixed and has no fixed-points over A. We define $\tilde{L}_{(B,A)}(f, U)$ to be the image of $\tilde{L}_{(B,A)}(i \circ f \circ r, r^{-1}U)$ under the map

$$\omega^0_{(B,A)}\{B \times S^0; (r^{-1}U)_{+B}\} \to \omega^0_{(B,A)}\{B \times S^0; U_{+B}\}$$

induced by the restriction of $r : r^{-1}U \to U$.

It is not immediately apparent that the index is independent of the choice of embedding of M as a neighbourhood retract in $B \times E$. We shall return to this point after stating the basic properties of the index and proving them for the special case in which M is a trivial vector bundle $B \times E$. The proof will appear as a consequence of the commutativity property of the fixed-point index. However, a direct proof is not hard, and we give such a proof in the next section when dealing with ANRs.

It is an exercise to verify the following properties in the special case.

Proposition 6.3 (Properties of the fixed-point index).
(i)   (Naturality). *Let $\alpha : (B', A') \to (B, A)$ be a map of compact ENR pairs. Then*

$$\tilde{L}_{(B',A')}(\alpha^* f, \alpha^* U) = \alpha^* \tilde{L}_{(B,A)}(f, U).$$

(ii)   (Localization). *Let $U' \subseteq U$ be an open subset with $\mathrm{Fix}(f) \subseteq U'$. Then $\tilde{L}_{(B,A)}(f \mid U', U')$ maps to $\tilde{L}_{(B,A)}(f, U)$ under the inclusion map*

$$\omega^0_{(B,A)}\{B \times S^0; U'_{+B}\} \to \omega^0_{(B,A)}\{B \times S^0; U_{+B}\}.$$

(iii)   (Additivity). *If U is the disjoint union of open sets U_1 and U_2, then*

$$\tilde{L}_{(B,A)}(f, U) = (i_1)_* \tilde{L}_{(B,A)}(f \mid U_1, U_1) + (i_2)_* \tilde{L}_{(B,A)}(f \mid U_2, U_2),$$

where $i_1 : U_1 \to U$ and $i_2 : U_2 \to U$ are the inclusion maps.

(iv)   (Multiplicativity). *Let U' be an open subset of a fibrewise ENR M' over B' and let $f' : U' \to M'$ be a compactly fixed map with no fixed-points over a compact sub-ENR A'. Then the product $f \times f' : U \times_B U' \to M \times_B M'$ is compactly fixed and has no fixed-points over $A \cup A'$ (assumed to be an ENR). Its index is*

$$\tilde{L}_{(B,A \cup A')}(f \times f', U \times_B U') = \tilde{L}_{(B,A)}(f, U) \wedge \tilde{L}_{(B,A')}(f', U').$$

Homotopy invariance of the fixed-point index is implicit in the fibrewise definition.

Proposition 6.4 (Homotopy invariance). *Let $f_t : U \to M$, $0 \le t \le 1$, be a family of fibrewise maps such that $\{(x, t) \in U \times [0, 1] \mid f_t(x) = x\}$ is compact and empty over $A \times [0, 1]$. Then*

$$\tilde{L}_{(B,A)}(f_t, U) \in \omega^0_{(B,A)}\{B \times S^0; U_{+B}\}$$

is independent of t.

The property of commutativity is more subtle.

Lemma 6.5 (Commutativity). *Let M and N be fibrewise ENRs over B, and let $U \subseteq M$ and $V \subseteq N$ be open subsets. Suppose that $f : V \to M$ and $g : U \to N$ are fibrewise maps such that the restriction of $g \circ f : f^{-1}U \to N$ is compactly fixed with no fixed-points over the closed sub-ENR A. Then $f \circ g : g^{-1}V \to M$ is also compactly fixed with no fixed-points over A, and*

$$(f \times 1)_* \tilde{L}_{(B,A)}(gf|, f^{-1}U) = (1 \times g)_* \tilde{L}_{(B,A)}(fg|, g^{-1}V)$$
$$\in \omega^0_{(B,A)}\{B \times S^0; (U \times_B V)_{+B}\}.$$

Notice at once that we have inverse fibrewise homeomorphisms

$$\text{Fix}(g \circ f) \underset{g}{\overset{f}{\rightleftarrows}} \text{Fix}(f \circ g).$$

Before looking at the proof of commutativity, we use the result, for the special case of open subspaces of $B \times E$, to show that the index is well-defined. Suppose, in Definition 6.2, that $i' : M \to W' \subseteq E'$, $r' : W' \to M$, is another representation of M as a retract of an open subspace of a trivial vector bundle. We apply Lemma 6.5 to the maps

$$i' \circ f \circ r : r^{-1}U \to B \times E' \quad \text{and} \quad i \circ r' : r'^{-1}U \to B \times E.$$

The composition $(i \circ r')(i' \circ f \circ r)$ is $i \circ f \circ r : r^{-1}(U \cap f^{-1}U) \to B \times E$, and the composition $(i' \circ f \circ r)(i \circ r')$ is $i' \circ f \circ r' : r'^{-1}U \to B \times E'$.

By the localization property, Proposition 6.3(ii), the inclusion of the subspace $r^{-1}(U \cap f^{-1}U)$ in $r^{-1}U$ sends $\tilde{L}_{(B,A)}(ifr, r^{-1}(U \cap f^{-1}U))$ to $\tilde{L}_{(B,A)}(ifr, r^{-1}U)$. Now from commutativity, ir' sends $\tilde{L}_{(B,A)}(i'fr', r'^{-1}U)$ to $\tilde{L}_{(B,A)}(ifr, r^{-1}U)$. Since $r(ir') = r'$, the image of $\tilde{L}_{(B,A)}(ifr, r^{-1}U)$ under r coincides with the image of $\tilde{L}_{(B,A)}(i'fr', r'^{-1}U)$ under r'. This completes the proof that the fixed-point index is well-defined.

It is implicit in the definition of the fixed-point index that it is topologically invariant, that is, if $g : M \to M'$ is a fibrewise homeomorphism between fibrewise ENRs over B, then

$$\tilde{L}_{(B,A)}(gfg^{-1}, gU) \in \omega^0_{(B,A)}\{B \times S^0; (gU)_{+B}\}$$

is the image under g of $\tilde{L}_{(B,A)}(f, U)$.

Another consequence of commutativity is:

Corollary 6.6 *Suppose that M is a fibrewise sub-ENR of a fibrewise ENR N and that $f : U \to M$ extends to a fibrewise map (denoted by the same symbol) $f : V \to M$ on an open subspace V of N such that $V \cap M = U$. Then*

$$\tilde{L}_{(B,A)}(f : U \to M, U) \in \omega^0_{(B,A)}\{B \times S^0; U_{+B}\}$$

maps by the inclusion to the index

$$\tilde{L}_{(B,A)}(f : V \to N, V) \in \omega^0_{(B,A)}\{B \times S^0; V_{+B}\}$$

of the composition (again denoted by f) $V \to M \to N$ of the extension f and the inclusion of the sub-ENR.

This follows easily from Lemma 6.5 applied to the extension $f : V \to M$ and the inclusion $g : U \to N$ of U as a subspace of N.

It remains to establish commutativity in the Euclidean case $M = B \times E$, $N = B \times F$, for some finite-dimensional vector spaces E and F. Consider the map h

$$(x, y) \mapsto (f(y), g(x)) : U \times_B V \to B \times (E \times F),$$

defined on the open subspace $U \times_B V$ of $B \times (E \times F)$. (We make the natural identification of $(B \times E) \times_B (B \times F)$ with $B \times (E \times F)$.) Clearly, (x, y) is a fixed-point of h if and only if $x = f(g(x))$ is a fixed-point of $f \circ g$ and $y = g(x)$ is a fixed-point of $g \circ f$, and we have a fibrewise homeomorphism $x \mapsto (x, g(x))$ from $\text{Fix}(f \circ g)$ to $\text{Fix}(h)$. The lemma will be proved by showing that the fixed-point index of h is equal to $(1 \times g)_* \tilde{L}_{(B,A)}(fg|, g^{-1}V)$. This will follow from homotopy invariance in two steps.

Consider first the linear homotopy $k_t : g^{-1}V \times_B V \to B \times (E \times F)$ given by

$$k_t(x, y) = ((1 - t)f(y) + tf(g(x)), g(x)), \quad 0 \le t \le 1.$$

If (x, y) is fixed by k_t, then $y = g(x)$ and $x = f(y)$. So $\text{Fix}(k_t) = \text{Fix}(h)$. By Lemma 6.4, the fixed-point index of $h = k_0$ is equal to the fixed-point index of $k_1 : (x, y) \mapsto (f(g(x)), g(x))$.

In the second step we shall deform the second component to a constant map. Some care is necessary in order to keep within $g^{-1}V \times_B V$. (It is easy to write down the deformation in $g^{-1}V \times F$.) Now we may choose an open neighbourhood U' of $\text{Fix}(f \circ g)$ in $g^{-1}V$ and a small open disc D centred at 0 in F such that

$$W := \{(x, g(x) + v) \mid x \in U', v \in D\}$$

is a neighbourhood of $\text{Fix}(h)$ in $U' \times_B V$. Let $i_t : U' \times D \to B \times (E \times F)$ be embeddings defined by

$$i_t(x, v) = (x, tg(x) + v), \quad 0 \le t \le 1.$$

Thus, i_1 is a homeomorphism to the subspace W. We think of each i_t as the inclusion of $U' \times D$ as a subspace of $B \times (E \times F)$ and define on that subspace a map $l_t : U' \times D \to B \times (E \times F)$ by

$$l_t(x, v) = (f(g(x)), tg(x)).$$

The fixed subspace of l_t is the subset

$$\{(x,v) \mid l_t(x,v) = i_t(x,v)\} = \text{Fix}(f \circ g) \times \{0\}.$$

Now the fixed-point index of l_1 corresponds (under the homeomorphism between $U' \times D$ and W) to that of k_1 and is homotopic to the fixed-point index of l_0. The index of l_0 is easily seen directly to coincide with the index of $f \circ g$. See also Proposition 6.7. This completes the verification of commutativity in the special case.

Having established the various properties in the Euclidean case and so justified the general definition of the fixed-point index, it is routine to deduce the properties in full generality.

Finally we record, as a special case of Lemma 6.5 or by inspection of the definition:

Proposition 6.7 *Let $s : B \to M$ be a section of the fibrewise ENR M and let U be an open neighbourhood of $s(B)$. Consider the fibrewise map $f : U \to M$ taking the constant value $s(b)$ in the fibre at $b \in B$. Then $\bar{L}_B(f, U)$ is equal to the stable class $s_* : B \times S^0 \to U_{+B}$ of the section.*

Remark 6.8. Suppose that M is embedded as a fibrewise retract of an open subspace of a finite-dimensional vector bundle ξ over B. Then the fixed-point index can be constructed by working in ξ, instead of the trivial bundle $B \times E$ used in Definition 6.2, as illustrated below in a special case. See also the related definition of the vector field index in Section 12, Remark 12.26.

Finite coverings

It is instructive to look at the fixed-point index for a map defined on a finite covering. This corresponds in the classical theory to looking at fixed-point theory for finite sets.

Let $M \to B$ be a finite d-fold covering, and suppose that we have a fibrewise embedding $M \to \xi$ into a finite-dimensional vector bundle ξ. (See Section 11 for a general discussion of such embeddings.) Let $f : M \to M$ be a fibrewise map. Then we can describe the construction of the fixed-point index quite concretely. Choose a Euclidean metric on ξ. Then, by the compactness of M, there is a real number $\epsilon > 0$ such that distinct points in any fibre of M are a distance greater than 2ϵ apart, so that the closed ϵ-discs in ξ centred on the points of M are disjoint. Let F denote the set of fixed-points of f; it is, of course, open in M. Let V be the set of points of ξ at a distance less than ϵ from M. Then $\bar{L}_B(f, F)$ is represented by the fibrewise map

$$q : \xi_B^+ \to \xi_B^+ \wedge_B F_{+B}$$

which maps the complement of V to the basepoint (in the fibre) and maps the open ϵ disc centred on $x \in M$ homeomorphically to the fibre of ξ, by the map c defined at the beginning of this section, with second component

x, when x is fixed by f, and collapses the open disc to the basepoint when $f(x) \neq x$.

ENR fixed-point sets

The fixed-point index gives, by the localization property, a system of compatible fibrewise stable maps $B \times S^0 \to V_{+B}$ for the family of open subsets V of U which contain the fixed-point set $\mathrm{Fix}(f)$. In general this fixed-point set, which again we denote temporarily by F, will be unpleasant as a topological space. Let us suppose that $F \to B$ is actually a fibrewise ENR.

Choose some open neighbourhood V of F in U which retracts onto F, say by $r : V \to F$. Then we define

$$M_{(B,A)}(f) \in \omega^0_{(B,A)}\{B \times S^0; \mathrm{Fix}(f)_{+B}\} \tag{6.9}$$

to be the image of $\tilde{L}_{(B,A)}(f \,|\, V, V)$ under r.

This class is independent of the choice of V and r. For, by Proposition 5.15, two retractions, extending the identity on F, will be fibrewise homotopic on some smaller open neighbourhood of F. Moreover, the class $M_{(B,A)}(f)$ determines the Lefschetz–Hopf index.

Proposition 6.10 *Suppose that* $\mathrm{Fix}(f)$ *is a fibrewise ENR over B. Then the inclusion map sends* $M_{(B,A)}(f)$ *to* $\tilde{L}_{(B,A)}(f, U)$:

$$\omega^0_{(B,A)}\{B \times S^0; \mathrm{Fix}(f)_{+B}\} \to \omega^0_{(B,A)}\{B \times S^0; U_{+B}\}.$$

For we can choose some smaller open neighbourhood W of F in V such that

$$W \xrightarrow{r|W} F \xrightarrow{i} V$$

is fibrewise homotopic to the inclusion of W in V.

The Nielsen–Reidemeister index

We return to the general situation of a compactly fixed map $f : U \to M$ defined on an open subset of a fibrewise ENR M over a compact ENR B and having no fixed-points over a closed sub-ENR A.

Definition 6.11 The *fibrewise homotopy fixed-point set* of f is the subspace h-$\mathrm{Fix}(f)$ of $\mathcal{P}_B U = \mathrm{map}_B(B \times [0,1], U)$ with fibre at $b \in B$ consisting of the paths $\omega : [0,1] \to U_b$ such that $\omega(1) = f_b(\omega(0))$, that is, paths joining a point of U_b to its image under f_b.

The homotopy fixed-point set h-$\mathrm{Fix}(f)$ contains $\mathrm{Fix}(f)$ as the space of constant paths, and evaluation at 0 defines a map $q : \text{h-}\mathrm{Fix}(f) \to U$. Using the fibrewise uniform local contractibility of M, we shall define the Nielsen–Reidemeister index of f as a class

$$N_{(B,A)}(f,U) \in \omega^0_{(B,A)}\{B \times S^0; \text{h-Fix}(f)_{+B}\} \qquad (6.12)$$

lifting $\tilde{L}_{(B,A)}(f,U)$.

Choose an open neighbourhood W of the diagonal in $M \times_B M$ and homotopy $H_t : W \to M$, as in Proposition 5.11, such that $H_0(x,y) = x$, $H_1(x,y) = y$ and $H_t(x,x) = x$, for $x, y \in M_b$. Let $V \subseteq U$ be the inverse image of W under the map $x \mapsto (x, f(x)) : U \to M \times_B M$. Then we have a fibrewise map $V \to \text{h-Fix}(f)$ taking x to the path $t \mapsto H_t(x, f(x))$ from x to $f(x)$. It extends the embedding of $\text{Fix}(f)$ in the homotopy fixed-point set. The usual arguments, following Proposition 5.15, show that two fibrewise maps constructed in this way will be homotopic on some neighbourhood of $\text{Fix}(f)$. Hence, the image of $\tilde{L}_{(B,A)}(f|V,V)$ under this map is independent of choices. We call it the *Nielsen–Reidemeister index* (or, more accurately, the Lefschetz–Hopf–Nielsen–Reidemeister index).

The classical indices

Let us look at the classical situation in which the base B is reduced to a point. Consider an ENR M and compactly-fixed map $f : U \to M$ defined on an open subspace U. Then $\tilde{L}(f,U)$ is an element of the stable homotopy group $\omega_0(U) = \omega^0\{S^0; U_+\}$, which is the free Abelian group generated by the path-components of U. The Lefschetz–Hopf index assigns an integer to each path-component, with only finitely many non-zero (since only finitely many path-components intersect the compact fixed-point set). The sum of the integers associated to the components is the classical Lefschetz fixed-point index. This is traditionally defined via integral homology as an element of the cohomology ring $H^0(*) = \mathbb{Z}$ of a point. (We refer again to [46].) From the fibrewise (or equivariant) viewpoint it is better regarded as an element of the stable cohomotopy ring $\omega^0(*)$. The Nielsen–Reidemeister index $N(f,U) \in \omega_0(\text{h-Fix}(f))$ likewise assigns an integer to each path-component of the homotopy-fixed-point set and is usually called the *Reidemeister trace*. The classical *Nielsen number* is the number of components for which the corresponding index is non-zero.

The Lefschetz index

In the classical situation, the distinction between the Lefschetz–Hopf index containing information about the location of the fixed-points and the Lefschetz index which is the sum of the local indices is of minor interest; it does become significant in the fibrewise theory. Returning now to the general setting we construct the Lefschetz index $L_{(B,A)}(f,U)$ from $\tilde{L}_{(B,A)}(f,U)$ as an element of the stable cohomotopy group $\omega^0(B, A)$ of the base by a procedure which reduces in the classical case to adding the local integer indices.

Definition 6.13 We define the *Lefschetz index*

$$L_{(B,A)}(f, U) \in \omega^0_{(B,A)}\{B \times S^0; B \times S^0\} = \omega^0(B, A)$$

as the image of $\tilde{L}_{(B,A)}(f, U)$ under the fibrewise map $U_{+B} \to B \times S^0 = B_{+B}$ projecting U to B.

When $A = \emptyset$ is empty, the Lefschetz index $L_B(f, U)$ lies in the stable cohomotopy ring $\omega^0(B)$. When B is non-compact we can define an index $_cL_B(f, U) \in {}_c\omega^0(B)$ in the compactly supported group.

The fixed-point transfer

Another specialization of the Lefschetz–Hopf index $\tilde{L}_{(B,A)}(f, U)$, which is a fibrewise stable map $B \times S^0 \to U_{+B}$ (which is zero over the subspace A), is obtained by collapsing the basepoints B and the subspace A to produce a stable map

$$B/A \to U_+,$$

which is called the *fixed-point transfer* of the map f.

In the special case when $\text{Fix}(f)$ is an ENR (not necessarily a fibrewise ENR over B), the transfer factors canonically through a stable map

$$B/A \to \text{Fix}(f)_+.$$

The reasoning is just as in the definition of $M_{(B,A)}(f, U)$ above. This construction has been exploited by Richter to give a stable splitting of $\Omega SU(n)$. See [35], p. 102.

The transfer construction is particularly important when M is compact and f is the identity map (and A is necessarily empty). In that case we refer simply to the *transfer* $B_+ \to M_+$ of the compact fibrewise ENR $M \to B$.

Collapsing basepoints to define the transfer does, however, lose information. We next examine more carefully the fibrewise nature of the construction.

The fibrewise transfer

Let $\alpha : B' \to B$ be a compact fibrewise ENR over B. The associated transfer is a stable map in the opposite direction $B_+ \to B'_+$. But the fixed-point index $\tilde{L}_B(1, B')$ is a *fibrewise* stable map

$$B \times S^0 \ (= B_{+B}) \to B'_{+B},$$

which we might call the *fibrewise transfer* of the map α. Let $A \subseteq B$ be a compact sub-ENR and write $A' := \alpha^{-1}(A)$. Consider fibrewise pointed spaces X and Y over B. We shall define a *transfer homomorphism*

$$\alpha_{\sharp} : \omega^*_{(B',A')}\{\alpha^*X; \alpha^*Y\} \to \omega^*_{(B,A)}\{X; Y\}. \tag{6.14}$$

We need a generalization of Proposition 3.4.

Lemma 6.15 *There is a natural isomorphism:*

$$\omega^*_{B'}\{B' \times_B X; B' \times_B Y\} \to \omega^*_B\{B'_{+B} \wedge_B X; Y\}.$$

The essential insight is in the evident equivalence between fibrewise maps $B' \times_B M \to B' \times_B N$ over B' and fibrewise maps $B' \times_B M \to N$ over B, for any fibrewise spaces M and N over B. This is a special case of the adjoint property of the direct image, (Proposition 5.37). For $B'_{+B} \wedge_B X$ is $\alpha_*(\alpha^*X)$.

Now compose with the fibrewise transfer smashed with the identity on X to get the required transfer homomorphism

$$\omega^*_{B'}\{B' \times_B X; B' \times_B Y\} \to \omega^*_B\{B'_{+B} \wedge_B X; Y\} \to \omega^*_B\{X; Y\}.$$

The relative case, using the identification of

$$\omega^*_{(B',A')}\{B' \times_B X; B' \times_B Y\} \quad \text{with} \quad \omega^*_{(B,A)}\{B'_{+B} \wedge_B X; Y\},$$

is scarcely more difficult.

Proposition 6.16 (Properties of the transfer homomorphism).
(i) (Functoriality). *Let* $\alpha : B' \to B$ *and* $\beta : B'' \to B'$ *be compact fibrewise ENRs, and let* A *be a closed sub-ENR of* B. *Write* $A' = \alpha^{-1}(A)$ *and* $A'' = \beta^{-1}(A')$. *Then* $(\alpha \circ \beta)_{\sharp} = \alpha_{\sharp} \circ \beta_{\sharp}$:

$$\omega^*_{(B'',A'')}\{\beta^*\alpha^*X; \beta^*\alpha^*Y\} \to \omega^*_{(B',A')}\{\alpha^*X; \alpha^*Y\} \to \omega^*_{(B,A)}\{X; Y\},$$

for any fibrewise pointed spaces X *and* Y *over* B.
(ii) (Frobenius reciprocity). *Let* $\alpha : B' \to B$ *be a compact fibrewise ENR,* A *a closed sub-ENR of* B *and* $A' = \alpha^{-1}(A)$.
    (a) *Consider fibrewise pointed spaces* X, Y, X' *and* Y' *over* B. *We have*

$$\alpha_{\sharp}(\alpha^*(x) \wedge y) = x \wedge \alpha_{\sharp}(y) \in \omega^*_{(B,A)}\{X \wedge_B X'; Y \wedge_B Y'\}$$

for $x \in \omega^*_{(B,A)}\{X; Y\}$, $y \in \omega^*_{B'}\{\alpha^*X; \alpha^*Y\}$.
    (b) *Consider fibrewise pointed spaces* X, Y *and* Z *over* B. *We have*

$$\alpha_{\sharp}(\alpha^*(x) \circ y) = x \circ \alpha_{\sharp}(y) \in \omega^*_{(B,A)}\{X; Z\}$$

for $x \in \omega^*_{(B,A)}\{Y; Z\}$, $y \in \omega^*_{B'}\{\alpha^*X; \alpha^*Y\}$.

In the statement of Frobenius reciprocity – the terminology is adapted from the representation theory of finite groups – we have formulated only one of several variants: in (b) the order of composition can be reversed, and in (a) and (b) we could multiply a class x over B by a class y over (B', A') (with $A' = \alpha^{-1}(A)$ as usual). Verification of (ii) is straightforward; it amounts to checking that the equivalence in Lemma 6.15 is compatible with multiplication by classes defined over B.

Functoriality, (i), requires more work. It follows from a special case (f the identity, $M = B''$) of the next result.

Proposition 6.17 *Let* $\alpha : B' \to B$ *be a compact fibrewise ENR, and let* $M \to B'$ *be a fibrewise ENR over* B'. *Suppose that* $f : U \to M$ *is a compactly fixed fibrewise map over* B' *defined on an open subset* U *of* M. *Let* A *be a closed sub-ENR of* B *such that* f *has no fixed-points over* $A' := \alpha^{-1}(A)$.

Then $\alpha_* f : \alpha_* U \to \alpha_* M$ *(that is,* $f : U \to M$ *considered as a map over* B *via* $\alpha : B' \to B$*) is compactly fixed and its index is given by*

$$\tilde{L}_{(B,A)}(\alpha_* f, \alpha_* U) = \alpha_\sharp(\tilde{L}_{(B',A')}(f, U)) \in \omega^0_{(B,A)}\{B \times S^0; (\alpha_* U)_{+B}\}.$$

The proof is just a matter of writing down the definitions of the two sides of the identity carefully and comparing the results. The case in which $B' \to B$ is a trivial bundle is straightforward, and is a special case of the multiplicativity of the index, Proposition 6.3(iv).

There is no real loss of generality, because of the way we defined the fixed-point index, in assuming that $M = B' \times F$ for some Euclidean space F. Choose a representation $i : B' \to W$, $r : W \to B'$, of B' as a fibrewise retract over B of an open subspace W of $B \times E$, for some Euclidean space E. It is convenient to suppose that i is the inclusion of B' as a subspace of E. Then the fibrewise transfer $\alpha_\sharp$ is given by the fixed-point index of r. For the purposes of exposition we lose very little by assuming that B is a point.

Choose an open neighbourhood T of $\mathrm{Fix}(f)$ in B' with compact closure $\overline{T}$ contained in U and an open neighbourhood V of B' in E with compact closure $\overline{V}$ contained in W. Then

$$S := \{(x, y) \in V \times F \mid (r(x), y) \in T\}$$

is an open neighbourhood of $\mathrm{Fix}(f) = \mathrm{Fix}(f \circ (r \times 1))$ in the intersection $(r \times 1)^{-1}(U) \cap (V \times F)$. And its closure, being contained in $\overline{V} \times K$, where K is the image of the projection $\overline{T} \to F$ to the second factor, is compact. Let us write f in components as: $f(x, y) = (x, f'(x, y))$. Using a subscript on the norm to indicate the vector space considered, choose $\epsilon > 0$ such that

$$\|x - r(x)\|_E \geq \epsilon \quad \text{for} \quad x \in \overline{V} - V,$$

$$\|y - f'(x, y)\|_F \geq \epsilon \quad \text{for} \quad (x, y) \in \overline{T} - T,$$

$$\text{and} \quad \|(x, y) - f(r(x), y)\|_{E \times F} \geq \epsilon \quad \text{for} \quad (x, y) \in \overline{S} - S.$$

We use the supremum norm on $E \times F$.

Now the index $\tilde{L}_{(B',A')}(f,U)$ is constructed as a map $B' \times F^+ \to U_{+B'}$ over B'. Collapsing the basepoints B' we obtain a map

$$B'_{+B} \wedge_B (B \times F^+) \to U_{+B} \wedge_B (B \times F^+)$$

given by

$$(x,y) \mapsto \begin{cases} * & \text{if } (x,y) \notin T, \\ [f(x,y), x, c_F(y - f'(x,y))] & \text{if } (x,y) \notin \overline{T}. \end{cases} \qquad (6.18)$$

Remember that we are omitting the base B from the notation, and so think of (x,y) as an element of $E \times F$. The subscript 'F' on c_F indicates the construction c of (6.1), given by the norm on F, in the second factor.

The transfer $\alpha_{\sharp}$ is determined by the map $B_+ \wedge E^+ \to B'_+ \wedge E^+$:

$$x \mapsto \begin{cases} * & \text{if } x \notin V, \\ [r(x), c_E(x - r(x))] & \text{if } x \in \overline{V}. \end{cases}$$

Smashing with the identity on F^+ and composing with (6.18), we obtain a map: $B_+ \wedge E^+ \wedge F^+ \to U_{+B} \wedge E^+ \wedge F^+$ given by

$$(x,y) \mapsto \begin{cases} * & \text{if } x \notin S, \text{ that is, } x \notin V \text{ or } (r(x),y) \notin T, \\ [f(r(x),y), c_E(x - r(x)), c_F(y - f'(r(x),y))] & \text{if } x \in \overline{S}. \end{cases}$$

Noting that $c_{E \times F}(u,v) = [c_E(u), c_F(v)]$, at least up to homotopy, we see that this gives the fixed-point index of f over B as asserted by the proposition.

7 Fixed-point theory for fibrewise ANRs

In this section we examine the extension of the Lefschetz fixed-point theory from finite to infinite dimensions, that is, from Euclidean to absolute neighbourhood retracts. Our primary source is the paper [67] of Granas. As usual, unless otherwise indicated, the base space B is an ENR.

Compactly fixed maps

Let $M \to B$ be a fibrewise ANR, U be an open subset of M and $f : U \to M$ be a fibrewise map.

Definition 7.1 We say that f is *compactly fixed* if the fixed-point set $\text{Fix}(f)$ is compact and there is an open neighbourhood V of $\text{Fix}(f)$ in U, such that $f(V)$ has compact closure in M.

We record a few elementary properties of such a compactly fixed map f and open set V. Clearly any open neighbourhood $V' \subseteq V$ also has the property that the closure of the image $f(V')$ is compact. There is, therefore, no loss of generality in supposing that $\overline{V} \subseteq U$ so that f is defined on $\overline{V}$. Now $f(\overline{V})$ and $f(V)$ have the same closure, and the projection $p(\overline{V})$ has compact closure in the base B.

This definition is consistent with our simpler definition for a fibrewise ENR: in that case $\overline{V}$ can be chosen to be itself compact.

For any metric d on M, the distance $d(x, f(x))$ is bounded away from 0 on $\overline{V} - V$. Indeed, suppose not. Then, by the compactness of the closure of $f(V)$, there is a sequence (x_n) in $\overline{V} - V$ such that $d(x_n, f(x_n)) \to 0$ and $f(x_n)$ converges to some point $y \in M$. It follows that $x_n \to y$ and so that $f(y) = y$. This contradicts the fact that f has no fixed-points in the closed set $\overline{V} - V$.

Finite-dimensional approximation

As in the fixed-point theory for ENRs, we begin with the case that M is a trivial vector bundle: $M = B \times E$ for some normed vector space E. Let U be an open subspace of $B \times E$ and let $f : U \to B \times E$ be a compactly fixed map over B. The index will be defined via finite-dimensional approximations to f.

Choose an open neighbourhood V of $\mathrm{Fix}(f)$ in U such that the closure $\overline{V}$ in $B \times E$ is contained in U and $f(V)$ has compact closure. From the discussion above, $\|f(x) - x\|$ (that is, to be precise, the difference in norms of the projections onto E) is bounded away from 0 on $\overline{V} - V$. Choose $\epsilon > 0$ such that $\|f(x) - x\| \geq \epsilon$ for all $x \in \overline{V} - V$.

By the compactness of the closure of $f(V)$, there exist finitely many points $y_1, \ldots, y_n$ in E such that

$$f(\overline{V}) \subseteq B \times \bigcup_{1 \leq j \leq n} B_\epsilon(y_j),$$

where $B_\epsilon(y)$ is as usual the open ball of radius ϵ in E with centre y. Since $\overline{V}$ is metrisable, it is easy to write down, in terms of a metric, a partition of unity subordinate to any finite open cover. Let $(\phi_j)_{1 \leq j \leq n}$ be a partition of unity subordinate to the cover by the open sets $f^{-1}(B \times B_\epsilon(y_j))$, $1 \leq j \leq n$.

Now let F be a (necessarily closed) finite-dimensional vector subspace of E containing $y_1, \ldots, y_n$. We define $h : \overline{V} \to F$ by

$$h(x) = \sum_{1 \leq j \leq n} \phi_j(x) y_j.$$

Because $\phi_j(x) = 0$ unless $\|f(x) - (b, y_j)\| < \epsilon$, we have

$$\|f(x) - (p(x), h(x))\| < \epsilon,$$

where p is, as usual, the projection.

Remark 7.2. Incidentally, h maps into the closed convex hull K of the points $y_1, \ldots, y_n$ in F, and K is a finite polyhedron.

We can now define the required finite-dimensional approximation to f. Let

$$g : \overline{V} \to B \times F$$

be the map $g(x) = (p(x), h(x))$. We have $\|f(x) - g(x)\| < \epsilon$ for all $x \in V$. And for all $x \in \overline{V} - V$, $\|x - g(x)\| > 0$ by the choice of ϵ so that $\|x - f(x)\| \geq \epsilon$. Hence the restriction of g to the open subset $V \cap (B \times F)$ of $B \times F$

$$g| : V \cap (B \times F) \to B \times F$$

is compactly fixed. (For the fixed-point set is closed and contained in the compact set $\overline{p(V)} \times K$.)

We define the fixed-point indices of $f : U \to B \times E$ in terms of the fixed-point indices of $g| : V \cap (B \times F) \to B \times F$. If B is compact and f has no fixed-points over a closed sub-ENR A we can choose V so that $\overline{V}_A$ is empty and define

$$\tilde{L}_{(B,A)}(f, U) = j_* \tilde{L}_{(B,A)}(g|, V \cap (B \times F)) \in \omega^0_{(B,A)}\{B \times S^0; U_{+B}\},$$

the image under the inclusion $j : V \cap (B \times F) \to U$. If B is not compact,

$$_c\tilde{L}_B(f, U) \in {_c\omega^0_B}\{B \times S^0; U_{+B}\}$$

is defined as the image under the inclusion map of $_c\tilde{L}_B(g|, V \cap (B \times F))$.

This definition has involved several choices. We must verify that the result is independent of the choice of the approximation h (and the vector space F) for given V, and then check that it is unchanged if we replace V by any open subset V' with $\mathrm{Fix}(f) \subseteq V' \subseteq V$.

Suppose first of all that we replace F by a finite-dimensional subspace F' of E containing F. Then the index defined by $g| : V \cap (B \times F') \to B \times F'$ is equal to that defined by $g| : V \cap (B \times F) \to B \times F$, by Corollary 6.6.

Next, suppose that $g' : \overline{V} \to B \times F$ is another approximation to f, satisfying $\|f(x) - g'(x)\| < \epsilon$ for all $x \in \overline{V}$. We have a linear homotopy $g_t = (1-t)g + tg'$, $0 \leq t \leq 1$, from g to g', and $\|g_t(x) - f(x)\| < \epsilon$, by convexity. Hence, $\|g_t(x) - x\| > 0$ for $x \in \overline{V} - V$. This means that we have a compactly fixed homotopy $g_t| : V \cap (B \times F) \to B \times F$, and the indices determined by $g_0|$ and $g_1|$ are equal, by the homotopy invariance (Proposition 6.4) of the Lefschetz index.

Consider, finally, an open neighbourhood V' of $\mathrm{Fix}(f)$ in V. For ϵ sufficiently small, we have $\|f(x) - x\| \geq \epsilon$ for all $x \in \overline{V} - V'$. Then we can compute the index on $V' \cap (B \times F)$.

This completes the definition of the fixed-point index for a map defined on an open subset of $B \times E$. We turn now to the general case.

The definition of the fixed-point index

Let $M \to B$ be a fibrewise ANR over B and $f : U \to M$ be a compactly fixed fibrewise map defined on an open subspace U. To define the fixed-point index of f we choose an embedding of M as a fibrewise retract of an open subset W of $B \times E$ for some normed vector space E: $i : M \to W$, $r : W \to M$, $r \circ i = 1$. The map $ifr : U' = r^{-1}(U) \to B \times E$ is compactly fixed. Indeed, $\text{Fix}(ifr) = i(\text{Fix}(f))$ is compact (so closed in $B \times E$), and, if $V \subseteq U$ is an open neighbourhood of $\text{Fix}(f)$ such that $f(V)$ has compact closure in M, then $r^{-1}(V)$ is an open neighbourhood of $\text{Fix}(ifr)$ such that $(ifr)(r^{-1}V)$ has compact closure in $B \times E$.

We define the *Lefschetz–Hopf fixed-point index* to be

$$\tilde{L}_{(B,A)}(f, U) = r_* \tilde{L}_{(B,A)}(ifr, r^{-1}(U)),$$

in the case where B is compact and f has no fixed-points over a closed sub-ENR A, and

$$_c\tilde{L}_B(f, U) = r_{*c}\tilde{L}_B(ifr, r^{-1}(U)),$$

where B is (not necessarily) compact.

Verification that the index is well-defined

In order to show that this is a good definition we need to check that the result is independent of the choice of the representation of M as a fibrewise retract of an open subspace of a trivial bundle $B \times E$. Homotopy invariance, the essential input, is intrinsic to our fibrewise construction.

We begin by observing that the index depends only on the embedding $i : M \to B \times E$, not on the choice of open neighbourhood W and retraction $r : W \to M$. First, if we replace W by an open subspace W' containing $i(M)$ and use the restriction of r to $W' \to M$, we can choose V above so that $\overline{V} \subseteq W'$, and the outcome is unchanged. Consider next another retraction $r' : W \to M$. By Proposition 5.15, since $i(M)$ is closed in W, the restrictions of r and r' to some open neighbourhood of $i(M)$ are homotopic. The construction of the index over $B \times [0, 1]$ establishes homotopy invariance: the indices defined using r and r' coincide.

Now let E' be a second normed vector space. We have a fibrewise embedding $i_0 : M \to B \times (E \times E')$ with first component i and second component the constant $0 \in E'$. As an open neighbourhood of $i_0(M)$ we can take $W \times E'$, with retraction r_0 given by the composition of the projection $W \times E' \to W$ and $r : W \to M$. The index defined using i_0 agrees with that defined using i, again by Corollary 6.6.

Finally, suppose that we have a second embedding $i' : M \to W'$ and retraction $r' : W' \to M$, where W' is an open subspace of $B \times E'$. Define a family of embeddings

$$i_t : M \to B \times (E \times E'), \qquad 0 \le t \le 1,$$

by $i_t(x) = ((1 - t)i(x), ti'(x))$. This determines a fibrewise embedding $j : M \times [0,1] \to (B \times [0,1]) \times (E \times E')$ over $B \times [0,1]$. If we can show that $j(M \times [0,1])$ is a fibrewise retract of an open neighbourhood, then we can construct the index over $B \times [0,1]$ and deduce that the index given by i_0, which according to the preceding paragraph coincides with that defined by i, is the same as that given by i_1, or i'. We cannot apply Proposition 5.10 directly to see that $j(M \times [0,1])$ is a neighbourhood retract in $B \times (E \times E')$, since it need not be a closed subspace. But it is closed in an open subspace, for example in the union of the open subspaces $\{((1 - t)x, y, t) \mid x \in W, y \in E', 0 \le t < 1\}$ and $\{(x, ty, t) \mid x \in E, y \in W', 0 < t \le 1\}$.

This completes the proof that the fixed-point indices are well-defined and so topologically invariant.

Properties of the fixed-point index

The basic properties of the Lefschetz–Hopf fixed-point index carry through by finite-dimensional approximation from the ENR theory to ANRs. We refer to Propositions 6.3 and 6.4 for precise statements.

Proposition 7.3 *The Lefschetz–Hopf fixed-point index*

$$\tilde{L}_{(B,A)}(f, U) \in \omega^0_{(B,A)}\{B \times S^0; U_{+B}\}$$

of a compactly fixed fibrewise map $f : U \to M$ defined on an open subset U of a fibrewise ANR M over a compact ENR B and having no fixed-points over the closed sub-ENR A enjoys the properties: (i) *Naturality,* (ii) *Localization,* (iii) *Additivity,* (iv) *Multiplicativity,* (v) *Homotopy invariance.*

The formulation of a commutativity property requires a little more care; the proof proceeds again by finite-dimensional approximation.

Proposition 7.4 (Commutativity). *Let M and N be fibrewise ANRs over B, and let $U \subseteq M$ and $V \subseteq N$ be open subsets. Suppose that $f : V \to M$ and $g : U \to N$ are fibrewise maps such that the map*

$$U \times_B V \to M \times N \quad : \quad (x, y) \mapsto (f(y), g(x))$$

is compactly fixed with no fixed-points over the closed sub-ENR A. Then

$$(f \times 1)_* \tilde{L}_{(B,A)}(gf|, f^{-1}U) = (1 \times g)_* \tilde{L}_{(B,A)}(fg|, g^{-1}V)$$
$$\in \omega^0_{(B,A)}\{B \times S^0; (U \times_B V)_{+B}\}.$$

The compactness condition can be reformulated as follows. The (homeomorphic) fixed-point sets $\mathrm{Fix}(g \circ f)$ of $g \circ f : f^{-1}U \to N$ and $\mathrm{Fix}(f \circ g)$ of

$f \circ g : g^{-1}V \to M$ are compact and admit neighbourhoods $P \subseteq f^{-1}(U)$ and $Q \subseteq g^{-1}(V)$ such that $f(P)$ and $g(Q)$ have compact closure in N and M respectively.

Observe that we cannot deduce from this commutativity property, as we did for ENRs in Section 6, that the index is well-defined, because the compactness condition is not, in general, satisfied by the inclusion maps which are involved. It is the following *contraction property* which lies behind the definition.

Proposition 7.5 *Let* $q : N \to M$ *be a fibrewise map between fibrewise ANRs over a compact ENR* B, *and let* $k : U \to N$ *be a fibrewise map defined on an open subset* U *of* M. *Write* $V = q^{-1}(U) \subseteq N$, $f = q \circ k : U \to M$ *and* $g = k \circ q| : V \to N$, *as illustrated in the diagram:*

$$
\begin{array}{ccccc}
q^{-1}U & = & V & \xrightarrow{\ g\ } & N \\
& & {\scriptstyle q|}\Big\downarrow & {\scriptstyle k}\nearrow & \Big\downarrow{\scriptstyle q} \\
& & U & \xrightarrow[\ f\ ]{} & M
\end{array}
$$

Suppose that the fixed-point set of f, *which is homeomorphic to the fixed-point set of* g, *is compact and empty over the closed sub-ENR* A. *Suppose that* k *maps a neighbourhood of* $\mathrm{Fix}(f)$ *into a compact subspace of* N, *so that both* f *and* g *are compactly fixed.*

Then $\tilde{L}_{(B,A)}(f, U)$ *is the image under* q *of* $\tilde{L}_{(B,A)}(g, V)$.

This leads to an effective description of the technique of finite-dimensional approximation for a globally defined fibrewise map.

Proposition 7.6 *Suppose that the base* B *is compact. Let* $f : M \to M$ *be a fibrewise self-map of a fibrewise ANR* M *such that* $f(M)$ *has compact closure in* M. *Then there exist a fibrewise ENR* N *over* B, *fibrewise maps* $q : N \to M$ *and* $k : M \to N$, *and a fibrewise homotopy* $h : M \times [0,1] \to M$ *from* f *to* $q \circ k$, *such that* $k(M)$ *has compact closure in* N *and* $h(M \times [0,1])$ *has compact closure in* M.

The maps f, $q \circ k$ and $g = k \circ q : N \to N$ in the statement of the proposition are automatically compactly fixed, because the fixed-point set is closed in a compact subspace. By Propositions 7.3 and 7.5, the fixed-point index of f is computed from the finite-dimensional approximation g as

$$
\tilde{L}_B(f, M) = q_* \tilde{L}_B(g, N).
$$

The fibrewise ENR N is constructed by adapting the definition of the index. Choose a representation $i : M \to W$, $r : W \to M$, $ri = 1$, of M as a fibrewise retract of an open subspace W of $B \times E$ for some normed space E.

We use a metric d_1 on $B \times E$ of the form $d_1((b, y), (b', y')) = d(b, b') + \|y - y'\|$ for b, $b' \in B$ and y, $y' \in E$, where d is a metric on the base.

By compactness, there is an $\epsilon > 0$ such that the closed ball in $B \times E$ with centre $i(f(x))$ and radius 2ϵ is contained in U for each $x \in M$. The set $i(f(M))$ is contained in the union of a finite number of open balls $B_\epsilon(i(f(x_j)))$, $1 \le j \le n$. Let (ϕ_j) be a partition of unity on M subordinate to the open cover $((if)^{-1}B_\epsilon((if)(x_j)))$.

We write $f(x_j) = (b_j, y_j) \in B \times E$. Let F be the finite-dimensional vector subspace of E generated by $y_1, \ldots, y_n$. We take N to be the open subspace $W \cap (B \times F)$ of $B \times F$ and $q : N \to M$ to be the restriction of r. Then the required map $k : M \to N$ is defined by

$$k(x) = (p(x), \sum_j \phi_j(x)y_j),$$

and the homotopy $h_t : M \to M$, $0 \le t \le 1$, by

$$h_t(x) = r((1 - t)(if)(x) + tk(x)),$$

the addition taking place in the vector bundle $B \times F$.

Finally, we verify that the set P of points $(1 - t)(if)(x) + tk(x)$, for $t \in [0, 1]$ and $x \in M$, lies in a compact subset of W. By construction, $k(x) \in B_\epsilon((if)(x))$ and so also $(1 - t)(if)(x) + tk(x) \in B_\epsilon((if)(x))$. Hence C is contained in the union of the open balls $B_{2\epsilon}((if)(x_j))$, and the closure of C in $B \times E$ is contained in W. Since C evidently lies in a compact subset of $B \times E$, we have shown that $\overline{C}$ is compact.

Remark 7.7. This result can be used to relate the Lefschetz fixed-point index for ANRs to the homology trace. Let $f : M \to M$ be a self-map of an ANR M, such that $f(M)$ has compact closure in M. (We are thus considering the case in which $B = *$ is reduced to a point.) Then the induced homomorphism $f_* : H_*(M; \mathbb{Q}) \to H_*(M; \mathbb{Q})$ has finite-dimensional image and its super-trace is equal to the Lefschetz index $L(f, M) \in \omega^0(*) = \mathbb{Z} \subseteq \mathbb{Q}$.

Let $g : N \to N$ be a self-map of an ENR as constructed above. We obtain the result from the commutative diagram:

$$
\begin{array}{ccc}
H_*(N) & \xrightarrow{g_*} & H_*(N) \\
{\scriptstyle q_*} \downarrow & {\scriptstyle k_*} \nearrow & \downarrow {\scriptstyle q_*} \\
H_*(M) & \xrightarrow{f_*} & H_*(M)
\end{array}
$$

The ENR N is an open subset of some Euclidean space F. Let $K \subseteq F$ be a finite polyhedron in F which contains $k(M)$ and is contained in N. By the contraction property again, the Lefschetz index of $g : N \to N$ is the same as that of $g| : K \to K$. So we reduce the result for g to the classical result, which we shall generalize in Proposition 15.37, for the self-map $g|$ of the finite polyhedron K.

The property stated in Proposition 6.17 also generalizes. Suppose that $\alpha : B' \to B$ is a compact fibrewise ENR and that $M \to B'$ is a fibrewise ANR over B'. Then $\alpha_* M = M \to B$ is a fibrewise ANR. If U is an open subset of M and $f : U \to M$ is compactly fixed over B, then $\alpha_*(f) : \alpha_* U \to \alpha_* M$ is compactly fixed over B, and its index, in the case that B is compact, is $\alpha_\sharp(\tilde{L}_B(f, U))$.

The Nielsen–Reidemeister index

For a compactly fixed map $f : U \to M$ over a compact ENR B, with no fixed-points over the closed sub-ENR A, we can define the Nielsen–Reidemeister index

$$N_{(B,A)}(f, U) \in \omega^0_{(B,A)}\{B \times S^0;\ \text{h-Fix}(f)_{+B}\}$$

just as in Section 6.

If the fixed-point set happens to be a fibrewise ANR then, again following the theory for ENRs, we have a class

$$M_{(B,A)}(f) \in \omega^0_{(B,A)}\{B \times S^0;\ \text{Fix}(f)_{+B}\}$$

lifting $\tilde{L}_{(B,A)}(f, U)$.

Example 7.8. For example, suppose that s is a section of a compact fibrewise ANR $M \to B$ over a compact ENR B. Regard the section, as in Proposition 6.7, as a fibrewise map $s : B \to M$ over B, and let $f : M \to M$ be the compactly fixed map $s \circ p$. Identifying the fixed-point set $s(B) = \text{Fix}(f)$ with B, we have

$$M_B(f) = 1 \in \omega^0_B\{B \times S^0;\ B \times S^0\}.$$

8 Virtual vector bundles and stable spaces

In this again rather technical section we look at the formal enlargement of two categories: the category of real vector bundles and stable isomorphisms over a base B (which in this section will always be a compact ENR), and the stable homotopy category over B.

Stable isomorphisms of vector bundles

Let α and β be, first of all, finite-dimensional real vector bundles (of the same dimension) over B. We shall define stable isomorphisms in much the same way as we defined fibrewise stable maps. A *stable vector bundle isomorphism*

$\alpha \to \beta$ will be given by a pair (f, ξ), where ξ is a finite-dimensional real vector bundle over B and f is a vector bundle isomorphism

$$f : \xi \oplus \alpha \to \xi \oplus \beta.$$

An equivalence relation $\sim$ on such pairs is generated by:

(i)   (Homotopy) If f and f' are homotopic (through vector bundle isomorphisms) then $(f, \xi) \sim (f', \xi)$.
(ii) (Stability) $(f, \xi) \sim (1 \oplus f, \zeta \oplus \xi)$ if ζ is a finite-dimensional vector bundle over B.
(iii) (Isomorphism) $(f, \xi) \sim (f', \xi')$ if $a : \xi \to \xi'$ is a vector bundle isomorphism, where f' is defined by commutativity of the square:

$$
\begin{array}{ccc}
\xi \oplus \alpha & \xrightarrow{\ f\ } & \xi \oplus \beta \\
{\scriptstyle a\oplus 1}\downarrow & & \downarrow{\scriptstyle a\oplus 1} \\
\xi' \oplus \alpha & \xrightarrow[\ f'\ ]{} & \xi' \oplus \beta
\end{array}
$$

A stable isomorphism $\alpha \to \beta$ is an equivalence class of pairs (f, ξ).

It is clear, just as in the discussion of fibrewise stable maps in Section 3, that we may take all the vector bundles (ξ and ξ') involved in the definition to be trivial.

Composition of stable isomorphisms is defined in the obvious way. Given representatives (f, ξ) and (g, η) of stable isomorphisms $\alpha \to \beta$ and $\beta \to \gamma$, the composition is represented by $(h, \xi \oplus \eta)$ where h is given by the middle row of the diagram:

$$
\begin{array}{ccc}
\xi \oplus (\eta \oplus \beta) & \xrightarrow{1 \oplus g} & \xi \oplus (\eta \oplus \gamma) \\
{\scriptstyle =}\uparrow & & \uparrow{\scriptstyle =} \\
(\xi \oplus \eta) \oplus \alpha \longrightarrow & (\xi \oplus \eta) \oplus \beta \longrightarrow & (\xi \oplus \eta) \oplus \gamma \\
{\scriptstyle =}\downarrow & \downarrow{\scriptstyle =} & \\
\eta \oplus (\xi \oplus \alpha) & \xrightarrow[1 \oplus f]{} \eta \oplus (\xi \oplus \beta) &
\end{array}
$$

Defined even more easily than composition is the direct sum of stable isomorphisms $\alpha \to \beta$ and $\alpha' \to \beta'$: it is obtained by forming the direct sum of representatives.

The J-homomorphism

In this framework the *J-homomorphism* is given by fibrewise one-point compactification. From a representative $f : \xi \oplus \alpha \to \xi \oplus \beta$ of a stable isomorphism $\alpha \to \beta$ we obtain a representative $f^+ : \xi_B^+ \wedge_B \alpha_B^+ \to \xi_B^+ \wedge_B \beta_B^+$ of a stable fibrewise homotopy equivalence $\alpha_B^+ \to \beta_B^+$:

$$[f^+] \in \omega_B^0\{\alpha_B^+; \beta_B^+\}.$$

In the category of vector bundles and stable isomorphisms the automorphism group of any object α is $KO^{-1}(B)$, that is, $[B_+; O(\infty)]$ (or $[B_+; O(N)]$ for large N). Indeed, the direct sum with the identity on α gives an isomorphism from the automorphism group, $KO^{-1}(B)$, of the zero-bundle to the automorphism group of α.

In the stable homotopy category over B, the automorphism group of the sphere-bundle α_B^+ is the group of units $\omega^0(B)^\times$ in the stable cohomotopy ring $\omega^0(B)$. And our construction specializes to the classical J-homomorphism

$$J : KO^{-1}(B) \to \omega^0(B)^\times.$$

Virtual vector bundles

Virtual vector bundles arise by application of the Grothendieck construction to the category of vector bundles and stable isomorphisms. The isomorphism classes of the category will form the Grothendieck group $KO^0(B)$ of finite-dimensional real vector bundles over B.

Definition 8.1 A *virtual real vector bundle* α over B is an ordered pair (α_0, α_1) of (finite-dimensional) real vector bundles; it is often denoted by the formal difference $\alpha_0 - \alpha_1$. The *dimension* of α is the difference $\dim \alpha_0 - \dim \alpha_1$.

A genuine vector bundle α will be identified with the virtual bundle $(\alpha, 0)$. We have to give a definition of stable isomorphism between two virtual bundles extending the notion already defined for genuine vector bundles. In fact we shall give two, equivalent, definitions. The first is succinct, but a little contrived.

Definition 8.2 (First version). A *stable isomorphism*

$$\alpha = \alpha_0 - \alpha_1 \to \beta = \beta_0 - \beta_1$$

is defined to be a stable isomorphism of vector bundles $\alpha_0 \oplus \beta_1 \to \beta_0 \oplus \alpha_1$.

The composition of morphisms $f : \alpha \to \beta$ and $g : \beta \to \gamma$ is given by the stable isomorphism $h : \alpha_0 \oplus \gamma_1 \to \gamma_0 \oplus \alpha_1$ determined by the commutative diagram:

$$
\begin{array}{ccc}
(\alpha_0 \oplus \gamma_1) \oplus \beta_1 & \xrightarrow{h \oplus 1} & (\gamma_0 \oplus \alpha_1) \oplus \beta_1 \\
{\scriptstyle =} \downarrow & & \uparrow {\scriptstyle =} \\
(\alpha_0 \oplus \beta_1) \oplus \gamma_1 & & \alpha_1 \oplus (\gamma_0 \oplus \beta_1) \\
{\scriptstyle f \oplus 1} \downarrow & & \uparrow {\scriptstyle 1 \oplus g} \\
(\beta_0 \oplus \alpha_1) \oplus \gamma_1 & \xrightarrow{=} & \alpha_1 \oplus (\beta_0 \oplus \gamma_1)
\end{array}
$$

The identity morphism $\alpha \to \alpha$ is given by the stable isomorphism $1 \oplus (-1) : \alpha_0 \oplus \alpha_1 \to \alpha_0 \oplus \alpha_1$. The reason for the sign is that, for any vector bundle ζ, the linear map $1 \oplus (-1) : \zeta \oplus \zeta \to \zeta \oplus \zeta$ is homotopic to the map which interchanges the factors. (See Lemma 3.2.)

The direct sum $\alpha \oplus \alpha'$ of virtual bundles α and α' is defined to be $(\alpha_0 \oplus \alpha_0', \alpha_1 \oplus \alpha_1')$. Direct sums of morphisms are defined in the obvious way.

Notice that any virtual bundle $\alpha_0 - \alpha_1$ is isomorphic to one in which α_1 is a trivial vector bundle.

Another description of stable isomorphisms is useful and, perhaps, more natural than that given by Definition 8.2. We specify a morphism $\alpha \to \beta$ by giving two finite-dimensional real vector bundles ρ and σ over B and two vector bundle isomorphisms $f_0 : \rho \oplus \alpha_0 \to \sigma \oplus \beta_0$ and $f_1 : \rho \oplus \alpha_1 \to \sigma \oplus \beta_1$. Now introduce the equivalence relation $\sim$ on such data (ρ, σ, f_0, f_1) generated by:

(i)   (Homotopy) $(\rho, \sigma, f_0, f_1) \sim (\rho, \sigma, f_0', f_1')$ if f_0 and f_0', f_1 and f_1' are, respectively, homotopic through vector bundle isomorphisms.

(ii) (Stability) $(\rho, \sigma, f_0, f_1) \sim (\xi \oplus \rho, \xi \oplus \sigma, 1 \oplus f_0, 1 \oplus f_1)$, for any vector bundle ξ.

(iii) (Isomorphism) If $a : \rho \to \rho'$ and $b : \sigma \to \sigma'$ are vector bundle isomorphisms, $(\rho, \sigma, f_0, f_1) \sim (\rho', \sigma', f_0', f_1')$, where f_0' and f_0' are given by the commutative diagrams:

$$
\begin{array}{ccc}
\rho \oplus \alpha_0 & \xrightarrow{\ f_0\ } & \sigma \oplus \beta_0 \\
{\scriptstyle a\oplus 1}\downarrow & & \downarrow{\scriptstyle b\oplus 1} \\
\rho' \oplus \alpha_0 & \xrightarrow{\ f_0'\ } & \sigma' \oplus \beta_0
\end{array}
\qquad
\begin{array}{ccc}
\rho \oplus \alpha_1 & \xrightarrow{\ f_1\ } & \sigma \oplus \beta_1 \\
{\scriptstyle a\oplus 1}\downarrow & & \downarrow{\scriptstyle b\oplus 1} \\
\rho' \oplus \alpha_1 & \xrightarrow{\ f_1'\ } & \sigma' \oplus \beta_1
\end{array}
$$

Definition 8.3 (Second version). A *stable isomorphism* $\alpha \to \beta$ of virtual bundles is an equivalence class of data (ρ, σ, f_0, f_1) as above.

The two versions (8.2 and 8.3) of the definition are easily seen to be consistent. Given a stable isomorphism $\alpha_0 \oplus \beta_1 \to \beta_0 \oplus \alpha_1$ represented by a vector bundle isomorphism $f : \xi \oplus \alpha_0 \oplus \beta_1 \to \xi \oplus \beta_0 \oplus \alpha_1$ for some vector bundle ξ, making canonical identifications we set $\rho = \xi \oplus \beta_1$, $\sigma = \xi \oplus \alpha_1$, $f_0 = f : \xi \oplus \beta_1 \oplus \alpha_0 \to \xi \oplus \alpha_1 \oplus \beta_0$ and $f_1 = 1 : \xi \oplus \beta_1 \oplus \alpha_1 \to \xi \oplus \alpha_1 \oplus \beta_1$.

In Definition 8.3 the identity map $1 : \alpha \to \alpha$ is given, more satisfactorily, by taking ρ and σ to be zero and f_0 and f_1 to be the identity maps.

Extending the stable category

In the development of ordinary stable homotopy theory, one soon finds it necessary to introduce formal desuspensions of pointed spaces. At an elementary level this can be done by considering pairs (E, m), where E is a pointed space and $m \geq 0$ is a non-negative integer. The space E is identified with the pair $(E, 0)$. A stable map from a pair (E, m) to a pair (F, n) is defined to be an element of $\omega^{m-n}\{E; F\}$, that is, $\omega^0\{\Sigma^{r-m}E; \Sigma^{r-n}F\}$ where $r \geq \max\{m, n\}$. Smash products are defined by $(E, m) \wedge (F, n) := (E \wedge F, m + n)$. Then $S^m \wedge (E, m)$ is canonically isomorphic to E: (E, m) is the m-fold desuspension of the pointed space E. At a later stage, this treatment is superseded by a full discussion of spectra. In this introductory account we shall be content to give the fibrewise version of the elementary theory.

Definition 8.4 A *fibrewise stable space* over B is a pair (X, λ), where X is a fibrewise pointed space and λ is a finite-dimensional real vector bundle over B. The pair (X, λ) will be written symbolically as $X \wedge_B (-\lambda)_B^+$. We identify the fibrewise space X with the pair $(X, 0)$.

We have to define the morphisms, that is the stable maps, between two such fibrewise stable spaces (X, λ) and (Y, μ). Following the definition (first version) for virtual bundles, we define:

$$\omega_B^0\{X \wedge_B (-\lambda)_B^+; Y \wedge_B (-\mu)_B^+\} := \omega_B^0\{X \wedge_B \mu_B^+; Y \wedge_B \lambda_B^+\}.$$

Composition is defined using the suspension isomorphisms again in essentially the same way as we defined composition of stable isomorphisms of virtual bundles. We note also that the identity map on (X, λ) is given by $1 \wedge (-1)^+ : X \wedge_B \lambda_B^+ \to X \wedge_B \lambda_B^+$.

The smash product $(X, \lambda) \wedge_B (X', \lambda')$ of two fibrewise stable spaces over B is defined to be $(X \wedge_B X', \lambda \oplus \lambda')$, and the smash product of stable maps is defined in the natural way. There is a canonical equivalence between the fibrewise suspension $(X, \lambda) \wedge_B \lambda_B^+$ and X. (To be precise, there are two canonical equivalences, a left and a right, differing by the stable map $(-1)^+ : \lambda_B^+ \to \lambda_B^+$.)

As in the case of virtual bundles there is a variant, second version, of the definition. We may specify a stable map $(X, \lambda) \to (Y, \mu)$ by giving two vector bundles ρ and σ over B together with an isomorphism $g : \rho \oplus \lambda \to \sigma \oplus \mu$. Then

$$\omega_B^0\{(X, \lambda); (Y, \mu)\} = \omega_B^0\{\rho_B^+ \wedge_B X; \sigma_B^+ \wedge_B Y\}. \tag{8.5}$$

(Following the definition of stable isomorphism a little more closely, we could describe a stable map by a fibrewise pointed map $f : \rho_B^+ \wedge_B X \to \sigma_B^+ \wedge_B Y$, together with the vector bundle isomorphism g, and introduce an equivalence relation on data (f, g, ρ, σ) in the now familiar way.)

Given a virtual real vector bundle $\alpha = (\alpha_0, \alpha_1)$, we have an associated stable fibrewise space $((\alpha_0)_B^+, \alpha_1)$, denoted according to our convention by $(\alpha_0)_B^+ \wedge_B (-\alpha_1)_B^+$; it will also be written simply as α_B^+.

Virtual bundles defined by Fredholm maps

We conclude this section with a digression on Fredholm maps. For simplicity we shall look only at Fredholm maps between *Hilbert bundles*, that is, locally trivial bundles of Hilbert spaces.

Let λ and μ be Hilbert bundles over B and let $d : \lambda \to \mu$ be a fibrewise (linear) Fredholm map. So d is continuous and its restriction to fibres is a Fredholm map between Hilbert spaces. We shall associate to d an essentially canonical virtual bundle. To be precise we shall give a (non-empty) subcategory of the category of virtual bundles over B having exactly one morphism between any two objects.

Consider a point $b \in B$ of the base. Let K_b be the (finite-dimensional) orthogonal complement in μ_b of the image of $d_b : \lambda_b \to \mu_b$, and let $i_b : K_b \to \mu_b$ be the inclusion. Then $d_b \oplus i_b : \lambda_b \oplus K_b \to \mu_b$ is surjective. Extend i_b to a fibrewise (continuous) linear map $\tilde{i}_b : B \times K_b \to \mu$. Then, because the set of invertible operators in a Hilbert space is open, $d \oplus \tilde{i}_b : \lambda \oplus (B \times K_b) \to \mu$ will be surjective over an open neighbourhood U_b of b in B.

By compactness of B there is a finite subset A such that the sets U_a, $a \in A$, cover B. Let E be the finite-dimensional vector space $\bigoplus_{a \in A} K_a$ and let $f : B \times E \to \mu$ be the direct sum of the linear maps $\tilde{i}_a$. Then $d \oplus f : \lambda \oplus (B \times E) \to \mu$ is surjective in each fibre and its kernel, α_0 say, is a finite-dimensional vector bundle. Let α_1 be the trivial bundle $B \times E$.

We associate to the Fredholm map d the virtual vector bundle $\alpha = (\alpha_0, \alpha_1)$.

The point is that this construction is essentially functorial. For suppose that we have another finite-dimensional vector space F and linear map $g : B \times F \to \mu$ such that $d \oplus g : \lambda \oplus (B \times F) \to \mu$ is surjective. Write β_0 for the kernel of this map and β_1 for $B \times F$. Then a natural stable isomorphism $\alpha \to \beta$ can be obtained as follows.

Consider the linear map $h_t = (1 - t)f \oplus tg : B \times (E \oplus F) \to \mu$ for $0 \leq t \leq 1$. Then $d \oplus h_t$ gives a surjective linear map $\lambda \oplus (B \times (E \oplus F)) \to \mu$ over $B \times [0, 1]$ with kernel a finite-dimensional vector bundle ξ, say. The bundle ξ over $B \times [0, 1]$ determines an isomorphism, up to homotopy, from the restriction $\xi_0 = \alpha_0 \oplus (B \times F)$ at $t = 0$ to the restriction $\xi_1 = \beta_0 \oplus (B \times E)$ at $t = 1$. This data, understood in the second version using the natural identification $\alpha_1 \oplus (B \times F) \to \beta_1 \oplus (B \times E)$, gives a stable isomorphism $\alpha \to \beta$.

By looking at a third construction and a 2-parameter family it is easy to see that we have associated to d a category of virtual vector bundles in which there is a unique isomorphism between any two objects. Such a category is, for practical purposes, the same thing as a virtual vector bundle. (We are used to talking about the field of real numbers $\mathbb{R}$ without worrying about the model!)

9 The Adams conjecture

The homotopy-theoretic proof of the Adams conjecture by Becker and Gottlieb [9] was a landmark in fibrewise stable homotopy theory. In this section we shall look at a closely related, but more elementary, proof which emerged soon afterwards in the work of Brown [20], Meyerhoff and Petrie [109] and Dold [48]. (The exposition is based in part on an unpublished manuscript 'Adams trivialization, $\mathrm{Im}(J)$-theory and the codegree of vector bundles' by M.C. Crabb and K. Knapp, which was quoted in [33].)

Throughout this section $p > 1$ will be a fixed prime and l will be an integer prime to p. The base space B will be a compact ENR.

For a real vector space E of dimension n we write $\sigma(E)$ for the 2-element set of orientations of E, understood as the sphere $S(\Lambda^n E)$ (so that, in particular, $\sigma(0) = S(\mathbb{R})$). More generally, $\sigma(\xi)$ will denote the orientation bundle of a real vector bundle ξ.

Fibre degree

Consider two real vector bundles ξ and η over B of the same dimension, and let $f : \xi_B^+ \to \eta_B^+$ be a fibrewise stable map. The restriction to fibres at $b \in B$ is a stable map of spheres $f_b : (\xi_b)^+ \to (\eta_b)^+$. A choice of isomorphism $\sigma(\xi_b) \to \sigma(\eta_b)$ would allow us to define the degree, in $\mathbb{Z}$, of f_b. In any case the degree is well-defined up to sign and is constant on each component of B. If the fibre degree is nowhere zero, then the orientation bundles of ξ and η are isomorphic and there is a unique isomorphism with respect to which the fibre degree is everywhere positive.

A fibrewise stable map $f : \xi_B^+ \to \eta_B^+$ is a stable fibrewise homotopy equivalence if and only if its degree in each fibre is ± 1. This follows at once from Dold's theorem; but in this special case it is easy to give an elementary proof by induction over cells of the base as in Proposition 9.1 below.

Localization at the prime p

In elementary stable homotopy theory localization is a purely algebraic construction. We shall call an element f of the localized stable homotopy group $\omega_B^0\{\xi_B^+; \eta_B^+\}_{(p)}$ a *p-local fibrewise stable map*. (These algebraically defined 'maps' can be realized as fibrewise maps of spaces by introducing fibrewise localization; see, for example, [105].) Such maps have a $\mathbb{Z}_{(p)}$-valued fibre degree, defined only up to sign.

A p-local fibrewise stable map f as above is said to be a *p-local equivalence* if there is an inverse map $g \in \omega_B^0\{\eta_B^+; \xi_B^+\}_{(p)}$ such that $f \circ g = 1$ and $g \circ f = 1$. The local equivalences are characterized by their fibre degree as follows.

Proposition 9.1 *A p-local fibrewise stable map* $f \in \omega_B^0\{\xi_B^+; \eta_B^+\}_{(p)}$ *is a p-local equivalence if and only if its degree (defined up to sign) in each fibre is prime to p.*

As usual we may assume that B is a connected finite complex. (In general, we can work over a finite polyhedron in which B is a retract.)

It is not hard to see that a stable map in $\omega^0(B)_{(p)} = \omega_B^0\{\xi_B^+; \xi_B^+\}_{(p)}$ is invertible if and only if its fibre degree (now defined as an element of $\mathbb{Z}_{(p)}$) is prime to p. For

$$\omega^0(B)_{(p)} = \mathbb{Z}_{(p)} \oplus \mathfrak{n}, \tag{9.2}$$

where $\mathfrak{n}$ is the nilradical, which is finite. (Compare Proposition 4.5.)

There is no loss of generality in assuming that f has fibre degree ± 1 and we can fix the isomorphism $\sigma(\xi) \to \sigma(\eta)$ for which f has degree 1.

Suppose that B is obtained by attaching an m-cell to a subcomplex A, $m \geq 1$, and that we have constructed a map g over A with $f \circ g = 1$. The obstruction x to extending g to B lies in the relative group $\omega_{(B,A)}^1\{\eta_B^+; \xi_B^+\}_{(p)}$. But $f \circ x = 0$ and the restriction of f to D is an equivalence. So $x = 0$ and g extends to a map $\tilde{g}$ over B. Multiplying by the inverse of $f \circ \tilde{g}$ we get the inverse of f.

An alternative proof may be given using a Mayer–Vietoris argument as in Proposition 9.4 below.

Remark 9.3. In more geometric terms, the proposition shows that: if f is a stable map $\xi_B^+ \to \eta_B^+$ with fibre degree prime to p, then there is an integer k prime to p and a stable map $g : \eta_B^+ \to \xi_B^+$ such that $f \circ g = k$ and $g \circ f = k$.

The Adams conjecture gives a precise criterion, in terms of Adams operations, for such sphere-bundles ξ_B^+ and η_B^+ to be stably fibrewise homotopy equivalent at the prime p. In the course of the proof we shall also need to take kth roots, as follows.

Proposition 9.4 *Let $k \geq 1$ be prime to p and let ξ and η be real vector bundles of the same dimension over B with isomorphic orientation bundles. We fix an isomorphism $a : \sigma(\xi) \to \sigma(\eta)$, which determines a kth power isomorphism $\sigma(k\xi) \to \sigma(k\eta)$. Suppose that $g \in \omega_B^0\{(k\xi)_B^+; (k\eta)_B^+\}_{(p)}$ is a p-local stable fibrewise homotopy equivalence with fibre degree 1 (with respect to a). Then there is a unique stable map $f \in \omega_B^0\{\xi_B^+; \eta_B^+\}_{(p)}$ with fibre degree 1 (with respect to a) such that $f^{\wedge k} = f \wedge \cdots \wedge f = g$.*

In the statement we are, of course, identifying $(k\xi)_B^+$, where $k\xi$ is the direct sum of k factors $\xi \oplus \cdots \oplus \xi$, with the k-fold smash product $\xi_B^+ \wedge_B \cdots \wedge_B \xi_B^+$. Notice that, if k is odd, the existence of a fibrewise homotopy equivalence g guarantees that the orientation bundles of ξ and η are isomorphic (and we can choose the isomorphism a with respect to which g has degree 1).

We first show uniqueness of the kth root. If f_1 and f_2 are both kth roots of g, then $f_2^{-1} \circ f_1$ is a kth root of the identity in $\omega_B^0\{\xi_B^+; \xi_B^+\}_{(p)} = \omega^0(B)_{(p)}$, with fibre degree 1. But, from (9.2), any element of degree 1 in $\omega^0(B)_{(p)}$ has order a power of p, and therefore has a unique kth root.

The existence proof is again by induction. If the base B is contractible, the result is clear. In general, we can express the connected finite complex B as a union of contractible closed sub-ENRs with each intersection an ENR and use a Mayer–Vietoris argument.

The inductive step is as follows. Suppose that B_1 and B_2 are subcomplexes of B with intersection the subcomplex A. Given kth roots f_1 over B_1 and f_2 over B_2, we show that there is a kth root over $B_1 \cup B_2$. For ease of notation suppose that $B = B_1 \cup B_2$. From the uniqueness clause already established, f_1 and f_2 agree on A. Let $h \in \omega_B^0\{\xi_B^+; \eta_B^+\}_{(p)}$ lift f_1 and f_2. Consider the kth power map:

$$
\begin{array}{ccc}
\omega_A^{-1}\{\xi_B^+; \eta_B^+\}_{(p)} & \longrightarrow & \omega_A^{-1}\{(k\xi)_B^+; (k\eta)_B^+\}_{(p)} \\[6pt]
\delta \downarrow & & \downarrow \delta \\[6pt]
\omega_B^0\{\xi_B^+; \eta_B^+\}_{(p)} & \xrightarrow{\ \wedge^k\ } & \omega_B^0\{(k\xi)_B^+; (k\eta)_B^+\}_{(p)}
\end{array}
$$

By hypothesis, we have $g = h^k + \delta y$ for some y. Now $(h + \delta x)^k = h^k + kh^{k-1}.\delta x$. Since k is prime to p and h is an equivalence, there is a unique class x such that $y = kh^{k-1}.x$. Then $f = h + \delta x$ is the required kth root.

Line bundles

Let λ be a complex line bundle over B. Then the lth power: $z \mapsto z \otimes z \otimes \cdots \otimes z$ (l factors) defines a fibrewise map of sphere (in fact, circle) -bundles

$$
\mathfrak{a}(\lambda) : S(\lambda) \to S(\lambda^{\otimes l}),
$$

which has degree l in each fibre. (We have already used this map in Example 4.6 to show that the Euler class is torsion if λ has finite order.) By extending the map radially (or, if $l \geq 1$, by taking the lth power on the whole of λ) we get an associated fibrewise pointed map

$$
\mathfrak{a}(\lambda) : \lambda_B^+ \to (\lambda^{\otimes l})_B^+.
$$

This determines a p-local equivalence, which we denote by the same symbol,

$$
\mathfrak{a}(\lambda) \in \omega_B^0\{\lambda_B^+; (\lambda^{\otimes l})_B^+\}_{(p)}.
$$

Adams operations

The Adams operation ψ^l is usually defined on elements of the complex K-group $K^0(B)$ in terms of the exterior powers. Here we require the operation at the level of virtual complex vector bundles over B. The account of virtual real vector bundles in Section 8 translates easily into the complex theory. More generally, we can introduce a compact Lie group G and define a category of virtual G-vector bundles over (a compact G-ENR) B. Given a complex vector bundle ξ of (complex) dimension n over B, we shall define a virtual bundle $\psi^l\xi$, of the same dimension n, in an (essentially) canonical way; that is, we shall define the virtual bundle up to unique isomorphism as we did in Section 8 when associating a virtual bundle to a Fredholm operator.

One way of doing this is to use G-equivariant theory, where G is a compact Lie group. The category of virtual complex G-modules has the pleasant feature that the automorphism group $K_G^{-1}(*)$ of any object is trivial. A virtual G-module is, therefore, defined up to unique isomorphism by its class in the representation ring $R(G) = K_G^0(*)$. So for any complex G-module E we have a well-defined virtual G-module $\psi^l E$, with character given by

$$\chi_{\psi^l E}(g) = \chi_E(g^l).$$

Now let us take G to be $U(n)$ and E to be the defining representation $\mathbb{C}^n$. The complex n-dimensional vector bundle ξ (with a choice of Hermitian inner product, which is immaterial when we work with homotopy classes) can be expressed as $P \times_G E$, where P is the associated principal G-bundle. We then define $\psi^l\xi$ to be the virtual bundle $P \times_G \psi^l E$.

Another, more explicit, method, due to Atiyah [3], is available when l is prime. The group $\mathbb{Z}/l$ acts on the l-fold tensor product $\bigotimes^l \xi$ by cyclic permutation, and the bundle splits as a direct sum of components α_i indexed by the characters of $\mathbb{Z}/l$: $j \mapsto \omega^{ij}$, where ω is a primitive lth root of unity. We can take $\psi^l\xi = (\alpha_0, \alpha_1)$. In general, one can write l as a product of primes and iterate this construction.

At the heart of the Adams conjecture is the construction of a natural p-local equivalence:

$$\mathfrak{a}(\xi) \in \omega_B^0\{\xi_B^+;\ (\psi^l\xi)_B^+\}_{(p)}.$$

We use transfer methods to reduce to a special case in which the equivalence can be written down explicitly using Adams' construction, described above, for line bundles. (For a line bundle λ we have, of course, $\psi^l\lambda = \lambda^{\otimes l}$.)

A reduction

Writing $G = U(n)$, let $T = \mathbb{T}^n \leq U(n)$ be the standard maximal torus with Weyl group $W = \mathfrak{S}_n$. Thus W is the quotient $N_G(T)/T$ of the normalizer $N_G(T)$ of T in G. Choose a Sylow p-subgroup W_p of W, and let $N_p = T \ltimes W_p \leq T \ltimes W = N_G(T)$.

Remark 9.5. Every finite p-subgroup P of G is conjugate to a subgroup of N_p. This is most easily seen using equivariant fixed-point theory: P acts on the space of maximal tori of G and the Euler characteristic of the fixed subspace, that is, the set of tori whose normalizer contains P, is congruent $(\bmod p)$ to the Euler characteristic of the space $G/N_G(T)$, and this is 1. See, for example, [43]. But it is not hard to give an elementary proof. We have to show that every finite-dimensional unitary representation E of P is monomial. We argue by induction on the dimension of E. We may suppose that P is a subgroup of the unitary group $U(E)$ and that E is an irreducible P-module. If P is Abelian, then E is 1-dimensional. If not, then there is a non-central element $h \in P$ such that the commutator $[g, h] \in Z(P)$ (the centre of P) for all $g \in P$. Let F be an eigenspace of h. Then G permutes the eigenspaces of h, and E is identified with the space of sections of the vector bundle $P \times_Q F \to P/Q$, where Q is the stabilizer of F. One applies the induction hypothesis to the Q-module F.

Now we have two complex representations of N_p:

$$E := \mathbb{C} \oplus \cdots \oplus \mathbb{C} \qquad (z_1, \ldots, z_n) \cdot (x_1, \ldots, x_n) = (z_1 x_1, \ldots, z_n x_n),$$
$$F := \mathbb{C} \oplus \cdots \oplus \mathbb{C} \qquad (z_1, \ldots, z_n) \cdot (x_1, \ldots, x_n) = (z_1^l x_1, \ldots, z_n^l x_n),$$

for $(z_1, \ldots, z_n) \in T$, $(x_1, \ldots, x_n) \in \mathbb{C}^n$, with $W_p \le \mathfrak{S}_n$ acting by permutation of the factors.

A check on characters, using the fact that l is prime to p, shows that $F = \psi^l E$. Here are the details. Write $g = (z_1, \ldots, z_n; w)$, with $w \in N_p$. We have to show that $\chi_F(g) = \chi_E(g^l)$. Since l is prime to p, the fixed-points of w and w^l coincide, and we have

$$\chi_F(g) = \sum_{w(i)=i} z_i^l = \sum_{w^l(i)=i} z_i^l = \chi_E(g^l).$$

(The simplest example, with $p = 2 = l$, shows why we require l and p to be coprime.)

Now the construction of Adams gives an N_p-equivariant map

$$\mathfrak{a}(E) : S(E) = S(\mathbb{C}) * \cdots * S(\mathbb{C}) \to S(F) = S(\mathbb{C}) * \cdots * S(\mathbb{C}),$$

obtained by taking the n-fold join of the lth power map $z \mapsto z^l : S(\mathbb{C}) \to S(\mathbb{C})$. It has (non-equivariant) degree l^n, prime to p.

This allows us to define, by the balanced product construction, a fibrewise stable map

$$\mathfrak{a}(\xi) \in \omega_B^0 \{\xi_B^+; (\psi^l \xi)_B^+\}_{(p)},$$

with fibre degree l^n, for any complex n-dimensional vector bundle ξ whose structure group is reduced from $U(n)$ to N_p.

The transfer argument

Consider now an arbitrary n-dimensional complex vector bundle ξ over B, and let $P \to B$ be the associated principal $U(n)$-bundle. We shall lift from B to $B' := P/N_p$ and write $\pi : B' \to B$ for the projection. So $\pi^*\xi = P \times_{N_p} E$ and $\pi^*\psi^l\xi = P \times_{N_p} F$. (The equality signs indicate canonical isomorphism in the category of virtual complex bundles over B'.)

Now the image $\pi_\sharp \mathfrak{a}(\pi^*\xi)$ of the class defined by $\mathfrak{a}(E)$ above under the transfer map

$$\pi_\sharp : \omega^0_{B'}\{(\pi^*\xi)^+_{B'}; (\pi^*\psi^l\xi)^+_{B'}\}_{(p)} \to \omega^0_B\{\xi^+_B; (\psi^l\xi)^+_B\}_{(p)}$$

has fibre degree $l^n \chi(G/N_p)$, which is prime to p. (Notice that up to this point we could have worked with integral groups; it is the next step which requires localization at p.)

Define

$$\mathfrak{a}(\xi) := (\pi_\sharp 1)^{-1}\pi_\sharp \mathfrak{a}(\pi^*\xi) \in \omega^0_B\{\xi^+_B; (\psi^l\xi)^+_B\}_{(p)}.$$

It is a fibrewise stable map (at p) with fibre degree l^n. For $\pi_\sharp(1) \in \omega^0(B)$ has degree $\chi(G/N_p)$. The existence of such a stable map is the essential solution of the Adams conjecture.

Proposition 9.6 *Let $p > 1$ be a prime and l be an integer prime to p. Then there is, for any n-dimensional complex vector bundle ξ over a compact ENR B a canonically defined p-local equivalence*

$$\mathfrak{a}(\xi) \in \omega^0_B\{\xi^+_B; (\psi^l\xi)^+_B\}_{(p)}$$

of fibre degree l^n.

Remark 9.7. It is not obvious that this definition of $\mathfrak{a}(\xi)$ is consistent with the earlier usage when the structure group of ξ can be reduced to N_p. This is, in fact, the case. Moreover, the construction $\mathfrak{a}$ is multiplicative:

$$\mathfrak{a}(\xi \oplus \xi') = \mathfrak{a}(\xi) \cdot \mathfrak{a}(\xi'),$$

for complex vector bundles ξ and ξ' over B. Another property is also important in the characterization of $\mathfrak{a}$. Let $\pi : B' \to B$ be a p-fold covering. Then

$$\mathfrak{a}(S^p_\pi \xi) = P^p_\pi(\mathfrak{a}(\xi)),$$

where the operations S^p_π and P^p_π are defined as follows. Let $P \to B$ be the principal $\mathfrak{S}_p$-bundle associated to the p-fold covering. The vector bundle $S^p_\pi \xi$ over B is the quotient

$$P \times_{\mathfrak{S}_p} (\xi \oplus \cdots \oplus \xi),$$

by the permutation action of $\mathfrak{S}_p$ on the p-fold direct sum $\xi \oplus \cdots \oplus \xi$ lifted to P. The power operation

$$P_\pi^p : \omega_B^0\{\xi_B^+; (\xi')_B^+\} \to \omega_{B'}^0\{(S_\pi^p\xi)_{B'}^+; (S_\pi^p\xi')_{B'}^+\}$$

is constructed by taking the pth power, with the permutation action of $\mathfrak{S}_p$, lifting from B to P, and then taking the quotient by the free group action. (See [25].)

Granted the multiplicativity of $\mathfrak{a}$, one can define a natural class of fibre degree $l^{\dim \gamma}$ for any virtual complex bundle $\gamma = (\gamma_0, \gamma_1)$ over B:

$$\mathfrak{a}(\gamma) \in \omega_B^0\{\gamma_B^+; (\psi^l\gamma)_B^+\}_{(p)}$$

as $\mathfrak{a}(\gamma_0) \cdot \mathfrak{a}(\gamma_1)^{-1}$.

For the remainder of this section we assume that l generates the p-adic units $\mathbb{Z}_p^\times$ if p is odd and is $\pm 3 \pmod 8$ if $p = 2$ (that is, generates $\mathbb{Z}_2^\times$ modulo ± 1). In his original work [1] on the conjecture which bears his name, Adams established:

Proposition 9.8 *Let α be a virtual complex bundle of dimension 0 over B such that α_B^+ is stably fibrewise homotopy trivial at p. Then there is a stable isomorphism*

$$v : k\alpha \xrightarrow{\cong} \gamma - \psi^l\gamma,$$

for some integer k prime to p and some virtual bundle γ of dimension 0 over B.

The proof is K-theoretic and we provide only an outline, using the language of fibrewise K-theory from Section 15. For more details see, for example, Section 5 of [34].

The Bott class u of the complex bundle α generates the K-theory $\tilde{K}^0(B^\alpha)$ of the Thom space of α as a free $K^0(B)$-module. We can think of this class as a K-theory map over B:

$$u \in K_B^0\{\alpha_B^+; B \times S^0\}.$$

Let us now localize the K-theory at p, so that ψ^l is defined as a stable operation. Then $\psi^l(u) = \rho^l(\alpha)u$ for some characteristic class $\rho^l(\alpha) \in K^0(B)_{(p)}$. Now if $f \in \omega_B^0\{\alpha_B^+; B \times S^0\}_{(p)}$ $(= \omega^0(B^\alpha)_{(p)})$ is a p-local equivalence, its Hurewicz image in K-theory will be fixed by the operation ψ^l. Writing the Hurewicz image of f as xu for some unit $x \in K^0(B)_{(p)}^\times$, we see that

$$\rho^l(\alpha) = \psi^l(x) \cdot x^{-1}.$$

Computations in K-theory show that this condition is equivalent to the existence of an element $y \in K^0(B)_{(p)}$ such that $[\alpha] = \psi^l y - y$.

The Adams conjecture asserts the converse to Proposition 9.8.

Proposition 9.9 (Adams conjecture). *Let α be a virtual bundle of dimension 0 over a compact ENR B such that $k\alpha$ is stably isomorphic to $\gamma - \psi^l\gamma$ for* *some integer k prime to p and some virtual bundle γ over B. Then α_B^+ is stably fibrewise homotopy trivial at p.*

Indeed, suppose that we have a stable isomorphism $v : k\alpha \to \gamma - \psi^l\gamma$. From this data we can actual construct a stable trivialization (or stable cohomotopy Thom class)

$$(v^*\mathfrak{a}(\gamma))^{1/k} \in \omega_B^0\{\alpha_B^+; \ B \times S^0\}_{(p)} = \tilde{\omega}^0(B^\alpha)_{(p)}, \qquad (9.10)$$

using the kth root provided by Proposition 9.4. Such a stable trivialization may be called an *Adams trivialization*. (See [32].)

10 Duality

The exposition in this section is strongly influenced by the important paper [49] of Dold and Puppe. The general proof of the existence of fibrewise duals for homotopy fibre bundles is due to Becker and Gottlieb [10]. We begin with a discussion of duality in a purely algebraic setting, taking as our motivating example the category of finite-dimensional super vector spaces.

Duality in the category of finite-dimensional super vector spaces

A *super vector space* E over a field k is a $\mathbb{Z}/2$-graded vector space (E_0, E_1). An element $x \in E_i$ has *degree* $\deg x = i$. A linear map

$$f : E = (E_0, E_1) \to F = (F_0, F_1)$$

of degree 0 is a pair of linear maps $f_0 : E_0 \to F_0$ and $f_1 : E_1 \to F_1$; these are the morphisms in the category of super vector spaces. A linear map $f : E \to F$ of degree 1 is a pair of linear maps $f_0 : E_0 \to F_1$ and $f_1 : E_1 \to F_0$. We write $\mathrm{Hom}_i(E, F)$ for the vector space of linear maps $E \to F$ of degree i; thus $\mathrm{Hom}_*(E, F) = (\mathrm{Hom}_0(E, F), \mathrm{Hom}_1(E, F))$ is a super vector space.

The direct sum and tensor product of super vector spaces E and E' are defined by:

$$E \oplus E' = (E_0 \oplus E_0', E_1 \oplus E_1'),$$
$$E \otimes E' = ((E_0 \otimes E_0') \oplus (E_1 \otimes E_1'), (E_0 \otimes E_1') \oplus (E_1 \otimes E_0')).$$

The direct sum of linear maps of the same degree is defined in the obvious way. In super algebra a minus sign is introduced with every transposition of elements of degree 1. Accordingly, the tensor product $f \otimes f'$ of linear maps $f : E \to F$ and $f' : E' \to F'$ of possibly different degree is defined by:

$$(f \otimes f')(x \otimes x') = (-1)^{(\deg f') \cdot (\deg x)} f(x) \otimes f'(x').$$

There is a canonical isomorphism $\mathbf{t}$ between $E \otimes E'$ and $E' \otimes E$ given by

$$E \otimes E' \to E' \otimes E : x \otimes x' \mapsto (-1)^{\deg x \cdot \deg x'} x' \otimes x. \qquad (10.1)$$

We can think of any vector space as a super vector space concentrated in degree 0, and then identify k with $(k, 0)$. There are natural isomorphisms:

$$k \otimes E \to E \quad \text{and} \quad E \otimes k \to E$$

given by multiplication by scalars.

From now on we restrict attention to finite-dimensional vector spaces. The *dual* E^* of E is defined to be $\text{Hom}_*(E, k) = (E_0^*, E_1^*)$. There is an isomorphism $\phi : E^* \otimes F \to \text{Hom}_*(E, F)$ given by

$$\phi(l \otimes y)(x) = (-1)^{\deg x \cdot \deg y} l(x) y, \quad \text{for } l \in E^*, x \in E \text{ and } y \in F.$$

There are two basic structure maps (that is, morphisms in the category):

$$k \overset{\mathbf{i}}{\longrightarrow} E^* \otimes E \overset{\mathbf{e}}{\longrightarrow} k, \qquad (10.2)$$

the first corresponding to the inclusion of the scalar multiples of the identity in $\text{Hom}_*(E, E)$ and the second given by evaluation: $l \otimes x \mapsto l(x)$. In terms of a basis (e_j) for E and dual basis (e_j^*) for E^*, with $e_j^*(e_j) = 1$, we have $\mathbf{i}(1) = \sum_j (-1)^{\deg e_j} e_j^* \otimes e_j$. The composition $\mathbf{e} \circ \mathbf{i} : k \to k$ is multiplication by $\dim E_0 - \dim E_1$.

It is an elementary exercise to verify the following two basic identities satisfied by $\mathbf{i}$ and $\mathbf{e}$, and involving the composition $\mathbf{e} \circ \mathbf{t} : E \otimes E^* \to E^* \otimes E \to k$.

Proposition 10.3 *The maps* $\mathbf{i}$ *and* $\mathbf{e}$ *have the following properties.*
(i) *The composition*

$$E^* = k \otimes E^* \overset{\mathbf{i} \otimes 1_{E^*}}{\longrightarrow} E^* \otimes E \otimes E^* \overset{1_{E^*} \otimes \mathbf{e} \circ \mathbf{t}}{\longrightarrow} E^* \otimes k = E^*$$

is the identity on E^*, *and*
(ii) *the composition*

$$E = E \otimes k \overset{1_E \otimes \mathbf{i}}{\longrightarrow} E \otimes E^* \otimes E \overset{\mathbf{e} \circ \mathbf{t} \otimes 1_E}{\longrightarrow} k \otimes E = E$$

is the identity on E.

In these two identities is contained the complete duality theory for finite-dimensional super vector spaces.

Corollary 10.4 *For any finite-dimensional super vector spaces D and F there are inverse isomorphisms*

$$\text{Hom}_*(D \otimes E, F) \underset{\psi}{\overset{\phi}{\rightleftarrows}} \text{Hom}_*(D, E^* \otimes F)$$

defined in terms of $\mathbf{i}$ *and* $\mathbf{e}$ *as follows. Let* $f \in \text{Hom}_*(D \otimes E, F)$ *and* $g \in \text{Hom}_*(D, E^* \otimes F)$. *Then* $\phi(f)$ *is the composition*

$$D = D \otimes k \overset{1 \otimes \mathbf{i}}{\longrightarrow} D \otimes E^* \otimes E \overset{\mathbf{t} \otimes 1}{\longrightarrow} E^* \otimes D \otimes E \overset{1 \otimes f}{\longrightarrow} E^* \otimes F$$

and $\psi(g)$ is the composition

$$D \otimes E \xrightarrow{g \otimes 1} E^* \otimes F \otimes E \xrightarrow{t \otimes 1} F \otimes E^* \otimes E \xrightarrow{1 \otimes e} F \otimes k = F.$$

We show that $\phi(\psi(g)) = g$ using the first identity in Proposition 10.3. It is clearly enough to take g to be the identity map $D = E^* \otimes F \to E^* \otimes F$. Then $\phi(\psi(g))$ reduces to the composition:

$E^* \otimes F$

$= \downarrow$

$E^* \otimes F \otimes k \xrightarrow{1 \otimes i} E^* \otimes F \otimes E^* \otimes E$

$\pi \downarrow$

$E^* \otimes F \otimes E^* \otimes E \xrightarrow{1 \otimes 1 \otimes e} E^* \otimes F \otimes k$

$\downarrow =$

$E^* \otimes F,$

where the isomorphism π interchanges the two factors E^*. This is the composition (i) of Proposition 10.3 tensored with the identity on F.

A similar argument using (ii) shows that $\psi(\phi(f)) = f$.

Remark 10.5. In the special case $D = k$, $F = E$,

$$\phi : \operatorname{Hom}_*(E, E) \to \operatorname{Hom}_*(k, E^* \otimes E)$$

maps $1 \in \operatorname{Hom}_0(E, E)$ to i. Since ϕ is the inverse of ψ, which is determined by e, the map i, given that it exists, is uniquely determined by e. Symmetrically, i determines e.

The construction of the algebraic dual is, of course, functorial. Let us see how the dual $f^* : F^* \to E^*$ of a linear map $f : E \to F$ can be described in terms of the structure maps i and e for the duals of E and F.

Lemma 10.6 *The dual f^* of $f : E \to F$ is equal to the composition:*

$$F^* = k \otimes F^* \xrightarrow{i \otimes 1} E^* \otimes E \otimes F^* \xrightarrow{1 \otimes f \otimes 1} E^* \otimes F \otimes F^* \xrightarrow{1 \otimes e \circ t} E^* \otimes k = E^*.$$

Expressed in terms of elements rather than arrows, the dual is given by the formula

$$(f^*m)(x) = (-1)^{(\deg f) \cdot (\deg m)} m(f(x)), \quad \text{for } m \in F^*, \ x \in E.$$

The verification is an elementary exercise. Before turning to stable homotopy we note the interpretation of the trace of an endomorphism in terms of the dual.

Proposition 10.7 *Let $f : E \to E$ be a linear endomorphism of degree 0. Then the composition*

$$k \xrightarrow{\ i\ } E^* \otimes E \xrightarrow{\ 1 \otimes f\ } E^* \otimes E \xrightarrow{\ e\ } k$$

is multiplication by the (super) trace of f, $\mathrm{tr}(f) = \mathrm{tr}(f_0) - \mathrm{tr}(f_1)$.

The duality structure which we have just described can be formalized in the abstract setting of a symmetric monoidal category; see [49]. We prefer not to do this, but take duality in the category of finite-dimensional super vector spaces as the model for duality in fibrewise stable homotopy theory.

Duality in stable homotopy theory

Throughout this section base spaces are understood to be compact. We discuss duality in the category $\mathfrak{C}_B$ whose objects are fibrewise stable spaces (X, λ) over B, as in Section 8, where X is a pointed homotopy fibre bundle with fibre of the homotopy type of a finite complex. The morphisms are the fibrewise stable maps. For convenience of notation, we shall denote objects of $\mathfrak{C}_B$ by single letters $X, Y, \ldots$ and often speak as if they are spaces rather than, possibly, stable spaces. We shall use $\mathbf{t}$ for the canonical identification of $X \wedge_B X'$ and $X' \wedge_B X$ for any fibrewise stable spaces X and X'.

Definition 10.8 Let X be a fibrewise stable space over B in the category $\mathfrak{C}_B$. A *dual* for X is a fibrewise stable space X^* in $\mathfrak{C}_B$ equipped with structure maps $\mathbf{i}$ and $\mathbf{e}$, which are fibrewise stable maps:

$$B \times S^0 \xrightarrow{\ \mathbf{i}\ } X^* \wedge_B X \xrightarrow{\ \mathbf{e}\ } B \times S^0,$$

(of degree 0) such that the compositions (i)

$$X^* = (B \times S^0) \wedge_B X^* \xrightarrow{\mathbf{i} \wedge 1_{X^*}} X^* \wedge_B X \wedge_B X^* \xrightarrow{1_{X^*} \wedge \mathbf{e} \circ \mathbf{t}} X^* \wedge_B (B \times S^0) = X^*$$

and (ii)

$$X = X \wedge_B (B \times S^0) \xrightarrow{1_X \wedge \mathbf{i}} X \wedge_B X^* \wedge_B X \xrightarrow{\mathbf{e} \circ \mathbf{t} \wedge 1_X} (B \times S^0) \wedge_B X = X$$

are, respectively, the identity on X^* and the identity on X.

The proof of the next proposition is exactly the same as that of the corresponding algebraic result (Corollary 10.4).

Proposition 10.9 *Let $(X^*; i, e)$ be a dual of a fibrewise stable space X in $\mathfrak{C}_B$ over B. Then for any fibrewise stable spaces W and Y in $\mathfrak{C}_B$ there are inverse isomorphisms*

$$\omega_B^*\{W \wedge_B X; Y\} \underset{\psi}{\overset{\phi}{\rightleftarrows}} \omega_B^*\{W; X^* \wedge_B Y\}$$

defined as follows.

For $f \in \omega_B^*\{W \wedge_B X; Y\}$, $\phi(f) = (1_{X^*} \wedge f) \circ (t \wedge 1_X) \circ (1_W \wedge i)$:

$$W = W \wedge_B (B \times S^0) \xrightarrow{1 \wedge i} W \wedge_B X^* \wedge_B X \xrightarrow{t \wedge 1} X^* \wedge_B W \wedge_B X \xrightarrow{1 \wedge f} X^* \wedge_B Y.$$

For $g \in \omega_B^*\{W; X^* \wedge_B Y\}$, $\psi(g) = (1_Y \wedge e) \circ (t \wedge 1_X) \circ (g \wedge 1_X)$:

$$W \wedge_B X \xrightarrow{g \wedge 1} X^* \wedge_B Y \wedge_B X \xrightarrow{t \wedge 1} Y \wedge_B X^* \wedge_B X \xrightarrow{1 \wedge e} Y \wedge_B (B \times S^0) = Y.$$

A dual, assuming that it exists, is unique.

Corollary 10.10 *Suppose that $(X^*; i, e)$ and $(X_0^*; i_0, e_0)$ are duals of X. Then there is a unique stable fibrewise homotopy equivalence $h : X_0^* \to X^*$ such that $i = (h \wedge 1_X) \circ i_0$ and $e_0 = e \circ (h \wedge 1_X)$.*

Indeed, we have an isomorphism

$$\phi : \omega_B^0\{X_0^* \wedge_B X; B \times S^0\} \to \omega_B^0\{X_0^*; X^*\}$$

determined by i. The equivalence h is $\phi(e_0)$. The verification is formal nonsense.

We may, therefore, talk about *the* dual $(X^*; i, e)$ of X. The dual of X^* is then, by the symmetry of the definition, $(X; t \circ i, e \circ t)$.

Remark 10.11. As in the algebraic case (Remark 10.5), specification of one component of the pair (i, e) determines the other.

It is clear from the definition that duality is compatible with pull-backs: the pull-back of an identity map is an identity map.

Proposition 10.12 *Let $\alpha : B' \to B$ be a map of compact ENRs. If $(X^*; i, e)$ is the dual of X over B, then $(\alpha^* X^*; \alpha^* i, \alpha^* e)$ is the dual of $\alpha^* X$ over B'.*

The existence of duals

We shall demonstrate the existence of duals in the category $\mathfrak{C}_B$ by a gluing construction.

Lemma 10.13 *Let B be a union of closed sub-ENRs B_1 and B_2 with intersection an ENR $A = B_1 \cap B_2$. Suppose that X is a fibrewise stable space in*

$\mathfrak{C}_B$. Writing X_j for the restriction of X to B_j, suppose that $(X_j^*; \mathbf{i}_j, \mathbf{e}_j)$ is a dual of X_j, for $j = 1, 2$. Then X has a dual.

Let $h : (X_1^*)_A \to (X_2^*)_A$ be the unique stable fibrewise homotopy equivalence, whose existence is guaranteed by Corollary 10.10 in the category $\mathfrak{C}_A$, such that

$$(\mathbf{i}_2)_A = (h \wedge 1) \circ (\mathbf{i}_1)_A \quad \text{and} \quad (\mathbf{e}_1)_A = (\mathbf{e}_2)_A \circ (h \wedge 1).$$

By the gluing construction (Proposition 1.30), there is an object X^* of $\mathfrak{C}_B$ equipped with stable fibrewise homotopy equivalences $f_j : X_j^* \to (X^*)_{B_j}$, $j = 1, 2$, compatible with h over A. Replacing X_j^* by the restriction of X^*, we may now assume that X_j^* is precisely the restriction of X^* to B_j and that h is the identity map. We have to construct structure maps $\mathbf{i}$ and $\mathbf{e}$ over B extending those on B_1 and B_2.

From the Mayer–Vietoris sequence (Proposition 3.13), there is a (not necessarily unique) class $\mathbf{e} \in \omega_B^0\{X \wedge_B X^*; B \times S^0\}$ extending $\mathbf{e}_1$ over B_1 and $\mathbf{e}_2$ over B_2, since they agree on the intersection A. By the five-lemma applied again to Mayer–Vietoris sequences, the map

$$\psi : \omega_B^*\{W; X^* \wedge_B Y\} \to \omega_B^*\{W \wedge_B X; Y\}$$

defined as in Proposition 10.9 in terms of $\mathbf{e}$ is an isomorphism. In particular, we have an isomorphism

$$\omega_B^0\{B \times S^0; X^* \wedge_B X\} \overset{\cong}{\longrightarrow} \omega_B^0\{X; X\}.$$

Let $\mathbf{i}$ be the class mapping to 1_X. It restricts to $\mathbf{i}_j$ on B_j and, by a Mayer–Vietoris argument again, defines as in Proposition 10.9 an isomorphism ϕ.

The verification that $(X^*; \mathbf{i}, \mathbf{e})$ is the required dual of X is a rather formal deduction.

The proof that duals exist in the category of finite complexes is due to Spanier and Whitehead. If $(F^*; \mathbf{i}, \mathbf{e})$ is a dual for a finite pointed complex F, then $(B \times F^*; 1 \times \mathbf{i}, 1 \times \mathbf{e})$ is a dual for the trivial bundle $B \times F$. Hence:

Corollary 10.14 *Duals exist in the category $\mathfrak{C}_B$ of locally trivial fibrewise stable spaces over B with fibre of the (stable) homotopy type of a finite complex.*

As in the algebraic model, the dual is functorial. From the uniqueness statement, Corollary 10.10, the dual $(X^*; \mathbf{i}, \mathbf{e})$ of X is a well-defined object of the category. Given a fibrewise stable map $f \in \omega_B^*\{X; Y\}$ between objects X and Y of $\mathfrak{C}_B$, we define the *dual* f^* of f to be the composition:

$$
\begin{array}{ccc}
Y^* = (B \times S^0) \wedge_B Y^* & & X^* \wedge_B (B \times S^0) = X^* \\[4pt]
{\scriptstyle \mathbf{i} \wedge 1} \big\downarrow & & \big\uparrow {\scriptstyle 1 \wedge \mathbf{e} \circ t} \\[4pt]
X^* \wedge_B X \wedge_B Y^* & \overset{1 \wedge f \wedge 1}{\longrightarrow} & X^* \wedge_B Y \wedge_B Y^*
\end{array}
\qquad (10.15)
$$

Then one can check that $(g \circ f)^* = (-1)^{ij} f^* \circ g^*$ for $f \in \omega_B^i\{X; Y\}$ and $g \in \omega_B^j\{Y; Z\}$; and that, under the identification of X^{**} with X and Y^{**} with Y, one has $f^{**} = f$.

The Lefschetz trace and transfer

The trace in the stable category is related to the Lefschetz fixed-point index.

Definition 10.16 Let X be an object of the category $\mathfrak{C}_B$ and $f \in \omega_B^*\{X; X\}$ a fibrewise stable map. Then the *Lefschetz trace* $T_B(f, X) \in \omega^*(B)$ is defined to be the composition:

$$B \times S^0 \xrightarrow{\ \mathbf{i}\ } X^* \wedge_B X \xrightarrow{1 \wedge f} X^* \wedge_B X \xrightarrow{\ \mathbf{e}\ } B \times S^0$$

over B.

We shall verify that this coincides with the fixed-point index defined in Section 6 when both definitions are applicable. The Lefschetz trace has properties analogous to those of the algebraic trace:

Proposition 10.17 (Properties of the Lefschetz trace). *Let X, X' and Y be objects of $\mathfrak{C}_B$.*

(i) $T_B(f, B \times S^0) = f \in \omega^*(B)$, *for a stable self-map* $f \in \omega_B^*\{B \times S^0; B \times S^0\}$.

(ii) (Multiplicativity). *Let* $f \in \omega_B^*\{X; X\}$ *and* $f' \in \omega_B^*\{X'; X'\}$. *Then*

$$T_B(f \wedge f', X \wedge_B X') = T_B(f, X) \cdot T_B(f', X') \in \omega^*(B).$$

(iii) (Commutativity). *Let* $f \in \omega_B^i\{Y; X\}$, $g \in \omega_B^j\{X; Y\}$. *Then*

$$T_B(f \circ g, X) = T_B((f \wedge g) \circ t, X \wedge_B Y) = (-1)^{ij} T_B(g \circ f, Y).$$

(iv) (Symmetry). *For* $f \in \omega_B^*\{X; X\}$ *with dual* $f^* \in \omega^*\{X^*; X^*\}$, *we have*

$$T_B(f^*, X^*) = T_B(f, X) \in \omega^*(B).$$

Taken together, (i) and (ii) show that $T_B : \omega_B^*\{X; X\} \to \omega^*(B)$ is an $\omega^*(B)$-module map.

To define the Lefschetz transfer we have to work with genuine fibrewise pointed spaces, rather than fibrewise stable spaces.

Definition 10.18 Let $X \to B$ be a pointed homotopy fibre bundle with fibre of the homotopy type of a finite complex. The diagonal map $\Delta : X \to X \wedge_B X$ enters crucially into the following definition. For $f \in \omega_B^*\{X; X\}$ we define the *Lefschetz transfer* $\tilde{T}_B(f, X) \in \omega_B^*\{B \times S^0; X\}$ to be the composition:

$$B \times S^0 \qquad\qquad\qquad (B \times S^0) \wedge_B X \;\; = X$$

$$\Big\downarrow {\scriptstyle i} \qquad\qquad\qquad\qquad\qquad \Big\uparrow {\scriptstyle e \wedge 1}$$

$$X^* \wedge_B X \;\xrightarrow{1 \wedge f}\; X^* \wedge_B X \;\xrightarrow{1 \wedge \Delta}\; X^* \wedge_B X \wedge_B X$$

Remark 10.19. The analogue in the category of finite-dimensional super vector spaces is defined for a linear self-map $f : K \to K$ of a finite-dimensional (super) commutative super algebra K. (To be precise, this is the dual of the situation in the stable homotopy category; for an exact analogy we should look at a co-algebra.) The multiplication $K \otimes K \to K$ replaces the diagonal Δ. Reversing arrows in the diagram above we obtain a linear map $K \to k$, which maps an element $x \in K$ to the trace of f composed with multiplication by x. When K is an extension field of k and f is the identity, this is familiar as the trace map in Galois theory.

To relate $T_B(f, X)$ and $\tilde{T}_B(f, X)$ we must specialize further. Suppose that $M \to B$ is a homotopy fibre bundle with fibre of the homotopy type of a finite complex. We now take $X = M_{+B}$. The unique fibrewise map $M \to B$ determines a fibrewise pointed map $p_+ : X = M_{+B} \to B_{+B} = B \times S^0$. In this situation we have:

$$T_B(f, M_{+B}) = p_+ \circ \tilde{T}_B(f, M_{+B}) \in \omega^*(B). \qquad (10.20)$$

(To pursue the algebraic analogy in Remark 10.19, one needs to consider an algebra K with identity, so that k is included as a subring in K. The transfer $K \to k$ restricts to multiplication by the trace: $k \to k$.)

Invertible fibrewise spaces

Hopkins has explained how many algebraic notions, including the idea of an invertible module, can be carried over to stable homotopy theory; see, for example, [75].

Definition 10.21 We say that an object X of $\mathfrak{C}_B$ is *invertible* if its duality structure maps $i : B \times S^0 \to X^* \wedge_B X$ and $e : X^* \wedge_B X \to B \times S^0$ are stable isomorphisms. Of course, if one of the pair (i, e) is an isomorphism, then so is the other.

In the classical situation that B is a point, it is an elementary exercise to see that the invertible objects are the (stable) spheres. From local homotopy triviality and Dold's theorem we deduce that X is invertible if and only if each fibre X_b is (stably) homotopy equivalent to a sphere. Traditionally such fibrewise spaces are called spherical fibrations; the term *pointed homotopy sphere-bundle* conforms better to our terminology.

Proposition 10.22 *An object X of $\mathfrak{C}_B$ is invertible if and only if its fibres have the stable homotopy type of spheres. In other words, the invertible fibrewise spaces are the pointed homotopy sphere-bundles.*

Duality for fibrewise ENRs

For the remainder of this section $M \to B$ will be a compact fibrewise ENR over the compact base B. Recall from Lemma 5.18 that such a fibrewise space is a homotopy fibre bundle. There is a geometric construction of the dual of $X = M_{+B}$, which we now describe. Choose a fibrewise embedding i of M as a closed subspace of ξ for some finite-dimensional real vector bundle ξ over B and a retraction r of some open neighbourhood W onto M:

$$M \overset{i}{\longrightarrow} W \overset{r}{\longrightarrow} M.$$

We identify M with the subspace $i(M) \subseteq \xi$. The construction will depend upon the technical results (5.23)–(5.30) from Section 5, and we use the notation employed there. We shall show that $Y := (C_B(\xi, \xi - M), \xi)$ with appropriate structure maps is the dual of $X := M_{+B}$. It is not obvious that $C_B(\xi, \xi - M)$ is stably locally fibre homotopy trivial (with fibres having the stable homotopy type of finite complexes). This will only emerge indirectly. We shall write down explicit maps

$$\xi_B^+ \overset{i}{\longrightarrow} C_B(\xi, \xi - M) \wedge_B M_{+B} \overset{e}{\longrightarrow} \xi_B^+,$$

determining, by desuspending, stable maps $B \times S^0 \to Y \wedge_B X \to B \times S^0$ satisfying the identities (i) and (ii) of Definition 10.8. Then, given that X has a dual X^* in the category $\mathfrak{C}_B$, we deduce from the argument (Corollary 10.10) on the uniqueness of the dual, that Y is stably fibrewise homotopy equivalent to X^*.

Recall (Proposition 5.26) that the inclusion of the open set W in M gives a fibrewise pointed homotopy equivalence: $C_B(W, W - M) \to C_B(\xi, \xi - M)$. We shall use this equivalence without comment.

The structure maps are defined as compositions:

$$\mathbf{i} : B \times S^0 \overset{p^*}{\longrightarrow} Y \overset{\Delta}{\longrightarrow} Y \wedge_B X,$$
$$\mathbf{e} : Y \wedge_B X \overset{\Delta^*}{\longrightarrow} X \overset{p}{\longrightarrow} B \times S^0.$$

The symbols p and Δ are used to suggest 'projection' and 'diagonal'; we shall see the reason when we look at duality for fibrewise manifolds in Section 12. The maps we write down will actually be suspensions, defined on the smash products with ξ_B^+.

The simplest is

$$p : \xi_B^+ \wedge_B (M_{+B}) \to \xi_B^+.$$

It is the map induced by the projection $M \to B$ (for which we have also used the symbol p).

To define p^* we scale the inner product on ξ so that M is contained in the closed unit disc-bundle $D(\xi)$. From Remark 5.28 we have a fibrewise homotopy equivalence $\xi_B^+ \to C_B(\xi, \xi - D(\xi))$. The composition with the fibrewise map $C_B(\xi, \xi - D(\xi)) \to C_B(\xi, \xi - M)$ induced by the inclusion is

$$p^* : \xi_B^+ \to C_B(\xi, \xi - M) \simeq C_B(W, W - M).$$

The linear map $(x, y) \mapsto (x - y, y) : \xi \times_B \xi \to \xi \times_B \xi$, defined by the difference in the vector bundle, maps the diagonal to the first factor $\xi \times_B (B \times \{0\})$ and restricts to a map of pairs

$$(\xi \times_B M, (\xi \times_B M) - (M \times M)) \to (\xi \times_B M, (\xi - (B \times \{0\}) \times_B M).$$

The composition of induced maps is

$$\Delta^* : C_B(\xi, \xi - M) \wedge_B (M_{+B}) \xrightarrow{\simeq} C_B(\xi \times_B M, (\xi \times_B M) - (M \times_B M))$$
$$\to C_B(\xi, \xi - (B \times \{0\})) \wedge_B C_B(M, \emptyset) \to \xi_B^+ \wedge_B (M_{+B}).$$

At the last step we have used the equivalence in Remark 5.28 and the obvious identification of $C_B(M, \emptyset)$ with M_{+B}.

It is only in the definition of Δ that we use the retraction r of W onto M. The map of pairs $(W, W - M) \to (W \times_B M, (W - M) \times_B M)$: $x \mapsto (x, r(x))$ induces a fibrewise map $C_B(W, W - M) \to C_B(W, W - M) \wedge_B (M_{+B})$, which gives

$$\Delta : C_B(\xi, \xi - M) \to C_B(\xi, \xi - M) \wedge_B (M_{+B}).$$

Having defined i and e, it is not difficult to check the two basic identities. The argument involves two further 'diagonal' maps:

$$\Delta : X \to X \wedge_B X \quad \text{and} \quad \Delta^* : Y \wedge_B Y \to Y.$$

The first is induced by the geometric diagonal map $M \to M \times_B M$. The second is the desuspension of the map

$$C_B(W, W - M) \wedge_B C_B(W, W - M) \to C_B(\xi, \xi - 0) \wedge_B C_B(W, W - M)$$

determined by the map of pairs from $(W \times_B W, W \times_B W - M \times_B M)$ to $(\xi \times_B W, \xi \times_B W - 0 \times_B M)$ taking (x, y) to $(x - y, y)$. It behaves like a geometric diagonal map in that it commutes with switching the factors.

Lemma 10.23 *In the notation above, we have*

$$\Delta^* \circ t = \Delta^* : Y \wedge_B Y \to Y.$$

Verification of the identities (i) and (ii) now reduces to establishing the commutativity of the following diagrams:

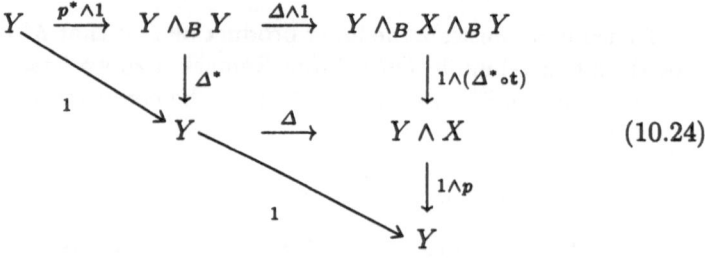

$$(10.24)$$

and

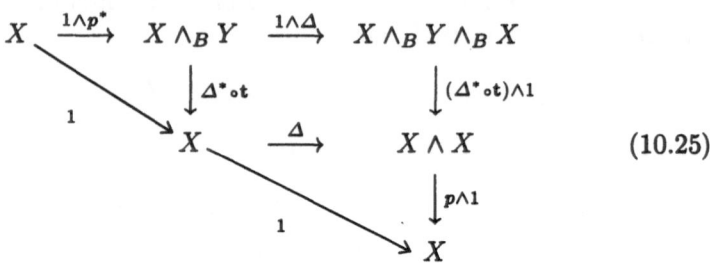

$$(10.25)$$

(Further details can be found in [49] and [99].)

In the classical case in which B is a point, one can verify, by using the fact that the homology of the stable dual constructed in this way is dual to the homology of the compact ENR M, that the stable dual has the stable homotopy type of a finite complex.

The fixed-point index

We conclude this discussion of duality theory by relating the fixed-point theory of Section 6 to the Lefschetz trace and transfer defined using duality.

Proposition 10.26 *Let $f : M \to M$ be a fibrewise self-map of a compact fibrewise ENR M over a compact ENR B. Then the Lefschetz–Hopf fixed-point index, $\tilde{L}_B(f, M)$, of f and the Lefschetz transfer, $\tilde{T}_B(f_+, M_{+B})$, of the induced fibrewise pointed self-map f_+ of M_{+B} coincide:*

$$\tilde{L}_B(f, M) = \tilde{T}_B(f_+, M_{+B}) \in \omega_B^0\{B \times S^0; M_{+B}\}.$$

As a consequence,

$$L_B(f, M) = T_B(f_+, M_{+B}) \in \omega^0(B).$$

The deduction of the equality of the fixed-point index and the trace is immediate, by (10.20). The main statement is proved by expanding the definition of the transfer $\tilde{T}_B(f_+, M_{+B})$ in terms of the explicit duality maps i and e constructed above.

Chapter 3. Manifold Theory

11 Fibrewise differential topology

In this section we take up the discussion of fibrewise differential topology begun in Part I, Section 22. As we observed there, fibrewise manifolds appeared in the work of Atiyah and Singer [5] on the Index Theorem for families of elliptic pseudodifferential operators and were there called *manifolds over a base*. The definition that we give here is rather more general than that given by Atiyah and Singer, although their definition covers most of the important examples. We broaden the definition in such a way that open subspaces of fibrewise manifolds are also regarded as fibrewise manifolds.

In this section and in the sequel, all manifolds are assumed to be smooth, finite-dimensional, Hausdorff, with a countable basis. Unless the contrary is indicated, manifolds are assumed to be without boundary.

Fibrewise manifolds and fibrewise smooth maps

Fibrewise manifolds over a base B will be modelled on open subsets

$$W \subseteq B \times E,$$

of a trivial bundle with fibre a finite-dimensional real vector space E. A fibrewise map $f : W \to B \times F$ to another such trivial vector bundle with fibre F is said to be *fibrewise smooth* if the following condition holds. The restriction of f to fibres at $b \in B$ is a map $f_b : W_b \to F$ from an open subset of the vector space E to the vector space F. We require that each map f_b be smooth (that is, C^∞), and that for each $k \geq 1$ the kth derivative should define a continuous map

$$W \to S^k(E; F) : (b, x) \mapsto D^k f_b(x)$$

to the vector space of symmetric k-multilinear maps from E to F.

We now define a *smooth structure* on a fibrewise space $M \to B$ over B in the classical way by a family of coordinate charts Φ covering M. Each chart $\phi \in \Phi$ is a fibrewise homeomorphism

$$\phi : W_\phi \to M_\phi \tag{11.1}$$

from an open subset W_ϕ of $B \times E_\phi$ for some (finite-dimensional) Euclidean space E_ϕ to an open subset M_ϕ of M. The transition functions are required to be fibrewise smooth. To spell this out, suppose that $\phi, \psi \in \Phi$. Then $\psi^{-1} \circ \phi : \phi^{-1}(M_\phi \cap M_\psi) \to W_\psi$, composed with the inclusion, is a fibrewise map from an open subset of $B \times E_\phi$ to $B \times E_\psi$. This map should be fibrewise smooth in the sense that we have defined above. Such a family of charts defines a smooth structure on the fibrewise space $M \to B$.

The general notion of fibrewise smooth map is defined in the obvious way. Let $M \to B$ and $N \to B$ be fibrewise spaces, each equipped with a smooth structure, and let $f : M \to N$ be a fibrewise map. Then f is *fibrewise smooth* if for each chart ϕ, defined on $W_\phi \subseteq B \times E$, for M and each chart ψ, defined on $W_\psi \subseteq B \times F$, for N the map

$$\psi^{-1} \circ f \circ \phi : W_\phi \cap f^{-1}(W_\psi) \to W_\psi \subseteq B \times F$$

is fibrewise smooth.

Definition 11.2 Let $M \to B$ be a fibrewise space (over an ENR B) admitting a smooth structure. We assume that the topological space M is Hausdorff and has a countable basis. As in the standard definition of a manifold we regard the smooth structures on M defined by two families of coordinate charts as equivalent if the identity map from M equipped with one smooth structure to the same space equipped with the other smooth structure is fibrewise smooth in the sense defined above. We call $M \to B$ with an equivalence class of smooth structures a *fibrewise (smooth) manifold* over B.

(The reader may consider it prudent to take the vector spaces E_ϕ allowed in the definition of a coordinate chart from some chosen set of vector spaces. In that case we may take the manifold structure to be given by a maximal atlas of charts.)

With this definition (which differs from that in Section 22 of Part I) any open subspace of a fibrewise manifold is naturally a fibrewise manifold.

Given a continuous map $\alpha : B' \to B$, the pull-back $\alpha^* M \to B'$ of a fibrewise manifold $M \to B$ has a natural structure as a fibrewise manifold: a chart $\phi : W_\phi \to M_\phi$ for M lifts to a chart

$$\alpha^* W_\phi \ (\subseteq B' \times E_\phi) \to \alpha^* M_\phi \ (\subseteq \alpha^* M)$$

for $\alpha^* M$. A fibrewise smooth map $f : M \to N$ over B pulls back to a fibrewise smooth map $\alpha^* f : \alpha^* M \to \alpha^* N$ over B'.

In particular, each fibre M_b of a fibrewise manifold $M \to B$ is a smooth manifold. If each fibre is of dimension n, that is, if the fibrewise manifold is modelled on open subsets of $B \times \mathbb{R}^n$, we say that $M \to B$ has (fibre) dimension n.

The term *fibrewise diffeomorphism* is used for an equivalence in the category of fibrewise manifolds and smooth maps over a given base. Products exist in the category: the topological product $M \times_B M'$ of fibrewise manifolds over B is given a smooth structure by forming the products of coordinate charts.

Example 11.3. As a simple example, let us look at dimension 0. The fibrewise space $p : M \to B$ is a 0-dimensional fibrewise manifold if each point of M has an open neighbourhood which maps homeomorphically under p to an open subset of B. The manifold structure is then unique. In this case it is not difficult to show that, if M is fibrewise compact, then $M \to B$ is a finite covering space; see Part I, Proposition 1.12. We shall generalize this result to higher dimensions in Proposition 11.13.

Example 11.4. Let M and B be smooth manifolds and let $p : M \to B$ be a smooth submersion. Then $M \to B$ has a natural structure as fibrewise manifold.

Indeed, let us say that a coordinate chart ϕ,

$$\phi : W_\phi \to M_\phi$$

in the notation of (11.1), is compatible with the smooth structures on M and B if it is a diffeomorphism from the open submanifold W_ϕ of $B \times E_\phi$ to the open submanifold M_ϕ of M. The collection of such compatible charts gives $M \to B$ the structure of a fibrewise manifold.

We say that $M \to B$ is a *smooth fibre bundle* if each point $b \in B$ has an open neighbourhood U such that there is a trivialization $U \times M_b \to M_U$ over B compatible with the smooth structures on the product $U \times M_b$ and the open submanifold M_U of M.

For the remainder of this section, $M \to B$ is supposed to be a fibrewise manifold.

A function $\rho : M \to \mathbb{R}$ defined on a fibrewise manifold is said to be *fibrewise smooth* if the associated fibrewise map $M \to B \times \mathbb{R}$ is fibrewise smooth over B. The assumptions on the topology of M ensure that M admits fibrewise smooth partitions of unity subordinate to any open covering. The proof proceeds just as in the case of an ordinary manifold.

Locally trivial fibrewise manifolds

The most important examples of fibrewise manifolds are locally trivial. If F is a (smooth, Hausdorff, with a countable basis) manifold, then the trivial bundle $B \times F \to B$, being the pull-back by the map $B \to *$, has a natural smooth structure. We call a fibrewise smooth manifold *trivial* if it is fibrewise diffeomorphic to such a manifold $B \times F \to B$ and *locally trivial* if each point of the base has an open neighbourhood over which the fibrewise manifold is trivial. We shall also refer to a locally trivial fibrewise manifold over B as a *fibrewise smooth fibre bundle* over B.

An important class of examples arises from fibre bundles with structure group a Lie group. Let G be a Lie group acting smoothly on a manifold F and let $P \to B$ be a principal G-bundle. Then $M := P \times_G F \to B$ has a natural structure as fibrewise smooth fibre bundle with local trivializations determined by local trivializations of $P \to B$. For example, any real vector bundle of dimension n over M is a fibrewise smooth fibre bundle of (fibre) dimension n.

In Section 5 we saw that a fibrewise compact ANR is a homotopy fibre bundle. We shall see below that a fibrewise compact manifold is a fibrewise smooth fibre bundle.

The fibrewise tangent bundle

A *tangent vector along the fibres* at a point $x \in M$, in the fibre over $b \in B$, is a tangent vector to the fibre M_b at x. The fibrewise tangent space $(\tau_B M)_x$ at $x \in M$ is the tangent space $(\tau M_b)_x$ to the fibre. These fibrewise tangent spaces are the fibres of a vector bundle $\tau_B M$ over M which we shall call the *fibrewise tangent bundle* of M (or, in more classical terminology, the *bundle of tangents along the fibres*).

The topology on $\tau_B M$ is defined by the coordinate charts as follows. A chart ϕ as in (11.1) gives a vector bundle isomorphism over B from $W_\phi \times E_\phi$ to the open subset $\tau_B M \mid M_\phi$ of $\tau_B M$:

$$
\begin{array}{ccc}
W_\phi \times E_\phi & \xrightarrow{\phi'} & \tau_B M \mid M_\phi \\
\downarrow & & \downarrow \\
W_\phi & \xrightarrow{\phi} & M_\phi
\end{array}
\tag{11.5}
$$

Considered as a fibrewise space over B, $\tau_B M \to B$ has a natural fibrewise smooth structure determined by requiring that the map ϕ' in the diagram above be fibrewise smooth over B when $W_\phi \times E_\phi$ is given the standard smooth structure as an open subspace of $B \times (E_\phi \oplus E_\phi)$. The given local trivialization (11.5) of $\tau_B M$ is then fibrewise smooth over B. We say that $\tau_B M$ is fibrewise smooth *as a vector bundle* over B.

The *derivative* of a fibrewise smooth map $f : M \to N$ over B, restricting on fibres at $b \in B$ to the usual derivative $f_b' : (\tau M)_b \to (\tau N)_b$, is a vector bundle homomorphism $f' : \tau_B M \to \tau_B N$ lifting $f : M \to N$. It is fibrewise smooth over B.

For the case of a smooth submersion, as discussed in Example 11.4, the fibrewise tangent bundle is the kernel:

$$
0 \to \tau_B M \longrightarrow \tau M \xrightarrow{p'} p^* \tau B \to 0
$$

of the derivative p' of the projection $p : M \to B$. When $M \to B$ is a smooth fibre bundle this is traditionally known as the bundle of tangents along the fibres of p and is sometimes written as $\tau(p)$.

Fibrewise submanifolds

Fibrewise submanifolds may be defined just as in the classical theory. Suppose that a fibrewise manifold $M \to B$ is a subspace of a fibrewise manifold $N \to B$. We say that M is a *fibrewise submanifold* of N if N admits a covering by coordinate charts of the form

$$\phi : W_\phi \to N_\phi,$$

where W_ϕ is an open subset of $B \times (E_\phi \oplus F_\phi)$ for some Euclidean spaces E_ϕ and F_ϕ, such that ϕ restricts to a coordinate chart

$$W_\phi \cap (B \times E_\phi) \to M \cap N_\phi$$

for M.

The *fibrewise normal bundle* of the embedding $i : M \hookrightarrow N$ is the quotient $\nu(i) := i^* \tau_B N / \tau_B M$. It is a vector bundle over M which is fibrewise smooth over B.

In talking about such a vector bundle ν over M there is a potential source of confusion in the notation for fibres: for $x \in M$, ν_x is the vector space fibre at x, whilst for $b \in B$, ν_b is the vector bundle over M_b obtained by restriction of ν to that subspace.

Immersions and embeddings

Definition 11.6 A fibrewise smooth map $f : M \to N$ is said to be (i) a *fibrewise immersion* if the derivative $f'_x : (\tau_B M)_x \to (\tau_B N)_{f(x)}$ is injective at each point $x \in M$; (ii) a *fibrewise embedding* if $f(M)$ is a fibrewise submanifold of N and f defines a fibrewise diffeomorphism $M \to f(M)$.

Remark 11.7. Suppose that f'_x is injective at each point x of a subspace $K \subseteq M$. Then, since injectivity of a linear map of finite-dimensional vector spaces is an open condition, there is an open neighbourhood W of K in M such that $f \,|\, W : W \to N$ is a fibrewise immersion.

As in the classical situation, an embedding is easily seen to be an immersion and with more effort one can show that an immersion is locally an embedding:

Lemma 11.8 *Let $f : M \to N$ be a fibrewise immersion over B, and let $x \in M$. Then there is an open neighbourhood W of x in M such that $f \,|\, W : W \to N$ is a fibrewise embedding.*

This is essentially a parametrized inverse/implicit function theorem. We omit the proof, which follows closely the familiar case in which B is a point.

Compact fibrewise manifolds can be embedded as submanifolds of a trivial vector bundle $B \times E$ by using a finite covering by coordinate charts and a partition of unity. More generally, we have:

Lemma 11.9 *Let K be a compact subspace of a fibrewise manifold M over B. Then there is an open neighbourhood W of K in M which embeds in $B \times E$ for some (finite-dimensional) Euclidean space E.*

Let Φ, in the notation that we have used before, in (11.1), be a finite family of coordinate charts of M covering K. From ϕ, defined on the open subset M_ϕ, we obtain a fibrewise smooth embedding $i_\phi : M_\phi \to B \times E_\phi$.

Now let W be an open neighbourhood of K with closure contained in $\bigcup M_\phi$. Then we may choose a partition of unity (ρ_ϕ) on W such that the closure of the support of ρ_ϕ is contained in M_ϕ. Take E to be the finite direct sum of the Euclidean spaces E_ϕ. Then

$$i := \sum_{\phi \in \Phi} \rho_\phi i_\phi : W \to B \times \bigoplus_{\phi \in \Phi} E_\phi = B \times E$$

is the required embedding. (The function $\rho_\phi i_\phi$, defined in the first place only on M_ϕ, is extended over M to be zero outside the support of ρ_ϕ.)

We shall need the following generalization of Lemma 11.8.

Lemma 11.10 *Let $f : M \to N$ be a fibrewise immersion, and let $K \subseteq M$ be a compact subspace such that the restriction of f to K is injective. Then there is an open neighbourhood W of K in M such that $f \mid W : W \to N$ is a fibrewise embedding.*

By Lemma 11.9, applied to K and $f(K)$, we may assume that both M and N are submanifolds of trivial vector bundles over B and so, in particular, are metrisable. We write d for the metric on both.

It follows from Lemma 11.8, by an elementary compactness argument, that there is a positive real number $\epsilon > 0$ such that the restriction of f to any open ball of radius 2ϵ in M centred on a point of K is a fibrewise embedding. From the compactness of K and the injectivity of f on K, there is a positive $\delta < \epsilon$ such that $d(f(x), f(y))$ is bounded away from 0 on the set of pairs

$$\{(x, y) \in M \times M \mid d(x, K) \leq \delta, d(y, K) \leq \delta, d(x, y) \geq \epsilon\}.$$

We may take W to be the union of the open balls of radius δ centred on points of K.

Fibrewise transversality

Let M, N and Q be fibrewise manifolds over B, and let $f : M \to N$ and $s : Q \to N$ be fibrewise smooth maps. We write

$$P := \{(x,y) \in Q \times_B M \mid s(x) = f(y)\} \subseteq Q \times_B N$$

for the fibre product $Q \times_N M$. The maps f and s are said to be *transverse* if for each $(x,y) \in P$, with $s(x) = z = f(y) \in N$, the sum of derivatives

$$s'_x \oplus f'_y : (\tau_B Q)_x \oplus (\tau_B M)_y \to (\tau_B N)_z \qquad (11.11)$$

is surjective.

The first assertion of the next lemma is local and is a consequence of the generalized inverse/implicit function theorem.

Lemma 11.12 *Suppose that the fibrewise smooth maps f and s as above are transverse. Then P has a natural structure as a fibrewise manifold over B and for each $(x,y) \in P$, there is a short exact sequence of vector spaces:*

$$0 \to (\tau_B P)_{(x,y)} \to (\tau_B Q)_x \oplus (\tau_B M)_y \xrightarrow{s'_x \oplus f'_y} (\tau_B N)_z \to 0,$$

where $s(x) = z = f(y)$.

If M is fibrewise compact and $s : Q \to N$ is proper, then P is a fibrewise compact submanifold of the product $Q \times_B M$.

The condition that the fibrewise map $s : Q \to N$ over B be proper means (in the present context of locally compact Hausdorff spaces) that the inverse image of a fibrewise compact subspace is fibrewise compact.

Fibrewise compactness

Proposition 11.13 *Let $M \to B$ be a fibrewise compact manifold over B. Then $M \to B$ is locally trivial as a fibrewise manifold.*

Since the problem is local, there is no loss of generality in assuming that B is compact. (Indeed, we could simply take B to be a closed disc.) The space M is then compact and, by Lemma 11.9, we may suppose that M is a fibrewise submanifold of $B \times E$ for some Euclidean space E. We may also assume that M is connected and so of constant dimension. Fix $a \in B$. We shall produce a trivialization in a neighbourhood of a.

Each fibre M_b is a closed submanifold of E. The normal bundle ν_b of $M_b \hookrightarrow E$ is the restriction of the fibrewise normal bundle ν over M of the embedding $M \hookrightarrow B \times E$. Using the Euclidean metric, we identify ν with the orthogonal complement of $\tau_B M$ in the trivial bundle $M \times E$.

We define $j : \nu \to B \times E$ as follows. A point $x \in M$ may be written as $(b,u) \in B \times E$, and an element y of ν_x is then described as $(x,v) \in M \times E$.

We define $j(y) := (b, u + v)$ and write the length of the normal vector as $\|y\| = \|v\|$. The map j is a fibrewise immersion: its derivative is essentially the identity map at each point. By Lemma 11.8, j embeds an open neighbourhood of M in ν into $B \times E$, and using compactness again we may take this neighbourhood to be of the form $B_\epsilon(\nu) := \{y \in \nu \mid \|y\| < \epsilon\}$ for some $\epsilon > 0$. We have thus constructed, rather explicitly, a fibrewise tubular neighbourhood of M in $B \times E$. (We shall look at the general construction of tubular neighbourhoods shortly.)

Let us write $i : M \hookrightarrow W$ for the inclusion of M in the open tubular neighbourhood $W := j(B_\epsilon(\nu))$ and $r : W \to M$ for the fibrewise retraction given by collapsing the disc-bundle to its zero-section: $r(j(y)) = x$ if $y \in \nu_x$, $x \in M$. We have established:

Corollary 11.14 *A fibrewise compact manifold over B is a fibrewise ENR.*

Consider the fibres M_b and W_b as subspaces of E. By compactness of M_a, there is an open neighbourhood U of a in B such that $M_a \subseteq W_b$ for $b \in U$. So over U we have an embedding of $U \times M_a$ in W_U. The composition with the retraction r:

$$U \times M_a \to W_U \overset{r}{\longrightarrow} M_U,$$

which is the identity on the fibre at a, will give the sought-for trivialization over a (possibly) smaller neighbourhood of a. By Remark 11.7 and Lemma 11.8 its restriction to a neighbourhood of a is a fibrewise embedding. Since the fibres are manifolds of the same dimension, M_a is embedded as a union of components of M_b for b close to a. To see that the map is surjective, near a, delete the image of the trivial bundle with fibre M_a to get a fibrewise compact manifold with empty fibre at a. Because of compactness fibres must be empty in a neighbourhood of a. This completes the proof of Proposition 11.13.

Fibrewise smooth Euclidean metrics

In our discussion of fibrewise tangent and normal bundles we have already met the idea of a fibrewise smooth vector bundle ξ over a fibrewise manifold $M \to B$. The (total) space ξ is a fibrewise manifold over B, and as a vector bundle over M admits, in a neighbourhood of any point, a local trivialization which is fibrewise smooth over B. We shall need to equip such a vector bundle with a *fibrewise smooth Euclidean metric*. The meaning should be clear. A metric is, in particular, a section of the fibrewise smooth vector bundle $(\xi \otimes \xi)^*$; it should be fibrewise smooth over B as a map $M \to (\xi \otimes \xi)^*$. Equivalently, the Euclidean vector bundle should admit fibrewise smooth local trivializations. The existence of fibrewise smooth metrics is established in the usual way with the help of a fibrewise smooth partition of unity on M.

Tubular neighbourhoods

For the rest of this section we shall take the base space B to be compact.

Let $M \to B$ be a compact fibrewise manifold and let $i : M \to N$ be a fibrewise embedding with fibrewise normal bundle $\nu = \nu(i)$. Choose a fibrewise smooth Euclidean metric g on ν and write the length of a normal vector y as $\|y\|$. Let $D_\epsilon(\nu) := \{y \in \nu \mid \|y\| \le \epsilon\}$ denote the disc-bundle of radius ϵ (> 0) in ν. The fibrewise manifold M is embedded in ν as the zero-section with normal bundle (naturally identified with) ν.

Definition 11.15 In the situation described above, a *tubular neighbourhood* of i is a fibrewise embedding

$$j : W \hookrightarrow N$$

of an open neighbourhood of the disc-bundle $D_\epsilon(\nu)$ of radius ϵ in ν, for some $\epsilon > 0$, which extends i on M and whose derivative j' induces the identity map from the normal bundle ν of M in $D_\epsilon(\nu)$ to the normal bundle ν of M in N.

In this definition we are only interested in the embedding of the disc-bundle $D_\epsilon(\nu)$, but require that j be defined on an open neighbourhood to provide differentiability on the bounding sphere-bundle $S_\epsilon(\nu) = \partial D_\epsilon(\nu)$.

For dimensional reasons, j gives a diffeomorphism between W and the open subset $j(W)$ of N.

The existence of tubular neighbourhoods is an essential ingredient in Differential Topology.

Proposition 11.16 *Let $i : M \to N$ be a fibrewise embedding of a compact fibrewise manifold M in a fibrewise manifold N over a compact ENR B. Then i admits a tubular neighbourhood.*

The standard construction, using a Riemannian metric on a neighbourhood of $i(M)$ in N and the exponential map which it defines, carries through with only minor modifications.

There is no loss of generality in assuming that M is a submanifold of N and that N is covered by a finite number of charts. Choose a fibrewise smooth Riemannian metric on N, that is, in the terminology above, a Euclidean metric on $\tau_B N$. We use the metric to identify $\nu = \nu(i)$ with the orthogonal complement of $\tau_B M$ in $(\tau_B N) \,|\, M$. The restriction of the exponential map, defined on a neighbourhood of the zero-section in $(\tau_B N) \,|\, M$, gives a fibrewise smooth map

$$j : W \to N$$

on an open neighbourhood of M in ν. And the derivative of j on M is, by construction, the identity map. By Lemma 11.8, we may replace W by a smaller open neighbourhood on which j is an embedding. Again by compactness W will contain the disc-bundle $D_\epsilon(\nu)$ for $\epsilon > 0$ sufficiently small.

This establishes the existence of tubular neighbourhoods. We also need the uniqueness up to homotopy.

Proposition 11.17 *Let $i : M \to N$ be a fibrewise embedding of a compact fibrewise manifold M in a fibrewise manifold N over a compact ENR B. Let j_0 and j_1 be tubular neighbourhoods of i of the same radius ϵ in possibly different metrics. Write $\pi : B \times [0,1] \to B$ for the projection. Then there is a tubular neighbourhood $j : W \to \pi^* N$ of the pull-back $\pi^* i : \pi^* M \to \pi^* N$ of $i : M \to N$ to $B \times [0,1]$ which restricts to j_0 and j_1 at the ends $B \times \{0\}$ and $B \times \{1\}$.*

In describing the tubular neighbourhoods j_0 and j_1 we use the notation g, $\|.\|$ and W of Definition 11.15 with distinguishing subscripts 0 and 1.

The basic idea is to deform j_0 linearly into j_1 by defining, for $0 \le t \le 1$, $j_t(y) = j_0((1-t)y + tj_0^{-1}(j_1(y)))$, where this makes sense. The addition takes place in the convex disc-bundle $D_\epsilon(\nu)$ in the g_0-metric.

To deal more carefully with the end-points, let $\rho : [0,1] \to [0,1]$ be a continuous function which takes the value 0 on $[0, 1/3]$ and the value 1 on $[2/3, 1]$. Write U for the open disc-bundle $\{y \in \nu \mid \|y\|_0 < \epsilon\}$. Then $V := U \cap j_1^{-1}(j_0 U)$ is an open neighbourhood of M in $W_0 \cap W_1 \subseteq \nu$. Let W' be the open subset $(W_0 \times [0, 1/3)) \cup (V \times [0,1]) \cup (W_1 \times (2/3, 1])$ of $\pi^* \nu$. We define a fibrewise map $j : W' \to N \times [0,1]$ over $B \times [0,1]$ by:

$$
j(y,t) = \begin{cases} j_0(y) & \text{for } 0 \le t \le \frac{1}{3}, \\ j_0((1 - \rho(t))y + \rho(t)j_0^{-1}(j_1(y))) & \text{for } y \in V, \\ j_1(y) & \text{for } \frac{2}{3} \le t \le 1. \end{cases}
$$

By the standard argument, there is an open neighbourhood W'' of $M \times [0,1]$ in W' on which j is a fibrewise embedding. And we may certainly arrange that W_t'' coincides with $W_t' = W_0$ for t in a neighbourhood of 0 and with $W_t' = W_1$ in a neighbourhood of 1. This allows us to rename W'' as the required open subset W of $\pi^* \nu$.

It remains to define a metric g on $\pi^* \nu$, extending g_0 and g_1, such that the closed ϵ-disc lies in W. This is easily done by taking $g_t = \alpha(t)g_0 + \beta(t)g_1$ for suitably chosen functions α, $\beta : [0,1] \to [0, \infty)$.

The diagonal embedding

An important example of a fibrewise embedding is the diagonal inclusion

$$
\Delta : M \to M \times_B M.
$$

The restriction of $\tau_B(M \times_B M)$ to M is the direct sum $\tau_B M \oplus \tau_B M$, into which the derivative Δ' includes $\tau_B M$ as the diagonal. The normal bundle $\nu(\Delta)$ can, therefore, be identified with $\tau_B M$ by the composition:

$$\tau_B M \xrightarrow{1 \oplus 0} \tau_B M \oplus \tau_B M \longrightarrow \nu. \tag{11.18}$$

We make this choice of isomorphism, taking the first rather than the second factor, once for all.

The group $\mathbb{Z}/2$ acts on $M \times_B M$ by interchanging the factors, with fixed subspace $\Delta(M)$. If we choose a Riemannian metric on M and use the exponential map defined by the product metric on $M \times_B M$ to construct a tubular neighbourhood of the diagonal, then everything will be automatically equivariant.

Let L denote the $\mathbb{Z}/2$-module $\mathbb{R}$ with the non-trivial action of the generator as -1. Then $L \otimes \tau_B M$ is just the vector bundle $\tau_B M$ with the antipodal action of $\mathbb{Z}/2$. The equivariant tubular neighbourhood embeds the disc-bundle

$$D_\epsilon(L \otimes \tau_B M) \hookrightarrow M \times_B M \tag{11.19}$$

into the square $M \times_B M$.

Local triviality of fibrewise embeddings

In the differentiable category we have the following analogue of the local homotopy triviality of maps between homotopy fibre bundles (Proposition 1.2).

Proposition 11.20 *Let $i : M \to N$ be a fibrewise embedding of a compact fibrewise manifold M into a fibrewise smooth fibre bundle N over a compact ENR B. Then for each $b \in B$ there exists an open neighbourhood U of b and fibrewise diffeomorphisms $\phi : M_U \to U \times M_b$ and $\psi : U \times N_b \to N_U$ over U such that $i_U = \psi \circ (1 \times i_b) \circ \phi$.*

The key idea is familiar as the statement that, for any $v \in \mathbb{R}^n$, there is a diffeomorphism of $\mathbb{R}^n$ which is the identity outside a compact set and maps 0 to v. It implies, for example, that the diffeomorphism group of a connected manifold acts transitively on the manifold. The construction generalizes easily in the following way.

Lemma 11.21 *Let ξ be a smooth vector bundle, equipped with a smooth Euclidean metric, over a manifold A. Let s be a smooth section of ξ with $\|s(b)\| < 1/2$ for all $b \in A$. Then there is a smooth fibrewise diffeomorphism $\theta : \xi \to \xi$ over A such that, for all $b \in A$, $\theta_b(0) = s(b)$ and $\theta_b(v) = v$ for $v \in \xi_b$ with $\|v\| \geq 1$.*

Choose a smooth bump function $\rho : [0, \infty) \to \mathbb{R}$ such that $0 \leq \rho(t) \leq 1$ for all t, while $\rho(t) = 1$ for $t \leq 1/4$ and $\rho(t) = 0$ for $t \geq 1$. For $b \in A$ consider the vector field v_b on the Euclidean space ξ_b given by

$$v_b(x) = \rho(\|x\|^2)s(b).$$

This has compact support contained in the unit disc $D(\xi_b)$ and $v_b(x) = s(b)$ for $\|x\| \leq 1/2$. Take θ_b to be $\Theta_b(1)$, where Θ_b is the flow determined by the vector field v_b. Since the section s is smooth, the diffeomorphism θ_b will vary smoothly with b.

This completes the proof of the lemma and we can proceed to the proof of Proposition 11.20. We may assume that $M = B \times F$ and $N = B \times F'$ are trivial. Embed F in F' by i_b with normal bundle ν. Let $D_2(\nu) \hookrightarrow F'$ be a tubular neighbourhood of radius 2 for some Euclidean metric on ν. We shall think of the disc-bundle as a subspace of F'. For a sufficiently close to b, say $a \in U$ where U is an open neighbourhood of b, i_a will map F into the open disc-bundle of radius $1/2$ and composition with the projection onto F:

$$F \xrightarrow{i_a} D_2(\nu) \longrightarrow F$$

will be a diffeomorphism ϕ_a. To simplify notation we may now suppose that the fibrewise diffeomorphism ϕ is the identity. The embedding i_a is then described by a section s_a of ν: $i_a(x) = (x, s_a(x))$, where $s_a(x) \in \nu_x$ and $\|s_a(x)\| < 1/2$.

We are going to use the lemma to construct a fibrewise diffeomorphism $\psi : U \times F' \to U \times F'$ such that ψ_a is the identity outside the tubular neighbourhood $D_1(\nu)$ of radius 1. To do this we can work entirely on the bundle ν.

Write A for the open disc-bundle of radius $1/2$ in ν and let ξ be the pull-back of ν to A. We take the canonical section of ξ to be s: $s(x) = (x, x)$. The required fibrewise diffeomorphism ψ of $U \times \nu$ is obtained by pulling back the diffeomorphism θ of Lemma 11.21 by the map

$$U \times F \to A : (a, x) \mapsto (x, s_a(x)).$$

Remark 11.22. As an example we apply the result to a special case mentioned earlier. Given a closed manifold B, consider the trivial smooth fibre bundle $p : N = B \times B \to B$, the projection onto the first factor. The diagonal map gives a fibrewise embedding $M = B \to B \times B = N$ over B. Then Proposition 11.20 establishes the result claimed in Proposition 1.13, at least for a smooth manifold. But it is easy to see that the proof in this case uses only the existence of a tubular neighbourhood of a point, and for that it suffices that B be a topological manifold.

Fibrewise manifolds with boundary

We shall not deal at length with fibrewise manifolds with boundary. In any case it is only fibrewise compact manifolds with boundary which we shall meet. We extend the class of coordinate charts to include homeomorphisms over B to open subsets of $B \times (E \times [0, \infty))$ for some Euclidean space E. Given a fibrewise space $M \to B$, a local trivialization $M_U \to U \times F$ where F

is a compact manifold with boundary supplies such a smooth structure over U. Local trivializations are compatible if the transition functions on charts are fibrewise smooth. A family of compatible local trivializations defines on $M \to B$ the structure of a *fibrewise manifold with boundary*. The boundary of M is a fibrewise manifold $\partial M \to B$ which is *closed*, that is, fibrewise compact and without boundary. By gluing two copies of M along their common boundary one obtains a fibrewise compact manifold $N \to B$ without boundary containing ∂M as a fibrewise submanifold of codimension 1 (with trivial normal bundle). A tubular neighbourhood $\partial M \times (-1, 1) \hookrightarrow N$ provides a collar neighbourhood of ∂M in M.

Existence and classification of fibrewise embeddings into vector bundles

We have already (in Lemma 11.9) established:

Proposition 11.23 *Let $M \to B$ be a fibrewise manifold over a compact ENR B. Suppose that M is compact. Then there is a fibrewise smooth embedding $M \hookrightarrow \xi$, where ξ is a finite-dimensional real vector bundle over B.*

In fact the vector bundle ξ that we constructed was trivial.

Remark 11.24. The fibrewise compactness of M is not essential. If M has constant (or bounded) fibre dimension one can still use a fibrewise smooth partition of unity to embed M in a finite-dimensional trivial bundle.

Given a compact fibrewise manifold M of constant (fibre) dimension m and a vector bundle ξ of dimension n over B, we now investigate the existence and classification of fibrewise embeddings as in Proposition 11.23 of M into $N = \xi$. (It is the affine structure of N, with the action of ξ by translation, which we shall actually use.)

Consider a fibrewise embedding $i : M \to N$. We define an associated map $\tilde{h}(i) : M \times_B M - M \to S(\xi)$ from the complement of the diagonal in $M \times_B M$ to the sphere-bundle of ξ, thought of as the quotient of the complement of the zero-section by the action of the positive reals (or, alternatively, as the unit sphere-bundle with respect to a chosen Euclidean metric), mapping (x, y) to the class of $i(x) - i(y)$. Interchanging x and y changes the sign of $i(x) - i(y)$. We regard $\tilde{h}(i)$ as a fibrewise $\mathbb{Z}/2$-equivariant map

$$\tilde{h}(i) : M \times_B M - M \to S(L \otimes \xi) \qquad (11.25)$$

over B. Here, as earlier in this section, the group $\mathbb{Z}/2$ permutes the factors of $M \times_B M$ and L is the $\mathbb{Z}/2$-module $\mathbb{R}$ with the antipodal action.

We observed above that one may choose a $\mathbb{Z}/2$-equivariant tubular neighbourhood

$$D(L \otimes \tau_B M) \hookrightarrow M \times_B M$$

of the diagonal. Let us write $\tilde{D}$ for the complement of the open disc-bundle. The inclusion of $\tilde{D}$ into $M \times_B M - M$ is clearly a $\mathbb{Z}/2$-equivariant fibrewise homotopy equivalence. For we can retract the complement of the zero-section in $D(L \otimes \tau_B M)$ onto the bounding sphere-bundle $S(L \otimes \tau_B M)$. The space $\tilde{D}$ has the advantage that it is compact; it is a fibrewise manifold with boundary: $\partial \tilde{D} = S(L \otimes \tau_B M)$.

Let D denote the quotient of $\tilde{D}$ by the free $\mathbb{Z}/2$-action. It is again a fibrewise manifold with boundary; its boundary is the real projective bundle $P(\tau_B M)$. The real line bundle over D associated to the double covering $\tilde{D}$ will be denoted by $\lambda = \tilde{D} \times_{\mathbb{Z}/2} L$. The $\mathbb{Z}/2$-map $\tilde{h}(i)$ determines, by passage to the orbit spaces, a section, denoted by $h(i)$, of the sphere-bundle $S(\lambda \otimes \xi)$ over D.

A necessary condition, therefore, for the existence of a fibrewise embedding $M \to N$ is that the vector bundle $\lambda \otimes \xi$ over D should admit a nowhere-zero cross-section. The theory developed by Haefliger [70] and Dax [39] shows that in a certain metastable range this condition is also sufficient.

To state their result we need to formulate an obstruction theory. Suppose that A is a closed sub-ENR of B and that i, now, is a fibrewise embedding $M_A \to N_A$ over A. The construction just explained gives a section $h(i)$ of the restriction of $S(\lambda \otimes \xi)$ to the subspace D_A of D. If the embedding i is to extend to an embedding over B, then the section $h(i)$ must extend from D_A to D. The relative Euler class

$$\gamma(\lambda \otimes \xi; h(i)) \in \omega^0_{(D,D_A)}\{D \times S^0; (\lambda \otimes \xi)^+_D\}, \tag{11.26}$$

of Definition 4.8, is a stable homotopy obstruction to extending i.

Proposition 11.27 *Let A be a subcomplex of a finite complex B. Let $M \to B$ be a compact fibrewise manifold of dimension m and $\xi \to B$ be a real vector bundle of dimension n, where the dimensions satisfy:*

$$\dim B + 3m < 2n - 2.$$

Suppose that $i : M_A \to \xi_A$ is a fibrewise embedding over A.

Then i extends to a fibrewise embedding $M \to \xi$ if and only if the associated section $h(i)$, as described above, of the sphere-bundle $S(\lambda \otimes \xi)$ over D_A extends to a section over D. This condition is equivalent to the vanishing of the relative Euler class $\gamma(\lambda \otimes \xi; h(i))$ in (11.26).

Since $\dim D = \dim B + 2m$ is less than $2(n-1)$, the equivalence of the two statements follows from Proposition 4.9.

The basic case in which B is a point (and A is empty) is the original theorem of Haefliger. The general case is a consequence of the work of Dax, although the translation is not immediate. We shall not describe the proof, but explain how to reduce the result to a statement about homotopy groups of spaces of embeddings.

Before doing so, we obtain at once from Proposition 11.27 a range of dimensions in which fibrewise embeddings exist, namely:

Corollary 11.28 *Suppose, in the situation set out in Proposition 11.27, that* $\dim B + 2m < n$. *Then* i *extends to an embedding* $M \to \xi$.

For in this case $\dim D < \dim(\lambda \otimes \xi)$.

By applying Proposition 11.27 to the pull-back to $B \times [0,1]$ and the subcomplex $B \times \{0,1\}$ we deduce an isotopy classification of fibrewise embeddings in the range $\dim B + 3m < 2n - 3$.

For a closed manifold F and Euclidean space E, let us write $\mathrm{emb}(F, E)$ for the space of smooth embeddings $F \to E$, topologized as a subspace of the Fréchet space of all smooth maps $F \to E$. Now the basic topological construction $\tilde{h}$ gives a map

$$\tilde{h} : \mathrm{emb}(F, E) \to \mathrm{map}^{\mathbf{Z}/2}(F \times F - F, S(L \otimes E))$$

from this space of embeddings to the space of equivariant maps. The results of Haefliger and Dax show that, if F has dimension m and E has dimension n, this map is a $(2n - 3m - 3)$-equivalence. (See [39], Section 3, Théorème B.)

For trivial bundles $M = B \times F$ and $N = B \times E$ over B we define $\mathrm{emb}_B(M, N)$ to be the trivial bundle $B \times \mathrm{emb}(F, E)$. This shows how to define the locally trivial bundle $\mathrm{emb}_B(M, N)$ in general and a fibrewise map

$$\tilde{h}_B : \mathrm{emb}_B(M, N) \to \mathrm{map}_B^{\mathbf{Z}/2}(M \times_B M - M, S(L \otimes \xi)).$$

Obstruction theory, Proposition 2.15, applied to $\tilde{h}_B$ leads to Proposition 11.27.

Dimension 0: finite coverings

Although the general theory of embeddings is difficult, the special case $m = 0$ can be treated by elementary methods. (See [72] for the existence result (Corollary 11.28).) Let F be a finite set, considered as a compact manifold of dimension 0, with cardinality $\#F = d$, and let E be a real vector space of dimension $n \geq 1$. We establish by induction on d:

Proposition 11.29 *For a finite set* F, *considered as a manifold of dimension* 0, *and vector space* E *of dimension* $n > 1$, *the map*

$$\tilde{h} : \mathrm{emb}(F, E) \to \mathrm{map}^{\mathbf{Z}/2}(F \times F - F, S(L \otimes E))$$

is a $(2n - 3)$-*equivalence (that is, to be precise, induces a monomorphism on homotopy groups up to dimension* $(2n - 4)$ *and an epimorphism up to dimension* $(2n - 3)$*).*

For the proof we introduce the abbreviations $\mathcal{A}(F) = \mathrm{emb}(F, E)$ and $\mathcal{B}(F) = \mathrm{map}^{Z/2}(F \times F - F, S(L \otimes E))$. The space $\mathcal{B}(F)$ is easily understood: it is just a product of $\binom{d}{2}$ spheres. When $d = 1$, $\mathcal{A}(F) = E$ is contractible and $\mathcal{B}(F)$ is a single point. When $d = 2$, we have a homeomorphism $\mathcal{A}(F) \to E \times \mathcal{B}(F)$, given by the mid-point and $\tilde{h}$.

Suppose that the result is true for F, where $d \geq 2$, and write F' for the disjoint union of F and a point $*$. Then $\mathcal{B}(F')$ is naturally identified with $\mathcal{B}(F) \times \mathrm{map}(F, S(E))$.

By restricting from F' to F we obtain the commutative diagram:

$$
\begin{array}{ccc}
\mathcal{A}(F') & \xrightarrow{\tilde{h}} & \mathcal{B}(F') \\
a \downarrow & & \downarrow b \\
\mathcal{A}(F) & \xrightarrow{\tilde{h}} & \mathcal{B}(F)
\end{array}
$$

The map b is a trivial fibration with fibre the product of d spheres $S(E)$; the map a is also a fibration, indeed a smooth fibre bundle, as can be seen from Proposition 11.20. (See also Section 23 of Part I.) Over $i \in \mathrm{emb}(F, E) = \mathcal{A}(F)$ the fibre of a is $E - i(F)$, which is homotopy equivalent to a wedge of d spheres $S(E)$ (with some chosen basepoint). The map $\tilde{h}$ on fibres maps the wedge product of d spheres S^{n-1} into the product in the standard way, and so induces an isomorphism of homotopy groups up to dimension $2n - 3$. The result follows from the inductive hypothesis, by the five-lemma.

12 The Pontrjagin–Thom construction

The fibrewise Pontrjagin–Thom construction, which is the subject of this section, is closely related to the construction of the fixed-point index that we described in Section 6. The fundamental concepts are already in the fourth paper [5] in the Atiyah–Singer series on the Index Theorem. After a preliminary definition, we begin by recalling the ordinary theory before proceeding to the fibrewise generalization. Initially, therefore, the underlying base space B is a point. Manifolds (and fibrewise manifolds) are understood to be (smooth) without boundary unless explicit reference is made to the boundary.

The fibrewise Thom space

Recall that if ξ is a (finite-dimensional real) vector bundle, admitting a Euclidean metric, over a space M, the *Thom space* M^ξ of ξ is usually defined to be the pointed space $D(\xi)/S(\xi)$. In fact, the Euclidean metric can be avoided by thinking of the sphere $S(\xi_x)$ as the space of oriented 1-dimensional subspaces of the fibre ξ_x at $x \in M$ and of the disc $D(\xi_x)$ as the cone on the sphere. When M is compact, M^ξ is just the one-point compactification of the total space of the bundle ξ. But in general the two concepts must be carefully distinguished.

We can also define the (stable) Thom space M^α of a virtual vector bundle α over M, provided that α can be written as $\xi - (M \times F)$ for some trivial vector bundle $M \times F$, as a desuspension of M^ξ. Recall that such a virtual bundle is said to be *of finite type*. In practice, α is likely to be the pull-back of a virtual bundle over a compact space.

As is often the case, the only difficulty in extending the definition to the fibrewise theory is notation.

Definition 12.1 Suppose that $M \to B$ is a fibrewise space over B and that ξ is a vector bundle over M. The *fibrewise Thom space* of ξ is the fibrewise pointed space

$$M_B^\xi := D(\xi)/_B S(\xi)$$

over B.

If M is fibrewise compact over B, then the fibrewise Thom space can be identified with the fibrewise one-point compactification ξ_B^+ over B or with the fibrewise quotient $(\xi_M^+)/_B M$. The definition can be extended, as in the classical case, to virtual bundles of finite type over M.

In general, the fibrewise Thom space will not be a homotopy fibre bundle. But if M is, say, a fibrewise compact ENR over B, so locally homotopy trivial with compact fibres, then M_B^ξ is also locally homotopy trivial. (Since the base B is an ENR, it suffices to look at the case in which $B = \mathbb{R}^n$ and M is a trivial bundle.)

The classical Gysin map

Let M and N be manifolds, with M closed (that is, compact without boundary), and let $f : M \to N$ be a smooth map. We write $\nu(f) = f^*\tau N - \tau M$ for the stable *normal bundle* over M; it is a virtual vector bundle (of finite type, since M is compact). Our immediate goal is to define the *Gysin map*, also known as the *Umkehr* or, when it acts contravariantly on stable cohomotopy or cohomology, as the *direct image* map,

$$f^! : N^+ \to M^{\nu(f)}$$

from the one-point compactification of the (possibly non-compact) manifold N to the Thom space of $\nu(f)$.

Suppose first that f is a smooth embedding of M as a submanifold of N with normal bundle $\nu = \nu(f)$ in the usual sense, and then we may choose a tubular neighbourhood

$$j : D(\nu) \hookrightarrow N$$

extending f. The Gysin map $f^! : N^+ \to D(\nu)/S(\nu)$ is defined by the Pontrjagin–Thom construction:

$$f^!(y) = \begin{cases} [x] & \text{if } y = j(x),\ x \in D(\nu), \\ * & \text{if } y \notin j(D\nu). \end{cases}$$

The homotopy class of $f^!$ is independent of the choice of tubular neighbourhood since different choices are isotopic.

Before proceeding to the general case we record an elementary property of this construction for embeddings.

Proposition 12.2 *Suppose that $f : M \hookrightarrow N$ is an embedding with normal bundle ν. Then the composition $f^! \circ f_*$:*

$$M_+ = M^+ \xrightarrow{\ f_*\ } N^+ \xrightarrow{\ f^!\ } M^\nu$$

is, up to homotopy, the map of Thom spaces $M^0 \hookrightarrow M^\nu$ given by the inclusion of the zero vector bundle over M into ν.

For the general case we choose a smooth embedding $i : M \hookrightarrow E$ of M into some Euclidean space E. Then we have an embedding $(i, f) : M \hookrightarrow E \times N$ with normal bundle ν to which we may apply the Pontrjagin–Thom construction above to obtain a map

$$E^+ \wedge N^+ = (E \times N)^+ \to D(\nu)/S(\nu) = M^\nu.$$

Identifying $\nu(f)$ with $\nu - (M \times E)$, we get a stable map

$$f^! : N^+ \to M^{\nu(f)}$$

represented by a map $E^+ \wedge N^+ \to E^+ \wedge M^{\nu(f)}$. We must check as we have done elsewhere (in Sections 6 and 7) that the stable map so defined is independent of the choice of embedding i.

Suppose that $i' : M \hookrightarrow E'$ is another embedding. Then we have a family of embeddings $i_t : M \hookrightarrow E \times E'$, $0 \le t \le 1$, given by the linear homotopy

$$i_t(x) = ((1 - t)i(x), ti'(x)).$$

It is easy to see that the construction above using i_0 rather than i produces the same stable Gysin map $f^!$. By performing the construction fibrewise over $[0, 1]$ to the fibrewise embedding

$$(x, t) \mapsto (i_t(x), f(x)) : M \times [0, 1] \to ((E \times E') \times N) \times [0, 1]$$

we see that the stable homotopy class of $f^!$ is independent of the choice of an embedding in Euclidean space.

A refinement of the Gysin map

The Gysin map $f^!$ can be lifted to a fibrewise map over N as follows. Let U be an open neighbourhood of the graph $\{(f(x), x) \mid x \in M\}$ of f in $N \times M$, which we regard as a fibrewise space over N by projection to the first factor. Let $q : U \to M$ be the projection onto the second factor. We shall define a fibrewise stable map

$$G(f, U) \in {}_c\omega_N^0\{N \times S^0; U_N^{q^*\nu(f)}\} \tag{12.3}$$

with compact supports. Composing with the inclusion, we get a fibrewise stable map with compact supports $N \times S^0 \to N \times M^{\nu(f)}$ over N, and this determines the Gysin map $f^! : N^+ \to M^{\nu(f)}$. (Compare this with the relation between the transfer and the fibrewise transfer in Section 6.)

To describe the construction we begin again with the case in which f is an embedding with normal bundle ν. We can certainly choose the tubular neighbourhood $j : D(\nu) \hookrightarrow N$ so that $(j(y), x) \in U$ for all $x \in M$, $y \in D(\nu_x)$. This gives us a map

$$
\begin{array}{ccc}
D(\nu) & \longrightarrow & U \quad : y \in D(\nu_x) \mapsto (j(y), x) \\
j \downarrow & & \downarrow \\
N & = & N
\end{array}
$$

over N. We specify $G(f, U)$ by a fibrewise map $N \to U_N^{q^*\nu}$ over N which is zero outside the compact subspace $D(\nu)$. On $D(\nu)$ it is given by the inclusion of $D(\nu)$ in U and the projection $D(\nu)/S(\nu) \to M^\nu$:

$$y \in D(\nu_x) \mapsto [(j(y), x), y].$$

In the general case we proceed again by choosing an embedding $i : M \hookrightarrow E$ and take a tubular neighbourhood $j : D(\nu) \hookrightarrow E \times N$ so small that $(j(y), x) \in E \times U$ for $x \in M$, $y \in D(\nu_x)$. The construction above gives a fibrewise map $E \times N \to E \times U_{E \times N}^{q^*\nu}$ with compact supports. We compose this with the projection maps $E \times N \to N$ and $E \times U \to U$ to define a fibrewise map

$$N \times E^+ \to U_N^{q^*\nu}$$

over N, and this gives $G(f, U)$.

The method already described for $f^!$ shows that the stable class $G(f, U)$ is well-defined. It is then clear, too, that if $U' \subseteq U$ is a smaller open neighbourhood of the graph of f in $N \times M$ then $G(f, U') \mapsto G(f, U)$ under the inclusion map. As in Lemma 1.5, let us replace $f : M \to N$ by the homotopy fibre bundle $M^\sharp \to N$, where

$$M^\sharp = \{(x, \omega) \in M \times \mathcal{P}N \mid f(x) = \omega(0)\} \quad : \quad (x, \omega) \mapsto \omega(1) \in N.$$

We write q again for the projection to the first factor $q : M^\natural \to M$. For U sufficiently small there is a natural fibrewise homotopy class $U \to M^\natural$ over N. In geometric terms, when N is equipped with a Riemannian metric we can map $(y, x) \in U \subseteq N \times M$ to the appropriately parametrized geodesic from $f(x)$ to y. When f is an embedding, this amounts to using the metric to construct a tubular neighbourhood $j : D(\nu) \hookrightarrow N$ and taking the line segment in ν_x to get a path from $f(x) = j(0)$ to $y = j(v)$, where $0, v \in \nu_x$. Alternatively, we may exploit the uniform local contractibility of N, as in the definition of the Nielsen–Reidemeister index (6.12). The image of $G(f, U)$ under the map $U \to M^\natural$ is independent of U; we denote it by

$$G(f) \in {}_c\omega_N^0\{N \times S^0; (M^\natural)_N^{q^*\nu}\}. \tag{12.4}$$

Functoriality of the Gysin construction

We suppose now that the manifold N, as well as M, is closed. Let $X \to N$ be a pointed homotopy fibre bundle with fibres of the homotopy type of finite complexes. The Gysin construction can be extended to define a Gysin map

$$f^! : X/N \to (f^*X \wedge_N \nu(f)_M^+)/M. \tag{12.5}$$

This may be done either by running through the original definition with the fibrewise space X as a sort of parameter space or, more illuminatingly, by taking the smash product with the identity on X of the refined Gysin map $G(f, U)$ defined over N. We shall return to this later; see Proposition 12.35. When ξ is a vector bundle over N and X is the pointed sphere-bundle ξ_B^+ the map (12.5) may be written more transparently as a stable map between Thom spaces

$$f^! : N^\xi \to M^{f^*\xi \oplus \nu(f)}.$$

Better still, if we replace ξ by $\alpha - \tau N$ where α is a virtual vector bundle over N, we obtain

$$f^! : N^{\alpha - \tau N} \to M^{f^*\alpha - \tau M}. \tag{12.6}$$

The proof of functoriality is now a straightforward exercise.

Proposition 12.7 (Functoriality). *Let $f : M \to N$ and $g : N \to P$ be smooth maps between closed manifolds, and let α be a virtual real vector bundle over P. Then*

$$(g \circ f)^! = f^! g^! : P^{\alpha - \tau P} \xrightarrow{g^!} N^{g^*\alpha - \tau N} \xrightarrow{f^!} M^{f^*g^*\alpha - \tau M}.$$

It is routine, too, to establish multiplicativity.

Proposition 12.8 (Multiplicativity). *Let $f : M \to N$ and $f' : M' \to N'$ be smooth maps between closed manifolds with product*

$$f \times f' : M \times M' \to N \times N'.$$

Then, for virtual vector bundles α over N and α' over N', we have $(f \times f')^! = f^! \wedge (f')^!$:

$$
\begin{array}{ccc}
(N \times N')^{(\alpha \oplus \alpha') - \tau(N \times N')} & = & N^{\alpha - \tau N} \wedge N'^{\alpha' - \tau N'} \\[2mm]
{\scriptstyle (f \times f')^!} \downarrow & & \downarrow {\scriptstyle f^! \wedge (f')^!} \\[2mm]
(M \times M')^{(f \times f')^*(\alpha \oplus \alpha') - \tau(M \times M')} & = & M^{f^* \alpha - \tau M} \wedge M'^{(f')^* \alpha' - \tau M'}
\end{array}
$$

The Frobenius property

We revert now to the original situation in which $f : M \to N$ is a (smooth) map from a closed manifold M to a possibly non-compact manifold N. Let Q be another (perhaps non-compact) manifold and let $s : Q \to N$ be a proper smooth map, such that f and s are transverse. Then

$$P := Q \times_N M = \{(x,y) \in Q \times M \mid s(x) = f(y)\}$$

is compact. We write r and g for the projections $P \to M$ and $P \to Q$ in the commutative square:

$$
\begin{array}{ccc}
P & \xrightarrow{\ r\ } & M \\[2mm]
{\scriptstyle g} \downarrow & & \downarrow {\scriptstyle f} \\[2mm]
Q & \xrightarrow[\ s\ ]{} & N
\end{array}
$$

By transversality, (11.11), the stable normal bundle $\nu(g)$ of g is the pull-back $r^* \nu(f)$ of the stable normal bundle of f.

Proposition 12.9 (Frobenius reciprocity). *In the situation described above one has a commutative square of stable maps:*

$$
\begin{array}{ccc}
Q^+ & \xrightarrow{\ s_*\ } & N^+ \\[2mm]
{\scriptstyle g^!} \downarrow & & \downarrow {\scriptstyle f^!} \\[2mm]
P^{\nu(g)} & \xrightarrow[\ r_*\ ]{} & M^{\nu(f)}
\end{array}
$$

Let us see, first of all, that we can reduce to the case in which f and so also g are embeddings. In general, we choose an embedding $i : M \hookrightarrow E$ into some Euclidean space E. Then we have a pull-back diagram:

$$
\begin{array}{ccc}
P & \xrightarrow{\ r\ } & M \\[2mm]
{\scriptstyle (j,g)} \downarrow {\scriptstyle \subseteq} & & {\scriptstyle \subseteq} \downarrow {\scriptstyle (i,f)} \\[2mm]
E \times Q & \xrightarrow[\ 1 \times s\ ]{} & E \times N
\end{array}
$$

in which $j(x, y) = (i(y), x)$, for $(x, y) \in P \subseteq Q \times M$.

Suppose, then, that f is an embedding with normal bundle $\nu = \nu(f)$. The normal bundle of the embedding g is the pull-back $r^*\nu$, and a tubular neighbourhood $D(\nu) \hookrightarrow N$ of M lifts to a tubular neighbourhood $r^*D(\nu) = D(r^*\nu) \hookrightarrow Q$ of P, by transversality. The Pontrjagin–Thom construction gives a diagram of maps

$$
\begin{array}{ccc}
Q^+ & \xrightarrow{\ s^+\ } & N^+ \\
\downarrow & & \downarrow \\
D(r^*\nu)/S(r^*\nu) & \xrightarrow[\ r\ ]{} & D(\nu)/S(\nu)
\end{array}
$$

which is genuinely commutative (not just up to homotopy). This completes the proof of Proposition 12.9.

Remark 12.10. There is also a reciprocity formula for the refined Gysin map $G(f, U)$ of (12.3). Let V be the pre-image of U under $s \times r : Q \times P \to N \times M$; it is an open neighbourhood of the graph of g. Then we have a pull-back diagram:

$$
\begin{array}{ccccc}
V & \longrightarrow & s^*U & \longrightarrow & U \\
& & \downarrow & & \downarrow \\
& & Q & \xrightarrow[\ s\ ]{} & N
\end{array}
$$

The induced map

$$
s^* : {}_c\omega_N^0\{N \times S^0; U_N^{\nu(f)}\} \to {}_c\omega_Q^0\{Q \times S^0; (s^*U)_Q^{\nu(f)}\}
$$

lifts $G(f, U)$ to the image of $G(g, V)$ under the map

$$
V_Q^{\nu(g)} \to (s^*U)_Q^{\nu(f)}
$$

determined by $1 \times r : Q \times P \to Q \times M$. (To ease the notation we have not distinguished between the stable normal bundles and their pull-backs under projection maps.)

Let us return to the case in which all the manifolds considered are compact. The proof of the following extension of the Frobenius property involves no new ideas.

Proposition 12.11 *Let $f : M \to N$ and $s : Q \to N$ be transverse smooth maps between closed manifolds, and let α be a virtual vector bundle over N. Writing $P := Q \times_N M$, $g : P \to Q$ and $r : P \to M$ for the projections, and β for the virtual bundle over Q defined by $\beta - \tau Q = s^*(\alpha - \tau N)$, we have a commutative diagram of stable maps:*

$$
\begin{array}{ccc}
Q^{\beta-\tau Q} & \xrightarrow{\;s_*\;} & N^{\alpha-\tau N} \\[4pt]
g^! \downarrow & & \downarrow f^! \\[4pt]
Pg^*\beta-\tau P & \xrightarrow{\;r_*\;} & Mf^*\alpha-\tau M
\end{array}
$$

In order to interpret the diagram one needs to use the identification $r^*(f^*\tau N - \tau M) = g^*\tau Q - \tau P$, that is, the equality $r^*\nu(f) = \nu(g)$.

Remark 12.12. A special case is worth noting. Given a smooth map $f : M \to N$ of closed manifolds we have a pull-back diagram:

$$
\begin{array}{ccc}
M & \xrightarrow{(f,1)} & N \times M \\[4pt]
f \downarrow & & \downarrow 1 \times f \\[4pt]
N & \xrightarrow[(1,1)]{} & N \times N
\end{array}
$$

to which we can apply Proposition 12.11. It follows, in particular, that the induced direct image map $(f^!)^*$, which we shall write as $f_!$, in stable cohomotopy:

$$
f_! : \tilde{\omega}^*(M^{f^*\alpha-\tau M}) \to \tilde{\omega}^*(N^{\alpha-\tau N}) \tag{12.13}
$$

obeys the familiar rule:

$$
f_!(f^*x \cdot y) = x \cdot f_!(y), \quad x \in \omega^*(N), \; y \in \tilde{\omega}^*(M^{f^*\alpha-\tau M}).
$$

The fibrewise theory

In the discussion of the fibrewise theory we shall work over a compact ENR B. There is, in fact, very little that needs to be done to extend the definitions and proofs from the classical theory. Let $M \to B$ and $N \to B$ be fibrewise manifolds, with M compact, and let $f : M \to N$ be a fibrewise smooth map. Then we have a stable Gysin map

$$
f^! : N_B^+ \to M_B^{\nu(f)} \tag{12.14}
$$

from the fibrewise one-point compactification of N to the fibrewise Thom space of the fibrewise stable normal bundle $\nu(f) = f^*\tau_B N - \tau_B M$. The construction proceeds by choosing a fibrewise embedding $i : M \hookrightarrow B \times E$ for some Euclidean space E and by applying the Pontrjagin–Thom construction fibrewise to a tubular neighbourhood of the embedding

$$
(i, f) : M \hookrightarrow (B \times E) \times_B N \quad (= E \times N).
$$

When N is compact and α is a virtual bundle over N, we have, more symmetrically, a Gysin map

$$f^! : N_B^{\alpha - \tau_B N} \to M_B^{f^* \alpha - \tau_B M}. \tag{12.15}$$

The fundamental properties: functoriality (Proposition 12.7); multiplicativity (Proposition 12.8); and Frobenius reciprocity (Propositions 12.9 and 12.11) all translate (by affixing a suffix 'B') to the fibrewise theory.

The construction is evidently compatible with the pull-back.

Proposition 12.16 *Let $\alpha : B' \to B$ be a map of compact ENRs. Then in the setting of (12.14) or (12.15)*

$$(\alpha^* f)^! = \alpha^* (f^!).$$

To amplify the formula, we have in the second setting (12.15), for the virtual bundle 0 (to avoid a clash of notation), the lifting

$$f^! \in \ \omega_B^0 \{ N_B^{-\tau_B N}; \ M_B^{-\tau_B M} \}$$

$$\downarrow \alpha^*$$

$$(\alpha^* f)^! \in \ \omega_{B'}^0 \{ (\alpha^* N)_{B'}^{-\tau_{B'} (\alpha^* N)}; \ (\alpha^* M)_B^{-\tau_{B'} (\alpha^* M)} \}$$

of $f^!$ to $(\alpha^* f)^!$.

As usual in the fibrewise theory, there is an implicit homotopy invariance property.

Proposition 12.17 (Homotopy invariance). *Let $f_t : M \to N$, $0 \le t \le 1$, be a homotopy between fibrewise smooth maps f_0, $f_1 : M \to N$, where M and N are fibrewise manifolds over B and M is compact. Using the homotopy to give a (homotopy class of) vector bundle isomorphism $f_0^* \tau_B N \to f_1^* \tau_B N$, we have*

$$f_0^! = f_1^! : N_B^+ \to M_B^{\nu(f_0)} = M_B^{\nu(f_1)}.$$

Since any continuous homotopy between smooth maps f_0 and f_1 is homotopic to a fibrewise smooth homotopy, there is no need to be precise about the nature of the homotopy in the statement above. Moreover, since any continuous fibrewise map is homotopic to a fibrewise smooth map, the homotopy invariance means that we can define the Gysin map $f^!$ without requiring f to be smooth. We shall see this again from a different point of view when we recognize $f^!$ as the dual of f_{+B} (when N is compact).

Duality for fibrewise manifolds

The stable homotopy-theoretic formulation of Poincaré duality is due to Atiyah [2]. (The canonical nature of the duality seems to have been 'folk-lore' for some time and was assumed to be generally understood when [25] was written, but the first textbook account of which we are aware is in [99]. See also [37].)

Proposition 12.18 *Let $M \to B$ be a compact fibrewise manifold (without boundary) over a compact ENR B. Then there are canonical duality maps:*

$$B \times S^0 \xrightarrow{\ i\ } M_B^{-\tau_B M} \wedge_B M_{+B} \xrightarrow{\ e\ } B \times S^0,$$

defined below. (To be precise, there are two canonical duality maps, a 'left' and a 'right'.)

The reason why there are two such maps can be seen at once. For the vector bundle $\tau_B M$ has a canonical antipodal involution, given by multiplication by -1 in each fibre, and this determines an involution of the Thom space $M_B^{-\tau_B M}$.

The duality structure maps appear as compositions of two factors defined in terms of the projection $p : M \to B$, which we regard as a fibrewise smooth map over B, and the diagonal embedding $\Delta : M \to M \times_B M$. From p we obtain first, by adjoining basepoints,

$$p_* : M_{+B} \to B \times S^0 (= B_{+B}).$$

The Gysin map defined by p is a stable map

$$p^! : B \times S^0 \to M_B^{-\tau_B M};$$

for the tangent bundle of the fibrewise manifold $B \to B$ is $B \times 0$, and the associated Thom space B_B^{-0} is $B \times S^0$.

In the description of the normal bundle ν of the embedding Δ we have a left/right choice. Writing τ_x and ν_x for the fibres of $\tau_B M$ and ν at a point $x \in M$, we have a short exact sequence:

$$0 \to \tau_x \xrightarrow{v \mapsto (v,v)} \tau_x \oplus \tau_x \xrightarrow{\ \pi\ } \nu_x \to 0.$$

There are, up to homotopy, two natural isomorphisms $\tau_x \to \nu_x$ given by $v \mapsto \pi(v, 0)$ and $v \mapsto \pi(0, v)$. Stably they differ by composition with $v \mapsto -v$; see Lemma 3.2. In (11.18) we chose the former. With this choice we have a Gysin map

$$\Delta^! : (M \times M)_B^{-\tau_B M \times 0} = M_B^{-\tau_B M} \wedge_B M_{+B} \to M_{+B} = M_B^0.$$

(In the notation of (12.15) the virtual bundle α is $0 \times \tau_B M$ over $M \times M$.) Identifying the restriction $\Delta^*(\tau_B M \times 0)$ with $\tau_B M$ in the obvious way, we have an inclusion map

$$\Delta_* : M_B^{-\tau_B M} \to M_B^{-\tau_B M} \wedge_B M_{+B}.$$

Having now assembled the ingredients we define $\mathbf{i}$ and $\mathbf{e}$ to be the compositions

$$\mathbf{i} : B \times S^0 \xrightarrow{p^!} M_B^{-\tau_B M} \xrightarrow{\Delta_*} M_B^{-\tau_B M} \wedge_B M_{+B},$$

$$\mathbf{e} : M_B^{-\tau_B M} \wedge_B M_{+B} \xrightarrow{\Delta^!} M_{+B} \xrightarrow{p_*} B \times S^0.$$

The verification of the two defining identities, (i) and (ii) in Definition 10.8, reduces to the functoriality and Frobenius property of the Gysin construction. We have to establish commutativity of the following diagrams, corresponding to (10.24) and (10.25). The fibrewise tangent bundle $\tau_B M$ is abbreviated to τ, and we use $\mathbf{t}$ again for the transposition of two factors.

$$ (12.19) $$

$$ (12.20) $$

The triangles commute by functoriality, and the squares by the Frobenius property applied to the diagrams:

$$
\begin{array}{ccc}
M \times M & \xrightarrow{\Delta \times 1} & M \times M \times M \\
{\scriptstyle \Delta} \uparrow & & \uparrow {\scriptstyle 1 \times \Delta} \\
M & \xrightarrow[\Delta]{} & M \times M
\end{array}
\qquad
\begin{array}{ccc}
M \times M & \xrightarrow{1 \times \Delta} & M \times M \times M \\
{\scriptstyle \Delta} \uparrow & & \uparrow {\scriptstyle \Delta \times 1} \\
M & \xrightarrow[\Delta]{} & M \times M
\end{array}
$$

Of course, $\mathbf{t} \circ \Delta = \Delta$. This completes the proof.

Proposition 12.21 Let $f : M \to N$ be a fibrewise map between compact (closed) fibrewise manifolds over B. Then, under the canonical duality isomorphisms,

$$f^! : N_B^{-\tau_B N} \to M_B^{-\tau_B M}$$

is the dual of $f_+ : M_{+B} \to N_{+B}$.

The proof is another commutative diagram. Let $\Phi : M \to M \times N$ be the map $x \mapsto (x, f(x))$. From the definition (10.15) we see that the dual of f_+ is the composition from top left to bottom right in the picture:

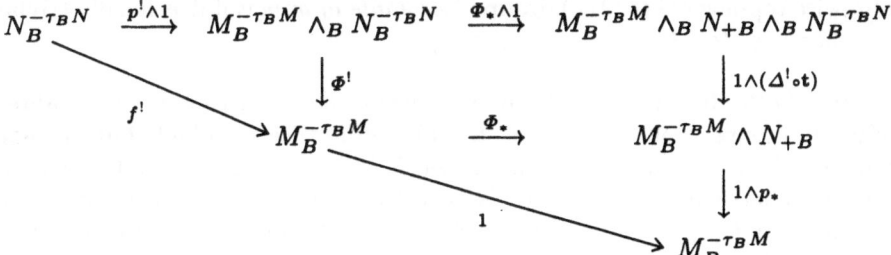

Again the triangles commute by functoriality. Commutativity of the square comes from the pull-back diagram:

$$
\begin{array}{ccc}
M \times N & \xrightarrow{\Phi \times 1} & M \times N \times N \\
\Phi \uparrow & & \uparrow 1 \times \Delta \\
M & \xrightarrow{\quad \Phi \quad} & M \times N
\end{array}
$$

Duality for a finite covering

A finite covering $M \to B$ is a 0-dimensional fibrewise manifold, with tangent bundle $\tau_B M = M \times 0$. As a special case of Proposition 12.18, we see that M_{+B} is *self-dual*. In fact, the Gysin maps $\Delta^!$ and $p^!$ can be constructed quite explicitly, for there is a natural fibrewise embedding of M into a vector bundle ξ defined as follows. Let $\xi = \eta^*$ be the vector bundle dual of $\eta := \mathrm{map}_B(M, B \times \mathbb{R})$, and embed ξ into M by evaluation. So the fibre ξ_b at $b \in B$ is the space of measures on the finite set M_b, containing M_b as the delta-functions. Alternatively, ξ_b is the free vector space on the set M_b.

Duality for a bundle of Lie groups

An important example, to which we shall return at the beginning of Section 13, is furnished by a bundle $M \to B$ of compact Lie groups.

We recall from Part I, Section 2, that a fibrewise space M over B is a *fibrewise topological group* if each fibre has a group structure such that the multiplication $(x, y) \mapsto xy \colon M \times_B M \to M$, the inverse $x \mapsto x^{-1} \colon M \to M$ and the inclusion $e : B \to M$ of the identity element 1 in each fibre are continuous fibrewise maps. If M is a fibrewise manifold over B and these maps are fibrewise smooth then we might call M a *fibrewise Lie group*. (Notice, in passing, that a fibrewise group can be thought of as a groupoid in which the base B is the space of objects of the category and the total space M is the space of morphisms: M_b is the automorphism group of b and distinct objects are non-isomorphic.)

Given a topological (or Lie) group G we can give the trivial bundle $B \times G$ a fibrewise group structure by identifying the fibre at each point with the group G. Fibrewise groups of this type are said to be *trivial*. We say that a fibrewise topological (or Lie) group is a *bundle of groups* if it is locally trivial as a fibrewise topological (or Lie) group.

Remark 12.22. In the world of nilpotent groups it is easy to write down fibre-wise Lie groups which are locally trivial as fibrewise manifolds but are not bundles of Lie groups. For example, let B be the (1-dimensional) space of skew-symmetric bilinear forms on $\mathbb{R}^2$ and let G_b be $\mathbb{R} \times \mathbb{R}^2$ with the multiplication $(x, y) \cdot (x', y') = (x + x' + b(y, y'), y + y')$. The group G_0 is Abelian, but G_b is non-Abelian (but nilpotent) if $b \neq 0$. However, compact Lie groups do not admit such deformations and a compact fibrewise Lie group is necessarily locally trivial. As explanation we offer the following heuristic argument. By Proposition 11.13 we certainly have local triviality as a fibrewise manifold. So it suffices to consider a fixed compact Lie group G and look at group multiplications $m : G \times G \to G$ close to the given group structure and having the same identity element 1. Now we can write m, in terms of the exponential map of G, as

$$m(x, y) = \exp(\alpha(x, y))xy,$$

where $\alpha : G \times G \to \mathfrak{g}$ is a smooth map with values in the Lie algebra $\mathfrak{g}$ of G. The conditions $m(x, 1) = x$ and $m(1, y) = y$ translate into: $\alpha(x, 1) = 0 = \alpha(1, y)$. To linearize the problem we treat α as an infinitesimal deformation. Then the associativity of the multiplication: $m(m(x, y), z) = m(x, m(y, z))$ becomes the cocycle condition:

$$\alpha(xy, z) + \alpha(x, y) = \alpha(x, yz) + x \cdot \alpha(y, z),$$

where the action of G is, of course, the adjoint. Now define $\beta : G \to \mathfrak{g}$ by integration with respect to the normalized Haar measure μ on G as:

$$\beta(x) = \int_G \alpha(x, z) \, d\mu(z).$$

It is then an elementary exercise to verify that

$$\alpha(x, y) = x \cdot \beta(y) - \beta(xy) + \beta(x).$$

This is a special case of the calculation which shows that the (smooth or continuous) cohomology $H^p(G; E)$ of the compact group G with coefficients in a finite-dimensional real G-module E is trivial for $p > 0$ (and equal to the invariant subspace E^G for $p = 0$). (See, for example, the early reference [117].) Now the coboundary condition above is the linearization of the identity

$$m(\exp(\beta(x))x, \exp(\beta(y))y) = \exp(\beta(xy))xy,$$

which shows that, at the infinitesimal level, $x \mapsto \exp(\beta(x))x$ is a diffeomorphism of G which relates m to the given group structure.

To return to our main subject, we suppose now that $M \to B$ is a bundle of compact Lie groups. Let $\zeta = e^* \tau_B M$ be the tangent space (or rather bundle) at the identity. Using (left) translation, as in the case of an ordinary Lie group, we can identify the fibrewise tangent bundle $\tau_B M$ with the bundle $p^* \zeta$, which is trivial on fibres. (We shall look at this in more detail at the beginning of the next section.) According to Proposition 12.16 the dual of M_{+B} is the fibrewise suspension (or, more accurately, desuspension) $(-\zeta)_B^+ \wedge_B M_{+B}$ of M_{+B}.

The duality structure maps admit another description involving the group structure. Let $n : M \times_B M \to M$ be the map $(x, y) \mapsto xy^{-1}$.

Proposition 12.23 *For a fibrewise compact Lie group $M \to N$, as described above, the canonical duality map* e $: (M_{+B})^* \wedge_B M_{+B} \to B \times S^0$ *can be expressed as the composition*

$$(-\zeta)_B^+ \wedge_B M_{+B} \wedge_B M_{+B} \xrightarrow{\; n_* \;} (-\zeta)_B^+ \wedge_B M_{+B} \xrightarrow{\; e^! \;} B \times S^0.$$

The Gysin map $e^!$ can be thought of as the fibrewise collapse onto the top cell; see Proposition 13.1.

For the proof of the proposition we use the Frobenius property once more. From the pull-back diagram:

$$
\begin{array}{ccc}
M \times_B M & \xrightarrow{\; n \;} & M \\
\Delta \uparrow & & \uparrow e \\
M & \xrightarrow[\; p \;]{} & B
\end{array}
$$

we obtain a commutative diagram of stable maps:

$$
\begin{array}{ccc}
(-\zeta)_B^+ \wedge_B M_{+B} \wedge_B M_{+B} & \xrightarrow{\; n_* \;} & M_{+B} \\
\Delta^! \downarrow & & \downarrow e^! \\
M_{+B} & \xrightarrow[\; p_* \;]{} & B \times S^0
\end{array}
$$

Duality for Thom spaces

Following Atiyah [2], we extend the duality theory to Thom spaces. For any virtual vector bundle α over M, the fibrewise Thom spaces M_B^α and $M_B^{-\alpha - \tau_B M}$ are canonically dual to each other. The structure maps involve Δ_* and $\Delta^!$:

$$M_B^{-\tau_B M} \xrightarrow{\; \Delta_* \;} M_B^{-\alpha - \tau_B M} \wedge_B M_B^\alpha \xrightarrow{\; \Delta^! \;} M_{+B},$$

whose definition depends upon the fact that the virtual bundle $(-\alpha) \times \alpha$ over $M \times M$ restricts to the zero bundle on the diagonal M.

The next proposition relates duality over M, in the special case of sphere-bundles, to duality of the associated Thom spaces over B.

Proposition 12.24 *Let α and β be virtual vector bundles over the compact fibrewise manifold M. Then there is a commutative square:*

$$
\begin{array}{ccc}
\omega_M^0\{\alpha_M^+; \beta_M^+\} & \xrightarrow[\cong]{D_M} & \omega_M^0\{(-\tau_B M - \alpha)_M^+; (-\tau_B M - \beta)_M^+\} \\
p_* \downarrow & & \downarrow p_* \\
\omega_B^0\{M_B^\alpha; M_B^\beta\} & \xrightarrow[D_B]{\cong} & \omega_B^0\{M_B^{-\tau_B M-\alpha}; M_B^{-\tau_B M-\beta}\}
\end{array}
$$

involving maps whose meaning is explained below.

The isomorphism D_B is duality over B: $f \mapsto f^*$, while D_M is the suspension isomorphism, which for sphere-bundles is duality over M. The vertical maps p_* are given by the quotient construction '$/_B M$'; a fibrewise map $a : \alpha_M^+ \to \beta_M^+$ gives

$$
p_*(a) : M_B^\alpha = (\alpha_M^+)/_B M \to M_B^\beta = (\beta_M^+)/_B M.
$$

Once more the proof is by inspection of a commutative diagram. Consider a fibrewise stable map $a : \alpha_M^+ \to \beta_M^+$ over M, and write a_* for the associated maps of Thom spaces $M_B^{\alpha-\beta} \to M_{+B}$ and $M_B^\alpha \to M_B^\beta$ in the following commutative diagram.

$$
\begin{array}{ccc}
M_B^{-\tau_B M-\beta} & & \\
\scriptstyle(1\times p)^! \downarrow & \searrow^{\;1} & \\
M_B^{-\tau_B M-\beta} \wedge_B M_B^{-\tau_B M} & \xrightarrow{\Delta^!} & M_B^{-\tau_B M-\beta} \\
\scriptstyle 1\wedge\Delta_* \downarrow & & \downarrow \scriptstyle\Delta_* \\
M_B^{-\tau_B M-\beta} \wedge_B M_B^\alpha \wedge_B M_B^{-\tau_B M-\alpha} & \xrightarrow{(\Delta\times 1)^!} & M_B^{\alpha-\beta} \wedge_B M_B^{-\tau_B M-\alpha} \\
\scriptstyle 1\wedge a\wedge 1 \downarrow & & \downarrow \scriptstyle a\wedge 1 \\
M_B^{-\tau_B M-\beta} \wedge_B M_B^\beta \wedge_B M_B^{-\tau_B M-\alpha} & \xrightarrow{(\Delta\times 1)^!} & M_{+B} \wedge_B M_B^{-\tau_B M-\alpha} \\
& & \downarrow \scriptstyle(p\times 1)_* \\
& & M_B^{-\tau_B M-\alpha}
\end{array}
$$

The composition from top left to bottom right is $D_B(p_* a) = p_* D_M(a)$.

Remark 12.25. The duality theory extends, with some notational discomfort, to fibrewise manifolds with boundary. Let $M \to B$ be such a fibrewise compact manifold over B with boundary $\partial M \to B$, and let α be a virtual vector bundle over M. Then the quotient $M_B^\alpha/_B \partial M_B^{\alpha|}$, which we might write as the *relative Thom space* $(M, \partial M)_B^\alpha$ is canonically dual to $M_B^{-\alpha - \tau_B M}$.

The Poincaré–Hopf vector-field index

The theorem of Hopf relating the zeros of a generic smooth vector field on a closed manifold to the Euler characteristic is one of the fundamental theorems of Algebraic Topology. Its generalization to fibrewise manifolds is our next subject. The Poincaré–Hopf index theory for vector fields which we shall describe is very closely related to the Lefschetz fixed-point theory, but there is an important conceptual distinction.

Let $M \to B$ be a fibrewise compact (closed) fibrewise manifold over an ENR B, and let $U \subseteq M$ be an open subset. A (continuous) section v of the fibrewise tangent bundle $\tau_B U = \tau_B M \,|\, U$ will be called a *vector field* on U; we refer to a fibrewise smooth section as a smooth vector field. The *zero-set* of v is the closed subspace

$$\text{Zero}(v) = \{x \in U \mid v(x) = 0\}$$

of U. This takes the place of the fixed-subspace in the Lefschetz theory. We suppose that $\text{Zero}(v)$ is compact. Then we shall define the *Poincaré–Hopf vector field index*

$$_c\tilde{I}_B(v, U) \in {}_c\omega_B^0\{B \times S^0; \, U_{+B}\}$$

in the general case when B is not necessarily compact, and

$$\tilde{I}_{(B,A)}(v, U) \in \omega_{(B,A)}^0\{B \times S^0; \, U_{+B}\} \tag{12.26}$$

when B is compact and v has no zeros over a closed sub-ENR $A \subseteq B$.

(In fact, the ambient manifold M does not play an essential rôle in the definition. It has been included in the description partly to emphasize the parallel with the fixed-point theory and partly because this is the situation that one normally encounters, in which M is given and the neighbourhood U is variable.)

As in the Lefschetz theory, the definition begins with the case in which M is a vector bundle. It would suffice to deal with a trivial bundle $B \times E$, but for the sake of variety let us treat directly the case that $M = \xi$ is a general vector bundle (of finite type). The fibrewise tangent bundle $\tau_B M$ is then naturally identified with the pull-back $p^* \xi$ of ξ to M. A vector field v on $U \subseteq M$ is thus a section of $p^* \xi$ over U, or simply a fibrewise map $v : U \to M = \xi$ over B. The zeros of v are the fixed-points of the fibrewise map

$$f : U \to M = \xi \quad : \quad x \mapsto x - v(x),$$

and we are going to define the vector field index of v to be the fixed-point index of f. It is worth repeating the explicit construction in Pontrjagin–Thom style.

We choose an open neighbourhood V of Zero(v) in U such that the closure $\overline{V}$ in U is compact, and choose $\epsilon > 0$ such that the norm in some Euclidean metric on ξ is bounded by $\|v(x)\| \geq \epsilon$ for $x \in \overline{V} - V$. Now define $q : \xi_B^+ \to \xi_B^+ \wedge_B U_{+B}$ in the usual way as:

$$q(x) := \begin{cases} * & \text{if } x \notin V, \\ [c(v(x)), x] & \text{if } x \in \overline{V}, \end{cases}$$

where $c : \xi \to \xi_B^+$ is the standard radial extension map which expands the open disc-bundle of radius ϵ to ξ and pushes the complement to the point at infinity (in each fibre). The map q represents the stable homotopy class $_c\tilde{I}_B(v, U)$ (or $\tilde{I}_{(B,A)}(v, U)$).

Turning to the general case, we choose an embedding $i : M \hookrightarrow \xi$ of the fibrewise manifold M in a vector bundle ξ, with normal bundle ν say, and choose a fibrewise tubular neighbourhood $j : D(\nu) \hookrightarrow \xi$. Let $W = j(B(\nu))$ be the image in ξ of the open unit disc-bundle, and let $r : W \to M$ be the retraction corresponding to the projection $B(\nu) \to M$. Then we can identify the tangent-bundle $\tau_B W$ with $r^*(\tau_B M \oplus \nu)$ and extend v to a field $\bar{v}$ on $r^{-1}(U)$, with the same zeros, by:

$$\bar{v}(j(y)) = (v(x), y) \in (\tau_B M)_x \oplus \nu_x, \quad \text{for } y \in \nu_x.$$

The vector-field index of (v, U) is defined as the composition of the index of $(\bar{v}, r^{-1}U)$ with the retraction map $r_+ : (r^{-1}U)_{+B} \to U_{+B}$.

Remark 12.27. In this general case the vector-field index is related to the fixed-point index as follows. Let V be an open neighbourhood of Zero(v) in U having compact closure. Choose a fibrewise Riemannian metric on M. Then, for all sufficiently small $\epsilon > 0$, the fixed-points of the associated exponential map

$$f : x \mapsto \exp_x(-\epsilon v(x)) : V \to M$$

are the zeros of v and the Lefschetz index of (f, V) coincides with the Poincaré–Hopf index of (v, V). (Of course, in the first case that we considered, when M was a Euclidean vector bundle ξ, the exponential map determined by the Euclidean metric was just translation in the vector bundle.)

The verification that the index is well-defined follows the familiar pattern, and the details are omitted. We summarize the basic properties for the index with compact supports.

Proposition 12.28 (Properties of the vector field index). *Let $M \to B$ be a fibrewise compact fibrewise manifold over an ENR B, and let v be a vector field with compact zero-set defined on an open subset U of M.*
(i)   (Naturality). *Let $\alpha : B' \to B$ be a proper map. Then*

$$_c\tilde{I}_{B'}(\alpha^* v, \alpha^* U) = \alpha^* {}_c\tilde{I}_B(v, U).$$

(ii)   (Localization). *Let $U' \subseteq U$ be an open subset with* $\mathrm{Zero}(v) \subseteq U'$. *Then* $_c\tilde{I}_B(v \,|\, U', U')$ *maps to* $_c\tilde{I}_B(v, U)$ *under the inclusion map*

$$_c\omega_B^0\{B \times S^0; U'_{+B}\} \to {}_c\omega_B^0\{B \times S^0; U_{+B}\}.$$

(iii)   (Additivity). *If U is the disjoint union of open sets U_1 and U_2, then*

$$_c\tilde{I}_B(v, U) = (i_1)_{+c}\tilde{I}_B(v \,|\, U_1, U_1) + (i_2)_{+c}\tilde{I}_B(v \,|\, U_2, U_2),$$

where $i_1 : U_1 \to U$ and $i_2 : U_2 \to U$ are the inclusion maps.

(iv)   (Multiplicativity). *Let (v', U', M') be as the data (v, U, M). Then the vector field $v \oplus v'$ defined on $U \times_B U' \subseteq M \times_B M'$ has compact zero-set* $\mathrm{Zero}(v) \times_B \mathrm{Zero}(v')$ *and its index is*

$$_c\tilde{I}_B(v, U) \wedge {}_c\tilde{I}_B(v', U') \in {}_c\omega_B^0\{B \times S^0; (U \times_B U')_{+B}\}.$$

(v)   (Homotopy invariance). *Let v_t, $0 \le t \le 1$, be a continuous family of vector fields on U such that $\{(x, t) \in U \times [0, 1] \,|\, v_t(x) = 0\}$ is compact. Then*

$$_c\tilde{I}_B(v_t, U) \in {}_c\omega_B^0\{B \times S^0; U_{+B}\}$$

is independent of t.

The Euler class of the tangent bundle

For the next part of the discussion we take the base B to be compact. Any vector field v on M is homotopic to the zero vector field 0 and hence, by Proposition 12.28(v), $\tilde{I}_B(v, M) = \tilde{I}_B(0, M) \in \omega_B^0\{B \times S^0; M_{+B}\}$. From Proposition 12.27 we see that this class coincides with the fixed-point index $\tilde{L}_B(1, M)$ of the identity map $1 : M \to M$. Recall from Section 4 that the Euler class $\gamma(\tau_B M)$ is the stable map $M \times S^0 \to (\tau_B M)_M^+$ over M given by the inclusion of the zero-section. We may, as in Proposition 3.4, make the identifications

$$\omega_M^0\{M \times S^0; (\tau_B M)_M^+\} = \omega_M^0\{(-\tau_B M)_M^+; M \times S^0\}$$
$$= \omega_B^0\{M_B^{-\tau_B M}; B \times S^0\},$$

and we use the same symbol and name for corresponding elements in any of the three Abelian groups. The associated map of Thom spaces, in the second interpretation, is the map

$$\gamma : M_B^{-\tau_B M} \to M_{+B} \tag{12.29}$$

induced by the 'inclusion' of $-\tau_B M$ in the zero bundle.

Proposition 12.30 *Let $M \to B$ be a compact fibrewise manifold over a compact ENR B. Then the Poincaré–Hopf index*

$$\tilde{I}_B(0, M) = \tilde{L}_B(1, M) \in \omega_B^0 \{B \times S^0; M_{+B}\}$$

is equal to the composition:

$$B \times S^0 \xrightarrow{p^!} M_B^{-\tau_B M} \xrightarrow{\gamma} M_{+B}.$$

It is, moreover, dual to the Euler class

$$\gamma(\tau_B M) \in \omega_B^0 \{M_B^{-\tau_B M}; B \times S^0\}$$

of the fibrewise tangent bundle of M.

To establish the first assertion it is probably simplest just to stare at the definitions. We choose a fibrewise embedding $i : M \hookrightarrow B \times E$ for some Euclidean space E, with normal bundle ν, and extend it to a tubular neighbourhood $D(\nu) \hookrightarrow B \times E$. The Gysin map $p^!$ is given, up to suspension, by the Pontrjagin–Thom construction $B \times E^+ \to D(\nu)/_B S(\nu) = M_B^\nu$, whereas the index class $\tilde{I}_B(0, M)$ is obtained by composition with the map

$$M_B^\nu \hookrightarrow M_B^{M \times E} = M_{+B} \wedge_B (B \times E^+)$$

given by including ν in the trivial bundle $M \times E$.

The second assertion follows by taking the dual of the composition $\gamma \circ p^!$. By Proposition 12.24, γ is self-dual, and by Proposition 12.21, $p^!$ is the dual of $p_+ : M_{+B} \to B \times S^0$. The composition $p_+ \circ \gamma$ defines the Euler class of the tangent bundle.

Isolated zeros

We develop next the fibrewise analogue of the classical computation of the index of a vector field with only isolated zeros on a closed manifold. Let $M \to B$ again be a compact fibrewise manifold over a compact base, and let v be a vector field defined on an open subset $U \subseteq M$. Suppose that $Z \subseteq U$ is a compact 0-dimensional submanifold, that is, a finite covering of B, containing all the zeros of v: $\mathrm{Zero}(v) \subseteq Z$. The normal bundle ν of the embedding $Z \hookrightarrow M$ is just the restriction $\tau_B M \mid Z$ of the tangent bundle of M. Let $j : D(\nu) \hookrightarrow U$ be a tubular neighbourhood. Pulling back the vector field v to $D(\nu)$ and identifying the tangent bundle of the disc-bundle with (the pull-back of) ν, we get a map $j^* v : D(\nu) \to \nu$, over Z, with zero-set in the zero-section $Z \subseteq D(\nu)$. The map $j^* v$ is non-zero on the sphere-bundle $S(\nu)$, and dividing by the norm we get a fibrewise map $S(\nu) \to S(\nu)$, that is,

$$S(\tau_B M \mid Z) \to S(\tau_B M \mid Z)$$

over Z. This extends radially to a pointed self-map of $(\tau_B M \mid Z)_Z^+$ giving a stable class

$$\mathfrak{o}(v) \in \omega^0(Z) \quad (= \omega_Z^0 \{(\tau_B M \mid Z)_Z^+; (\tau_B M \mid Z)_Z^+\}),$$

which we might call a *local index*.

Looking at a single fibre over $b \in B$ we are in the classical situation: Z_b is a finite set of points, $\omega^0(Z_b)$ is a direct sum of copies of $\mathbb{Z}$, one for each point, and the component of $\mathfrak{o}(v_b)$ at a point x is given by the degree of a self-map of the sphere $S((\tau Z_b)_x)$.

The next lemma reduces to the localization property, Proposition 12.28(ii), of the vector field index. Duality for the 0-dimensional manifold Z identifies

$$\omega^0(Z) = \omega_B^0\{Z_{+B}; B \times S^0\} \quad \text{with} \quad \omega_B^0\{B \times S^0; Z_{+B}\}.$$

Lemma 12.31 *Suppose, as in the text, that the zeros of the vector field v are contained in the compact 0-dimensional submanifold Z of U. Then the inclusion $z : Z \to U$ maps the local index, defined above, to the Poincaré–Hopf vector field index:*

$$\mathfrak{o}(v) \mapsto \tilde{I}_B(v, U) \; : \; \omega^0(Z) = \omega_B^0\{B \times S^0; Z_{+B}\} \to \omega_B^0\{B \times S^0; U_{+B}\}.$$

This leads to a fibrewise Poincaré–Hopf theorem. As in the Lefschetz fixed-point theory we associate to the index $\tilde{I}_B(v, U)$ a class

$$I_B(v, U) \in \omega^0(B), \tag{12.32}$$

by projecting U to B. In particular, $I_B(0, M) \in \omega^0(B)$ is the *fibrewise Euler characteristic*, $\chi_B(M)$, of $M \to B$.

Proposition 12.33 *Let v be a vector field on a compact fibrewise manifold M over a compact base B with zero-set contained in a compact 0-dimensional submanifold $Z \subseteq M$. Then, writing $p : Z \to B$ for the projection map of the finite covering and $p_! = (p^!)^*$ for the direct image map $\omega^0(Z) \to \omega^0(B)$, we have*

$$p_!(\mathfrak{o}(v)) = \chi_B(M) \in \omega^0(B).$$

Non-degenerate zeros

Suppose that v is a globally defined smooth vector field on M with *non-degenerate* zeros, in the sense that for each $x \in \text{Zero}(v)$ the derivative $Dv_x : (\tau_B M)_x \to (\tau_B M)_x$ is non-singular. (As this condition is by no means generic when we move out of the classical domain in which B is a point, the terminology is not perfect.) By transversality, $Z := \text{Zero}(v)$ is then a 0-dimensional submanifold of M. The local index is determined by the automorphism Dv of $\tau_B M \,|\, Z$, which defines an element of the real K-theory group $KO^{-1}(Z)$. (In the classical case we take the sign of the determinant in $KO^{-1}(*) = \{\pm 1\}$.) The class $\mathfrak{o}(v)$ in stable cohomotopy is its image under the J-homomorphism

$$J : KO^{-1}(Z) \to \omega^0(Z)^\times \subseteq \omega^0(Z).$$

For an application of these ideas see the work of Haibao and Rees [71].

The Lefschetz index of an isometry

The discussion of isolated singularities has exhibited one situation in which the vector field index is concentrated on the zero-set of the vector field. Another situation is that of a Killing vector field, that is, of an infinitesimal isometry. We state first a fixed-point version of the result.

Proposition 12.34 *Let $M \to B$ be a compact fibrewise manifold over a compact base B equipped with a (fibrewise) Riemannian metric. Suppose that $f : M \to M$ is a fibrewise smooth isometry with fixed-point set a submanifold. Then the fixed-point index $L_B(f, M) \in \omega^0(B)$ is equal to the fibrewise Euler characteristic $\chi_B(\mathrm{Fix}(f))$ of the fixed-point set $\mathrm{Fix}(f)$.*

More precisely, the fixed-point index in $\omega_B^0\{B \times S^0; \mathrm{Fix}(f)_+\}$ of the identity map on $\mathrm{Fix}(f)$ maps, under the inclusion, to the fixed-point index $\tilde{L}_B(f, M)$ of f in $\omega_B^0\{B \times S^0; M_{+B}\}$.

In the case $B = *$, when $\mathrm{Fix}(f)$ is discrete, this is a classical result, depending upon the fact that the derivative of an isometry at an isolated fixed-point has no eigenvalue equal to 1. When B is a point, $\mathrm{Fix}(f)$ is necessarily a submanifold. For the behaviour of f near a fixed-point x is determined by the derivative $(Df)_x : (\tau M)_x \to (\tau M)_x$, because the exponential map commutes with the action of the isometry f (on the tangent space and on M). So the $+1$-eigenspace of $(Df)_x$ is the tangent space $(\tau Z)_x$ of the fixed submanifold. This fact supplies the essential input for the case of a general base B, although we need to include the requirement that $\mathrm{Fix}(f)$ is a submanifold. (Consider, for example, the isometry $(g, x) \mapsto (g, gx)$ of the trivial bundle $O(n) \times S(\mathbb{R}^n) \to O(n)$ to see that this is not automatically the case.)

We shall give the proof for the vector field version, which we state next.

Proposition 12.35 *Let $M \to B$ be a compact fibrewise manifold with a Riemannian metric over a compact ENR B. Suppose that v is a Killing vector field, with zero-set a compact (closed) submanifold Z, defined on an open subset $U \subseteq M$. Then the Poincaré–Hopf index $I_B(v, U) \in \omega^0(B)$ is equal to the fibrewise Euler characteristic $\chi_B(Z)$ of the zero-set.*

If v is a Killing vector field (or infinitesimal isometry) then at a zero $x \in Z = \mathrm{Zero}(v)$ the derivative $(Dv)_x : (\tau_B M)_x \to (\tau_B M)_x$ is skew-symmetric with kernel equal to the tangent space $(\tau_B Z)_x$. On a small tubular neighbourhood $D(\nu)$ of Z the vector field is, under the usual identification of the tangent bundle of $D(\nu)$ with ν, homotopic to the (restriction of) the identity map $D(\nu) \to \nu$. The index of this new vector field is, essentially by definition, the Euler characteristic of Z.

Alternatively, if we look at the associated isometry f defined, as in Remark 12.27, by $x \mapsto \exp_x(-\epsilon v(x))$, we see, from the local description of the isometry on the tangent bundle, that f is homotopic, through maps with the same fixed-point set Z, to the projection of the tubular neighbourhood $D(\nu) \to Z$. Then the result follows from the contraction property (Corollary 6.6) of the fixed-point index.

Change of base

We turn now to differentiability conditions on the base instead of the fibre. Suppose that B and B' are compact ENRs and that $\alpha : B' \to B$ is a fibrewise manifold over B. Let us revert, in this context, to the standard notation $\tau(\alpha)$ for the bundle of tangents along the fibres $\tau_B B'$.

For any pointed homotopy fibre bundles $X \to B$ and $Y \to B$ with fibres of the homotopy type of finite complexes, we shall construct a Gysin (or direct image) map:

$$\alpha_! : \omega_{B'}^*\{\alpha^* X; \tau(\alpha)_{B'}^+ \wedge_{B'} \alpha^* Y\} \to \omega_B^*\{X; Y\}. \qquad (12.36)$$

Consider α as a fibrewise smooth map $B' \to B$ over B. The associated Gysin map $\alpha^!$ is a stable map $B \times S^0 \to B'^{-\tau(\alpha)}_B$, that is, an element

$$\alpha^! \in \omega_B^0\{B \times S^0; B'^{-\tau(\alpha)}_B\}.$$

Now we make the identifications, by desuspension and as in Lemma 6.15:

$$\omega_{B'}^*\{\alpha^* X; \tau(\alpha)_{B'}^+ \wedge_{B'} \alpha^* Y\} = \omega_{B'}^*\{(-\tau(\alpha))_{B'}^+ \wedge_{B'} \alpha^* X; \alpha^* Y\}$$
$$= \omega_B^*\{X \wedge_B B'^{-\tau(\alpha)}_B; Y\}.$$

Finally, composition with the class $\alpha^!$ maps this group to $\omega_B^*\{X; Y\}$, and so the direct image map $\alpha_!$ is defined.

Proposition 12.37 *The direct image map $\alpha_!$, (12.36), and transfer, (6.14), are related by multiplication by the Euler class of $\tau(\alpha)$:*

$$\alpha_\#(x) = \alpha_!(\gamma(\tau(\alpha)) \cdot x).$$

Thus the transfer $\alpha_\#$ is the composition:

$$\omega_{B'}^*\{\alpha^* X; \alpha^* Y\} \xrightarrow{\cdot \gamma(\tau(\alpha))} \omega_{B'}^*\{\alpha^* X; \tau(\alpha)_{B'}^+ \wedge_{B'} \alpha^* Y\} \xrightarrow{\alpha_!} \omega_B^*\{X; Y\}.$$

As for Proposition 12.30, this is best seen by looking directly at the definitions.

The properties of the direct image map correspond to those of the transfer (Proposition 6.16). We state only the simplest of the several variants of the Frobenius property.

Proposition 12.38 (Properties of the direct image map). *Let X and Y be fibrewise pointed spaces, as in the text, over B.*
(i)   (Functoriality). *Let $\alpha : B' \to B$ and $\beta : B'' \to B'$ be compact (closed) fibrewise manifolds, with B a compact ENR. Then $\alpha \circ \beta : B'' \to B$ is a compact fibrewise manifold with tangent bundle $\tau(\alpha \circ \beta)$ canonically isomorphic (up to homotopy) to $\beta^*\tau(\alpha) \oplus \tau(\beta)$, and $(\alpha \circ \beta)_! = \alpha_! \circ \beta_!$:*

$$\omega^*_{B''}\{(\alpha \circ \beta)^*X;\ \tau(\alpha \circ \beta)^+_{B''} \wedge_{B''} (\alpha \circ \beta)^*Y\}$$
$$\xrightarrow{\ \beta_!\ } \omega^*_{B'}\{\alpha^*X;\ \tau(\alpha)^+_{B'} \wedge_{B'} \alpha^*Y\} \xrightarrow{\ \alpha_!\ } \omega^*_B\{X;\ Y\}.$$

(ii)   (Frobenius reciprocity). *The direct image map $\alpha_!$ is $\omega^*(B)$-linear:*

$$\alpha_!(\alpha^*(x) \cdot y) = x \cdot \alpha_!(y) \quad \text{for } x \in \omega^*(B),\ y \in \omega^*_{B'}\{\alpha^*X;\ \tau(\alpha)^+_{B'} \wedge_{B'} \alpha^*Y\}.$$

When both B and B' are closed manifolds and α is a smooth submersion (so that $B' \to B$ is a smooth fibre bundle), we obtain by smashing Y with $(\tau B)^+_B$ the more egalitarian:

$$\alpha_! : \omega^*_{B'}\{\alpha^*X;\ (\tau B')^+_{B'} \wedge_{B'} \alpha^*Y\} \to \omega^*_B\{X;\ (\tau B)^+_B \wedge_B Y\}.$$

We should observe, too, that the two Gysin maps:

$$\alpha^! \in \omega^0_B\{B \times S^0;\ B'^{\tau(\alpha)}_B\} \quad \text{and} \quad \alpha^! \in \omega^0\{B_+;\ B'^{\tau(\alpha)}\},$$

the first defined by regarding α as a fibrewise smooth map $B' \to B$ over B and the second defined by thinking of α as a smooth map between manifolds, are compatible; the first determines the second by factoring out the fibrewise basepoints by the construction '$/B$'.

Let us return to the original case in which $\alpha : B' \to B$ is a fibrewise manifold over a compact ENR. We change notation and let X now be a pointed homotopy fibre bundle (with fibre of the homotopy type of a finite complex) over B', instead of B; the symbol Y denotes, as before, a space over B. Recall that the direct image α_*X (Definition 5.36) is the fibrewise pointed space

$$\alpha_*X = X/_B B' \tag{12.39}$$

over B. It is a pointed homotopy fibre bundle (with fibre of the homotopy type of a finite complex). We have already used, in special cases, the fact that the direct image is the left adjoint of the pull-back in the stable homotopy category.

Proposition 12.40 *Let* $\alpha : B' \to B$ *be a map between compact ENRs, and let* $X \to B'$ *and* $Y \to B$ *be fibrewise pointed spaces as above. Then there is a natural isomorphism:*

$$\omega^*_{B'}\{X; \alpha^*Y\} \overset{\cong}{\longrightarrow} \omega^*_B\{\alpha_*X; Y\}.$$

We are concerned now with a dual isomorphism.

Proposition 12.41 *Let* $\alpha : B' \to B$ *be a compact fibrewise manifold over a compact ENR, and let* $X \to B'$ *and* $Y \to B$ *be fibrewise pointed spaces as specified in the text. Then there is a natural isomorphism:*

$$\omega^*_{B'}\{\alpha^*Y; \tau(\alpha)^+_{B'} \wedge_{B'} X\} \overset{\cong}{\longrightarrow} \omega^*_B\{Y; \alpha_*X\}.$$

Corollary 12.42 *In the situation described above, if* X^* *is the dual of the fibrewise pointed space* X *over* B', *then the dual of* α_*X *over* B *is*

$$(\alpha_*X)^* = \alpha_*((-\tau(\alpha))^+_{B'} \wedge_{B'} X^*).$$

In order to see where these results come from, let us look at the special case in which $X = M_{+B'}$ for some smooth compact fibrewise manifold $M \to B'$. Then $\alpha_*M = M \to B$ is a fibrewise manifold too, $\alpha_*X = (\alpha_*M)_{+B}$, and there is a natural isomorphism (up to homotopy) between $\tau_B M$ (that is, $\tau_B \alpha_* M$) and $\tau(\alpha) \oplus \tau_{B'} M$. The dual X^* of X over B' is the fibrewise Thom space $M_{B'}^{-\tau_{B'} M}$ and the dual of $\alpha_*X = M_{+B}$ over B is

$$M_B^{-\tau_B M} = M_B^{-\tau(\alpha)-\tau_{B'} M} = \alpha_*((-\tau(\alpha))^+_{B'} \wedge_{B'} X^*).$$

Having identified the dual in this special case, we can deduce Proposition 12.40 from the chain of isomorphisms:

$$\omega^*_B\{Y; \alpha_*X\} = \omega^*_B\{Y \wedge_B \alpha_*((-\tau(\alpha))^+_{B'} \wedge_{B'} X^*); B \times S^0\}$$
$$= \omega^*_{B'}\{\alpha^*Y \wedge_{B'} ((-\tau(\alpha))^+_{B'} \wedge_{B'} X^*); B' \times S^0\}$$
$$= \omega^*_{B'}\{\alpha^*Y; \tau(\alpha)^+_{B'} \wedge_{B'} X\},$$

the first and last by duality over B and B', the middle isomorphism by (12.39). The same chain of reasoning in reverse order establishes Proposition 12.41, in general, as a corollary of Proposition 12.40.

The isomorphism in Proposition 12.40 is defined in the following way as a composition, λ say:

$$\omega^*_{B'}\{\alpha^*Y; \tau(\alpha)^+_{B'} \wedge_{B'} X\} \overset{\cong}{\longrightarrow} \omega^*_{B'}\{(-\tau(\alpha))^+_{B'} \wedge_{B'} \alpha^*Y; X\}$$
$$\overset{\alpha_*}{\longrightarrow} \omega^*_B\{B'^{-\tau(\alpha)}_B \wedge_B Y; \alpha_*X\}$$
$$\overset{\alpha_!}{\longrightarrow} \omega^*_B\{Y; \alpha_*X\}.$$

Here the first isomorphism is desuspension, the map α_* collapses B' to B, and the map $\alpha_!$ is composition with the Gysin map of $\alpha : B' \to B$ over B. We show that λ is an isomorphism by constructing an inverse μ of the form $x \mapsto u \circ \alpha^*(x)$, where

$$u \in \omega_{B'}^0 \{\alpha^*(\alpha_* X); \tau(\alpha)_{B'}^+, \wedge_{B'} X\}.$$

Now the pull-back $\alpha^*(\alpha_* X)$ of $\alpha_* X$ can be written as the fibrewise quotient $(B' \times_B X)/_{B'}(B' \times_B B')$, where $B' \times_B X$ and $B' \times_B B'$ are considered as fibrewise spaces over B' by projection onto the first factor. The diagonal

$$B' \xrightarrow{\Delta} B' \times_B B'$$
$$\searrow_1 \quad \swarrow$$
$$B'$$

is a fibrewise smooth embedding over B' with normal bundle $\tau(\alpha)$. We use the construction (12.5), extended to fibrewise manifolds, for the fibrewise pointed space $B' \times_B X \to B' \times_B B'$ to define a Gysin map:

$$(B' \times_B X)/_{B'}(B' \times_B B') \to \tau(\alpha)_{B'}^+.$$

This is the required element u. We omit what is once again a rather formal verification that the maps λ and μ are inverse to one another.

A manifold as base

Let us specialize the discussion now to the case in which $B = *$ is a point and B' is a closed manifold. To streamline the notation we revert to our standard 'B' for the base. Then Proposition 12.41 becomes:

Proposition 12.43 *Let B be a closed manifold, let $X \to B$ be a pointed homotopy fibre bundle with fibres of the homotopy type of finite complexes, and let F be a pointed finite complex. Then there is a natural isomorphism*

$$\omega_B^* \{B \times F; (\tau B)_B^+ \wedge_B X\} \xrightarrow{\cong} \omega^* \{F; X/B\}.$$

This result generalizes (as does the fibrewise version) to a manifold with boundary, as follows. The details of the proof, which builds on the argument given for a closed manifold, are omitted.

Proposition 12.44 *Let B be a compact manifold with boundary $A := \partial B$. Then, for $X \to B$ and F as in the statement of Proposition 12.43, there is a natural identification of two long exact sequences:*

the relative exact sequence of the pair (B, A) on the left,

and the exact sequence of the triple $(X, X_A \cup B, B)$ on the right,

in the following diagram.

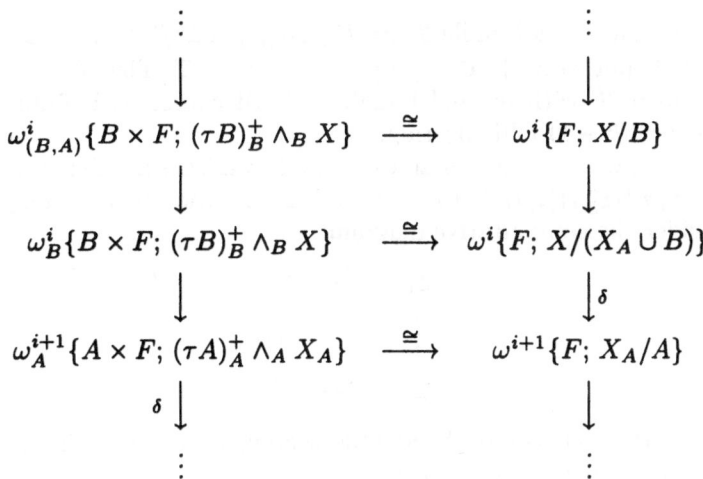

In writing the relative sequence in the form above we are using the identification of $\tau B \mid A$ with $\tau A \oplus \mathbb{R}$.

Finite coverings: an example of Sullivan

As a concrete illustration of the methods of this section we present an elegant result due to Sullivan [129]. We start with an elementary transfer argument.

Lemma 12.45 *Let $\pi : \tilde{B} \to B$ be a finite d-fold covering of a compact ENR B, and let ξ be a finite-dimensional real vector bundle over B whose pull-back $\tilde{\xi} := \pi^*\xi$ over $\tilde{B}$ has vanishing stable cohomotopy Euler class $\gamma(\tilde{\xi})$. Then $d^N \gamma(\xi) = 0$ for $N \gg 1$ sufficiently large.*

For the composition

$$\omega_B^0\{0_B^+;\, \xi_B^+\} \xrightarrow{\ \pi^*\ } \omega_{\tilde{B}}^0\{0_{\tilde{B}}^+;\, \tilde{\xi}_{\tilde{B}}^+\} \xrightarrow{\ \pi_!\ } \omega_B^0\{0_B^+;\, \xi_B^+\}$$

is multiplication by $\pi_!(1) \in \omega^0(B)$, by the Frobenius property (Proposition 12.38(ii)). If B is a point, then $\pi_!(1) = d$. In general, $\pi_!(1)$ has fibre degree equal to d and is equal to $d + x$, where x is nilpotent.

The proof is completed like that of Proposition 4.5. Since $\pi^*\gamma(\xi) = \gamma(\pi^*\xi)$, we have $d\gamma(\xi) = -x \cdot \gamma(\xi)$. If $x^N = 0$, then $d^N \gamma(\xi) = 0$ as claimed.

Proposition 12.46 *Let $P \to B$ be a principal $GL(n, \mathbb{Z})$-bundle over a compact ENR B, and let ξ denote the associated flat n-dimensional real vector bundle $P \times_{GL(n,\mathbb{Z})} \mathbb{R}^n \to B$. Then there exists a finite covering $\pi : \tilde{B} \to B$ such that the stable cohomotopy Euler class $\gamma(\pi^*\xi)$ of the pull-back of ξ vanishes.*

Consider the torus bundle $T := P \times_{GL(n,\mathbf{Z})} (\mathbb{R}^n/\mathbb{Z}^n)$ over B. For each integer $l > 1$, put $S_l := \{z \in T \mid z \neq 1,\ z^l = 1\} \subseteq T$. Thus S_l is a fibrewise submanifold of T with normal bundle $\pi^*\xi$. Of course, it is 0-dimensional; $S_l \to B$ is an $(l^n - 1)$-fold covering.

Inverting l, we show first of all that $\gamma(\pi^*\xi)$ vanishes in the localized group $\omega_{S_l}^0\{S_l \times S^0;\ \pi^*(\xi_B^+)\}[1/l]$. Let $m_l : T \to T$ denote the lth power map: $z \mapsto z^l$. Now consider the commutative diagram

$$
\begin{array}{ccc}
S_l & \overset{i}{\underset{\subset}{\longrightarrow}} & T \\[4pt]
{\scriptstyle \pi} \downarrow & & \downarrow {\scriptstyle m_l} \\[4pt]
B & \underset{e}{\overset{\subset}{\longrightarrow}} & T
\end{array}
$$

in which e is the inclusion of the identity section, and look at the corresponding diagram of direct image maps:

$$
\begin{array}{ccccc}
\omega^0(S_l) & = & \omega_B^0\{(S_l)_{+B};\ B \times S^0\} & \overset{i_!}{\longrightarrow} & \omega_B^0\{T_{+B};\ \xi_B^+\} \\[6pt]
 & & {\scriptstyle \pi_!} \downarrow & & \downarrow {\scriptstyle (m_l)_!} \\[6pt]
\omega^0(B) & = & \omega_B^0\{B \times S^0;\ B \times S^0\} & \underset{e_!}{\longrightarrow} & \omega_B^0\{T_{+B};\ \xi_B^+\}
\end{array}
$$

Because $m_l \circ e = e$, we have $(m_l)_! e_!(1) = e_!(1)$. Also $e_!(\pi_!(1)) = \pi_!(1).e_!(1)$, since $e_!$ is an $\omega^0(B)$-module homomorphism. This allows us to deduce that:

$$(m_l)_!(i_!(1) - \pi_!(1).e_!(1)) = 0.$$

We claim that $(m_l)_! : \omega_B^0\{T_{+B};\ \xi_B^+\}[1/l] \to \omega_B^0\{T_{+B};\ \xi_B^+\}[1/l]$ is bijective. As the $\mathbb{Z}[1/l]$-module is finitely generated, it will suffice to show that the map is surjective. Now $(m_l)_!(m_l)^*$ is multiplication by $(m_l)_!(1)$, according to Proposition 12.38(ii), where $(m_l)_!$ is the direct image map $\omega^0(T) = \omega_B^0\{T_{+B};\ B \times S^0\} \to \omega^0(T)$. So we just have to show that $(m_l)_!(1)$ is invertible in $\omega^0(T)[1/l]$. We can certainly assume that B is connected, and then $\omega^0(T)$ is the direct sum $\mathbb{Z}1 \oplus \mathfrak{n}$, where $\mathfrak{n}$ is a nilpotent ideal. Since $m_l : T \to T$ is an l^n-fold covering, $(m_l)_!(1)$ is equal to l^n modulo torsion.

We have established that: $i_!(1) = \pi_!(1).e_!(1) \in \omega_B^0\{T_{+B};\ \xi_B^+\}[1/l]$. But $i^* i_!(1)$ is the Euler class $\gamma(\xi) \in \omega_{S_l}^0\{S_l \times S^0;\ \pi^*(\xi_B^+)\}$, by Proposition 12.2, and $i^* e_!(1) = 0$, by Frobenius reciprocity, because S_l and $i(B)$ are disjoint.

Remark 12.47. It follows from Lemma 12.45 that $\gamma(\xi)$ is torsion, and, more precisely, that its order divides $l^M(l^n - 1)^N$ for M and N sufficiently large. Since this is true for any integer $l > 1$, a prime $p > 1$ divides the order of $\gamma(\xi)$ only if $p - 1$ divides n.

If one works in cohomology with integer coefficients, as did Sullivan, one sees that the cohomology Euler class $e(\xi)$, with coefficients twisted by the determinant bundle $P \times_{GL(n,\mathbf{Z})} \mathbb{Z}$, has order dividing $l^n(l^n - 1)$ for each $l > 1$. Equivalently, by elementary number theory, in terms of

$$m(n) = \prod_{p \,:\, p-1 \,|\, n} p^{\nu_p(n)},$$

where ν_p is the p-adic valuation, $p > 1$ prime, the order of $e(\xi)$ divides $m(n)$ if n is odd, $4m(n)$ if n is even.

After this digression we can complete the proof of Proposition 12.46 by taking $\tilde{B}$ to be, for example, the fibrewise product $S_2 \times_B S_3$.

13 Miller's stable splitting of $U(n)$

The theorem of Miller discussed in this section appeared in [111]. The method of proof, using fibrewise techniques, is taken from [26]. A fibrewise generalization of the theorem was exploited in [29].

We work throughout over a fixed compact base B.

Bundles of groups

Let G be a compact Lie group and H a Lie group acting on G by group automorphisms. For example, H might be the group of all automorphisms of G or H might be the group G acting on itself by inner automorphisms. We write $\mathfrak{g}$ for the Lie algebra of G, the tangent space $\tau_1 G$ at the identity $1 \in G$. The tangent bundle of G is trivialized by left translation: the derivative at 1 of left multiplication $G \to G$ by an element $g \in G$ gives an isomorphism $\mathfrak{g} = \tau_1 G \to \tau_g G$.

Associated to a principal H-bundle $P \to B$ over the compact ENR B there is the bundle of groups $M := P \times_H G \to B$. Let ζ be the Lie algebra bundle $P \times_H \mathfrak{g} \to B$.

By left translation on fibres, the fibrewise tangent bundle $\tau_B M$ is identified with the pull-back of ζ from B to M: the bundle of Lie groups is fibrewise parallelizable.

Just as we can split off the 'top cell' of a parallelizable closed manifold, we can split off the pointed sphere-bundle ζ_B^+. Let $e : B \to M$ be the inclusion of the identity in each fibre. We could use this section to make $M \to B$ a fibrewise pointed space, but it is more natural in the present context to add a disjoint basepoint and consider M_{+B}. The fibrewise normal bundle of the embedding e is ζ. Choosing an invariant inner product on $\mathfrak{g}$ and using the exponential map $\exp : \mathfrak{g} \to G$, which will embed a small disc in $\mathfrak{g}$ into G, we get an explicit fibrewise tubular neighbourhood of B in M. By scaling the metric we may arrange that the unit disc-bundle $D(\zeta)$ embeds in M. The fibrewise Gysin map $e^! : M_{+B} \to \zeta_B^+$ is the fibrewise collapse

$$M_{+B} \to M/_B(M - B(\zeta)) = D(\zeta)/_B S(\zeta)$$

onto the 'top cell'. (Here $B(\zeta)$ denotes the open unit disc-bundle.) The projection map $p : M \to B$ splits $e : p \circ e = 1_B$, and so $p^! : \zeta_B^+ \to M_{+B}$ splits $e^! : e^! \circ p^! = 1$.

Proposition 13.1 *There is a canonical fibrewise stable splitting*

$$M_{+B} \simeq_B \zeta_B^+ \vee_B N_{+B},$$

where $N = M - B(\zeta)$.

Notice that N is a fibrewise manifold with boundary $\partial N = S(\zeta)$. The fibrewise quotient $N/_B\partial N$ is fibrewise homeomorphic to M equipped with the basepoint e. The stable splitting just constructed is dual to the splitting of M_{+B} as $(B \times S^0) \vee_B M$ given by $e_* : B \times S^0 \to M_{+B}$ and the pointed map $M_{+B} \to M$ which is the identity on $M \subseteq M_{+B}$.

The geometry of $U(n)$

Let E be an n-dimensional $\mathbb{C}$-vector space equipped with a Hermitian inner product. We write $U(E)$ for the unitary group of E. The Lie algebra $\mathfrak{u}(E)$ of $U(E)$ is the space of skew-Hermitian endomorphisms of E.

Instead of using the exponential map to construct a tubular neighbourhood we shall use the Cayley transform.

Lemma 13.2 *The Cayley transform*

$$\psi : \mathfrak{u}(E) \to \{g \in U(E) \mid 1 - g \text{ is invertible}\}$$
$$\psi(v) = (v/2 - 1)(v/2 + 1)^{-1}$$

is a diffeomorphism mapping 0 *to* -1.

The inverse of ψ can be written as:

$$g \mapsto 2(1 - g)^{-1}(1 + g) = 2(1 - g^*)^{-1}(g - g^*)(1 - g)^{-1}.$$

Given a line (that is, a 1-dimensional complex subspace) $L \subseteq E$ and complex number $\lambda \in \mathbb{T}$ of modulus 1, we write $\rho_L(\lambda) \in U(E)$ for the *complex reflection* which is the identity on the orthogonal complement $L^\perp$ of L and multiplication by λ on L. There is a filtration

$$\{1\} = R^0(E) \subseteq \cdots \subseteq R^k(E) \subseteq \cdots \subseteq R^n(E) = U(E)$$

of $U(E)$ by the subspaces $R^k(E)$ consisting of the unitary transformations which can be expressed as a product of k complex reflections. By elementary linear algebra we see that:

Lemma 13.3 *The subspace $R^k(E)$ consists of those group elements $g \in U(E)$ with $\dim(\ker(1-g))^{\perp} \le k$.*

We shall need to know that the inclusion $R^{k-1}(E) \hookrightarrow R^k(E)$ is a closed cofibration. This follows from:

Lemma 13.4 *The space $R^k(E)$ is a compact ENR.*

One could quote here the general fact that a real algebraic variety is an ENR, but it is easy to give an explicit retraction of a neighbourhood of $R^k(E)$ in $U(E)$ onto the subspace. Given $g \in U(E)$ we may list its eigenvalues $\lambda_1, \dots, \lambda_n$ so that $|\lambda_1 - 1| \le |\lambda_2 - 1| \le \dots \le |\lambda_n - 1|$. Write $\lambda_{n-k} = e^{i\alpha}$, where $-\pi < \alpha \le \pi$, and define $\phi_k(g) = |\alpha| \in [0, \pi]$. The function $\phi_k : U(E) \to [0, \pi]$ so constructed is continuous and has zero-set $R^k(E)$. Continuity is clear from another description. For any vector subspace F of E, let $\|(1-g)_F\|$ denote the operator (supremum) norm of the restriction of $1 - g : F \to E$. Then

$$2|\sin(\alpha/2)| = \inf_{\dim F = n-k} \|(1-g)_F\|,$$

the infimum over the subspaces $F \subseteq E$ of dimension $n - k$.

The open subspace $\{g \in U(E) \mid \phi_k(g) < \pi\}$ can be retracted onto $R^k(E)$ by pushing eigenvalues towards 1. More precisely, let $f : \mathbb{T} \times [0, \pi) \to \mathbb{T}$ be a continuous function such that $f(e^{i\alpha}, t) = 1$ for $|\alpha| \le t$ and $f(\lambda, 0) = \lambda$. By the functional calculus for unitary operators, f extends to a homotopy $U(E) \times [0, \pi) \to U(E)$. We retract g to $f(g, \phi_k(g))$.

The Cayley transform identifies $R^n(E) - R^{n-1}(E)$ with the vector space $\mathfrak{u}(E)$, and thus the quotient $R^n(E)/R^{n-1}(E)$ is identified with the sphere $\mathfrak{u}(E)^+$.

A fibrewise stable splitting

Consider, more generally, an n-dimensional complex vector bundle ξ over the compact base B, with a Hermitian inner product. Then we can form the bundle $U(\xi) \to B$ of unitary groups over B, with fibre at $b \in B$ the unitary group $U(\xi_b)$ of the fibre of ξ. The bundle is filtered by sub-bundles

$$B = R^0(\xi) \subseteq \dots \subseteq R^k(\xi) \subseteq \dots \subseteq R^n(\xi) = U(\xi), \qquad (13.5)$$

the fibre of $R^k(\xi)$ at b being the subspace $R^k(\xi_b)$. As a bundle with each fibre an ENR, each space $R^k(\xi)$ is a fibrewise ENR, by Definition 5.5. (But the proof of Lemma 13.4 evidently carries through to give an explicit fibrewise neighbourhood retraction.)

The fibrewise quotient $R^n(\xi)/_B R^{n-1}(\xi)$ is identifed by the Cayley transform with the fibrewise one-point compactification $\mathfrak{u}(\xi)_B^+$ of the Lie algebra bundle $\mathfrak{u}(\xi)$. We think of the complement $R^n(\xi) - R^{n-1}(\xi)$ as the 'top cell' of $U(\xi)$, centred on -1 in each fibre. As in Proposition 13.1, the collapse map $U(\xi)_{+B} \to \xi_B^+$ is the Pontrjagin–Thom construction $i^!$ on the inclusion $i : B \to U(\xi)$ of -1 in each fibre, and it is split stably by

$$p^! : u(\xi)_B^+ \to U(\xi)_{+B}, \tag{13.6}$$

(where $p : U(\xi) \to B$, over B, is as usual the projection).

We shall exploit this discussion of the fibrewise case to split $U(E)$. Let $G_k(E)$, for $0 \le k \le n$, denote the Grassmann manifold of k-dimensional vector subspaces of E, and let η_k denote the canonical k-dimensional vector bundle over $G_k(E)$; the fibre of η_k at a point $F \in G_k(E)$ is the vector space F. The Hermitian structure on E induces an inner product on η_k. Now consider the bundle $U(\eta_k) \to G_k(E)$. Elements of $U(\eta_k)$ are pairs (F, h), where $F \subseteq E$ is a k-dimensional subspace and $h \in U(F)$. We have a surjective map

$$q : U(\eta_k) \to R^k(E) \qquad (F, h) \mapsto h \oplus 1 : E = F \oplus F^\perp \to F \oplus F^\perp = E.$$

This map is injective on the top cell $R^k(\eta_k) - R^{k-1}(\eta_k)$ and restricts to a fibrewise diffeomorphism

$$u(\eta_k) = (R^k(\eta_k) - R^{k-1}(\eta_k)) \to (R^k(E) - R^{k-1}(E))$$

over $G_k(E)$. The space $R^k(E) - R^{k-1}(E)$ fibres over $G_k(E)$ by the map: $g \mapsto (\ker(1 - g))^\perp$. Since the quotient $R^k(E)/R^{k-1}(E)$ is the one-point compactification of $R^k(E) - R^{k-1}(E)$ we have established:

Proposition 13.7 *There is a natural homeomorphism*

$$R^k(E)/R^{k-1}(E) \to G_k(E)^{u(\eta_k)}$$

between the kth quotient in the filtration of $U(E)$ and the Thom space (or one-point compactification) of $u(\eta_k)$.

We shall show:

Lemma 13.8 *The projection*

$$R^k(E)_+ \to R^k(E)/R^{k-1}(E) \to G_k(E)^{u(\eta_k)}$$

has a stable splitting.

From this lemma Miller's theorem follows immediately.

Proposition 13.9 (Miller's stable splitting of $U(n)$). *There is a stable decomposition*

$$\bigvee_{0 \le k \le n} G_k(E)^{u(\eta_k)} \xrightarrow{\simeq} U(E)_+.$$

The lemma itself is a consequence of (13.6), which provides a fibrewise stable splitting

$$(\mathsf{u}(\eta_k))^+_{G_k(E)} \to R^k(\eta_k)_{+G_k(E)}$$

over $G_k(E)$. Collapsing the fibrewise basepoints to a point and composing with $q : R^k(\eta_k) \to R^k(E)$ we obtain the required stable splitting map

$$G_k(E)^{\mathsf{u}(\eta_k)} \to R^k(E)_+.$$

The proof may be conveniently summarized in the diagram:

$$
\begin{array}{ccccc}
\mathsf{u}(\eta_k)^+_{G_k(E)} & \longrightarrow & R^k(\eta_k)_{+G_k(E)} & \longrightarrow & \mathsf{u}(\eta_k)^+_{G_k(E)} \\
& & & & \\
G_k(E)^{\mathsf{u}(\eta_k)} & \longrightarrow & R^k(\eta_k)_+ & \longrightarrow & G_k(E)^{\mathsf{u}(\eta_k)} \\
\Big\downarrow = & & \Big\downarrow {\scriptstyle q_*} & & \Big\downarrow = \\
G_k(E)^{\mathsf{u}(\eta_k)} & \longrightarrow & R^k(E)_+ & \longrightarrow & G_k(E)^{\mathsf{u}(\eta_k)}
\end{array}
$$

The top line is a sequence of fibrewise stable maps over $G_k(E)$ with composition equal to the identity. The second line is the sequence of stable maps obtained from the first by collapsing the fibrewise basepoints; the composition is then clearly the identity. From commutativity of the two squares it follows that the composition in the bottom line is the identity, and this gives us the stable splitting.

A fibrewise version

We have used fibrewise methods to establish Miller's theorem. The theorem itself has a natural generalization giving a fibrewise stable splitting of a bundle of unitary groups. Given a complex vector bundle ξ over B, we can form the Grassmann bundle $G_k(\xi)$, whose fibre over $b \in B$ is the Grassmann manifold $G_k(\xi_b)$ of k-dimensional subspaces of the fibre ξ_b, and over $G_k(\xi)$ we have a canonical k-dimensional complex vector bundle η_k (which, incidentally, is fibrewise smooth over B).

Proposition 13.10 *Let ξ be an n-dimensional complex vector bundle, with a Hermitian inner product, over the compact ENR B. There is a fibrewise stable decomposition*

$$\bigvee_{\substack{B \\ 0 \leq k \leq n}} G_k(\xi)^{\mathsf{u}(\eta_k)}_B \xrightarrow{\simeq} U(\xi)_{+B}.$$

By collapsing the basepoints in the fibres to a point, we obtain

Corollary 13.11 *There is a stable splitting of spaces:*

$$\bigvee_{0 \leq k \leq n} G_k(\xi)^{\mathsf{u}(\eta_k)} \xrightarrow{\simeq} U(\xi)_+.$$

The proposition may be established by following through the proof of Proposition 13.9 at the fibrewise level, a procedure which involves the systematic consideration of fibrewise fibre bundles. Alternatively, one may use equivariant methods, deducing Proposition 13.10 directly from the $U(E)$-equivariant extension of Proposition 13.9, which is proved by doing little more than inserting the word 'equivariant' at appropriate points.

The fibrewise Lusternik–Schnirelmann category of $U(\xi)$

It is interesting to note, in passing, a generalization of a result of Singhof [123, 124] on the category of the unitary group, which provides a good illustration of the concepts discussed in Section 19 of Part I.

For $\lambda \in \mathbb{T}$, a complex number of modulus 1, let W_λ denote the subspace of $U(\xi)$ consisting, in each fibre, of the unitary transformations with no eigenvalue equal to λ. The Cayley transform gives a fibrewise diffeomorphism from $\mathfrak{u}(\xi)$ to the open subspace W_1 of $U(\xi)$. By multiplying by λ, we obtain a fibrewise diffeomorphism $\mathfrak{u}(\xi) \to W_\lambda$ mapping the trivial summand $B \times i\mathbb{R}$ to $B \times (\mathbb{T} - \{\lambda\})$. So W_λ is fibrewise contractible, and if $\lambda \neq 1$ it can be contracted to the basepoint 1 in each fibre. Since any $n + 1$ of these sets cover $U(\xi)$, for distinct values of λ, we have established most of:

Proposition 13.12 *The fibrewise pointed Lusternik–Schnirelmann category of the bundle of unitary groups $U(\xi)$, with basepoint 1 in each fibre, of an n-dimensional complex Hermitian bundle ξ over B is $n + 1$. More precisely, we have*

$$\mathrm{cat}_B^* U(\xi) = \mathrm{wcat}_B^* U(\xi) = n + 1.$$

Recall that the fibrewise pointed category cat_B^* and weak category wcat_B^* (denoted by cat_B^B and wcat_B^B in Section 19 of Part I) are defined as follows. We consider fibrewise pointed open subsets W of $U(\xi)$ such that the inclusion $W \to U(\xi)$ is fibrewise null-homotopic; $\mathrm{cat}_B^* U(\xi)$ is the least number of sets in a covering of $U(\xi)$ by such open sets. The least integer k such that the diagonal map

$$U(\xi) \to \textstyle\bigwedge_B^k U(\xi)$$

is fibrewise null-homotopic is $\mathrm{wcat}_B^* U(\xi)$, and $\mathrm{wcat}_B^* U(\xi) \leq \mathrm{cat}_B^* U(\xi)$ (by Part I, (19.8)). The upper bound on $\mathrm{cat}_B^* U(\xi)$ is supplied by the geometric discussion and the lower bound on the weak category is provided by the cohomology of a fibre: $H^*(U(n); \mathbb{Z})$ is an exterior algebra on n generators in degrees $2i - 1$, $1 \leq i \leq n$.

Stiefel manifolds

The methods of this section can be generalized from the unitary group to the complex Stiefel manifolds. Let E and F be two finite-dimensional complex Hilbert spaces. We consider the Stiefel manifold $U(E; F)$ of isometric linear maps $E \to E \oplus F$. Elements can be written as pairs (g, h) with $g \in \text{End}(E)$, $h \in \text{Hom}(E, F)$ such that $g^*g + h^*h = 1$. The Cayley transform describes the open subspace consisting of the pairs (g, h) with $1 - g$ invertible.

Lemma 13.13 *There is a diffeomorphism*

$$\psi : \mathfrak{u}(E) \oplus \text{Hom}(E, F) \to \{(g, h) \in U(E; F) \mid 1 - g \text{ is invertible}\}$$

given by $\psi(v, w) = (g, h)$, *where*

$$b = (1 + w^*w/4)^{1/2}, \qquad a = bvb/2 + w^*w/4,$$
$$g = (a - 1)(a + 1)^{-1}, \qquad h = w(1 - g)/2.$$

Notice that $v = 2b^{-1}(1 - g^*)^{-1}(g - g^*)(1 - g)^{-1}b^{-1}$. The transformation $a + 1$ is invertible because $(a + 1) + (a + 1)^*$ is positive-definite.

The formulation and proof of the generalizations to Stiefel manifolds of Propositions 13.9, 13.10 and 13.12 are left as an exercise for the reader.

14 Configuration spaces and splittings

In this section the base B will be a compact ENR and all the fibrewise spaces will be locally homotopy trivial. But we begin with a review of the classical theory in which B is reduced to a point (and omitted from the notation). The generalization to the fibrewise theory turns out to be rather routine.

The classical theory

Let M be a (non-empty) smooth manifold. For a natural number $k \geq 0$, we define the *configuration space* of k indistinguishable particles in M to be the set

$$\mathcal{C}^k(M) := \{Q \subseteq M \mid \#Q = k\} \tag{14.1}$$

topologized as the quotient $\mathcal{F}^k(M)/\mathfrak{S}_k$ of the space

$$\mathcal{F}^k(M) := \{(x_1, \dots, x_k) \in M^k \mid x_i \neq x_j \text{ for } i \neq j\}$$

of ordered configurations (already studied in Part I, Section 23) by the free permutation action of the symmetric group $\mathfrak{S}_k$. (The special case $\mathcal{C}^0(M)$ is understood to consist of a single point, the empty subset of M.) As we have seen in Part I, Proposition 23.2, the space $\mathcal{F}^k(M)$ is a smooth manifold, and so, too, is the quotient $\mathcal{C}^k(M)$.

Example 14.2. The basic example is the manifold $M = \mathbb{R}$. We can order any k-element subset $Q \subseteq \mathbb{R}$ as: $x_1 < x_2 < \ldots < x_k$ and describe $C^k(\mathbb{R})$ as the contractible space $\mathbb{R} \times (0, \infty)^{k-1}$ by mapping Q to $(\bar{x}; x_2 - x_1, \ldots, x_k - x_{k-1})$, where $\bar{x}$ is the centre of mass $(x_1 + \ldots + x_k)/k \in \mathbb{R}$. The ordered configuration space $\mathcal{F}^k(\mathbb{R})$ is correspondingly identified with $\mathfrak{S}_k \times (\mathbb{R} \times (0, \infty)^{k-1})$, where $\mathfrak{S}_k$ acts by multiplication on the first factor and trivially on the bracketed second factor.

Example 14.3. Let M be a finite-dimensional Euclidean space E, and consider the case $k = 2$. There is a homeomorphism

$$\mathcal{F}^2(E) = E \times E - \Delta(E) \to E \times (0, \infty) \times S(E)$$

given by $(x, y) \mapsto (u, \|v\|, v/\|v\|)$, where u is the centre of mass $(x + y)/2$ and $v = (x - y)/2$ (say). This map is $\mathfrak{S}_2$-equivariant if the action on the sphere $S(E)$ is by the antipodal involution and the group acts trivially on the first two factors. We obtain a homeomorphism: $C^2(E) \to E \times (0, \infty) \times P(E)$ and, by projection, a homotopy equivalence $C^2(E) \to P(E)$ to the real projective space of E.

These first two examples have been important since configuration space models first appeared in homotopy theory; the third example came with the advent of cyclic homology.

Example 14.4. Let M be the circle S^1, which we regard as $\mathbb{R}/\mathbb{Z}$. Consider first an ordered configuration $(x_1, \ldots, x_k)$ in which $x_1, \ldots, x_k$ go anticlockwise around the circle. Let $(t_i)_{i \in \mathbb{Z}}$ be the monotone increasing infinite sequence

$$\ldots < t_i < t_{i+1} < \ldots \qquad \text{in } \mathbb{R}$$

with $x_i = t_i \pmod{\mathbb{Z}}$, $t_{i+k} = t_i + 1$, and, to fix the indexing, $0 \leq t_1 < 1$. Write $s_i = t_{i+1} - t_i$, so that $s_i > 0$ and $\sum_{1 \leq i \leq k} s_i = 1$. The normalized first term

$$\bar{x}_1 := \frac{1}{k}(t_1 + t_2 + \cdots + t_k) - \frac{k-1}{2k} \pmod{\mathbb{Z}} \in \mathbb{R}/\mathbb{Z}$$

is then well-defined. (If the distances s_i are all equal, then $\bar{x}_1$ will be equal to x_1.) Writing A for the subspace of $\mathcal{F}^k(S^1)$ consisting of all such ordered configurations $(x_1, \ldots, x_k)$ and Δ^o for the open simplex

$$\{(s_1, \ldots, s_k) \in \mathbb{R}^k \mid s_i > 0, \ s_1 + \cdots + s_k = 1\},$$

we obtain a homeomorphism:

$$A \to S^1 \times \Delta^o : (x_1, \ldots, x_k) \mapsto (\bar{x}_1; s_1, \ldots, s_k).$$

This homeomorphism is $\mathbb{Z}/k$-equivariant, where the generator 1 of the cyclic group acts on A and on Δ^o by cyclic permutation: $(x_1, \ldots, x_k) \mapsto (x_2, \ldots, x_1)$, $(s_1, \ldots, s_k) \mapsto (s_2, \ldots, s_1)$, and on $S^1 = \mathbb{R}/\mathbb{Z}$ by adding $1/k$.

The action of $\mathfrak{S}_k$ on $\mathcal{F}^k(S^1)$ leads to the description:

$$\mathcal{F}^k(S^1) \xrightarrow{\cong} (S^1 \times \Delta^o) \times_{\mathbb{Z}/k} \mathfrak{S}_k \xrightarrow{\simeq} S^1 \times_{\mathbb{Z}/k} \mathfrak{S}_k,$$

in which the first arrow is a homeomorphism and the second, given by projection, is a homotopy equivalence. The configuration space $C^k(S^1)$ itself is thus homotopy equivalent to the circle $S^1/(\mathbb{Z}/k)$. Geometrically, the homotopy equivalence pushes a configuration of k particles on the circle until the distances between adjacent particles are equal.

We denote the disjoint union of the configuration spaces for $k \geq 0$ by

$$C(M) := \coprod_{k \geq 0} C^k(M).$$

It is a pointed space, with basepoint in $C^0(M)$.

For a compact manifold M with boundary the inclusion of the open submanifold $M - \partial M \hookrightarrow M$ induces a homotopy equivalence $C^k(M - \partial M) \to C^k(M)$, as is easily checked using a collar neighbourhood $\partial M \times [0, 1) \subseteq M$ of the boundary. In establishing the basic results one also needs a pointed space $C(M, \partial M)$ defined by allowing particles to appear and disappear on the boundary. An equivalence relation $\sim$ is defined on $C(M)$ by:

$$Q \sim Q' \quad \text{if and only if} \quad Q \cap (M - \partial M) = Q' \cap (M - \partial M),$$

and $C(M, \partial M)$ is the topological quotient $C(M)/\sim$. The space is filtered by the subspaces $C^{\leq k}(M, \partial M)$ given by subsets $Q \subseteq M$ with $\#Q \leq k$. The following example provides a good illustration of the basic ideas.

Example 14.5. Let M be the closed unit disc $D(E)$ in the finite-dimensional Euclidean space E. Then $C(M, \partial M)$ is homotopy equivalent to the one-point compactification E^+.

The assertion is verified by producing a deformation retraction of the configuration space $C(M, \partial M)$ onto the subspace $C^{\leq 1}(M, \partial M)$ given by subsets with at most one particle. This subspace is clearly the quotient $D(E)/S(E)$, which is identified with E^+ in the usual way. The deformation is defined in the following way. Given a finite subset $Q \subseteq D(E)$, let $r > 0$ be the greatest real number ≤ 1 such that the open disc of radius r and centre the origin contains at most one element of Q. Now push Q out to $\frac{1}{r}Q$ so that excess particles fall off the edge of the disc.

Configuration spaces and spaces of sections

Let E again be a finite-dimensional Euclidean space. Using the Pontrjagin–Thom construction we shall define a map

$$\phi : C^k(E) \to \text{map}^*(E^+, E^+)$$

which is canonical up to homotopy. A finite subset $Q \in C^k(E)$ of E is a 0-dimensional compact submanifold. A tubular neighbourhood of Q is a disjoint union of discs centred on the points of Q. To be precise, we can take the discs to be of radius

$$\epsilon(Q) := \tfrac{1}{4} \min \{\|x - y\| \mid x, \, y \in Q, \, x \neq y\}.$$

The Pontrjagin–Thom construction, as described in Section 12, gives a pointed map $\phi(Q) : E^+ \to E^+$ varying continuously with Q.

This is clearly already a fibrewise construction (even in the ordinary case, $B = *$, that we are considering). Over the (generally non-compact) ENR $C^k(M)$ we have a fibrewise manifold $\{(Q, x) \mid x \in Q\}$ embedded in the trivial vector bundle $C^k(M) \times E$. The map ϕ is given by a fibrewise Pontrjagin–Thom construction.

We now generalize. Let M be a compact manifold with (possibly empty) boundary. Recall that $(\tau M)^+_M$ denotes the fibrewise one-point compactification (over M) of the tangent bundle of M with basepoint in each fibre the point at infinity. We write $\Gamma(M, \partial M; (\tau M)^+_M)$ for the space of sections of this bundle which are null (that is, take the value ∞) over the subspace ∂M. It contains as a subspace the space $_c\Gamma(M - \partial M; (\tau M)^+_M)$ of sections with compact support, that is, sections which are null outside a compact subspace of $M - \partial M$. Using a collar neighbourhood of the boundary again, it is easy to see that the inclusion

$$_c\Gamma(M - \partial M; (\tau M)^+_M) \hookrightarrow \Gamma(M, \partial M; (\tau M)^+_M)$$

is a homotopy equivalence.

We construct a canonical homotopy class

$$\phi : C^k(M - \partial M) \to {}_c\Gamma(M - \partial M; (\tau M)^+_M) \tag{14.6}$$

by a procedure familiar from Section 12. Consider a finite subset Q of $M - \partial M$, that is, a compact submanifold of dimension 0. The normal bundle ν of the embedding $Q \hookrightarrow M$ is the restriction $(\tau M)|Q$. A tubular neighbourhood allows us to identify a neighbourhood Q in $M - \partial M$ with the disc-bundle $D := D(\nu)$, of which the tangent bundle τD is just (the pull-back of) ν. The projection $D(\nu_x) \to D(\nu_x)/S(\nu_x)$, identified in the usual way, up to homotopy, with $(\nu_x)^+$ for $x \in Q$, gives a section of $(\tau D)^+_D$ over D which is null on the boundary $\partial D = S(\nu)$. Now extend this section to $M - \partial M$ to be null outside D. The map ϕ is defined by performing this construction fibrewise over $C^k(M - \partial M)$. This can be carried out quite explicitly by choosing a Riemannian metric on M and using the exponential map.

Remark 14.7. Suppose that the manifold M is connected. Then the components of $_c\Gamma(M - \partial M; (\tau M)^+_M)$ are indexed by the cohomological degree, and ϕ maps the configurations $C^k(M - \partial M)$ of k particles into the component of degree k. Indeed, the set of components can be identified with $\omega^0_{(M,\partial M)}\{M \times S^0; (\tau M)^+_M\}$, and this group is dual to $\omega_0(M) = \mathbb{Z}$.

Internal structure

Let Y, while we continue the discussion of the classical case in which the base B is a point, be a pointed compact ENR. To pursue the physical analogy, we will now suppose that our particles have some sort of internal structure (usually called a 'label') specified by a point of Y. The structure given by the basepoint of Y means non-existence. More formally, we consider pairs (Q, f) consisting of a finite subset Q of M and a map $f : Q \to Y$ and introduce the equivalence relation $\sim$:

$$(Q, f) \sim (Q', f') \quad \text{if and only if} \quad \begin{array}{c} f|(Q \cap Q') = f'|(Q \cap Q') \\ \text{and} \\ f^{-1}(Y - \{*\}) = f'^{-1}(Y - \{*\}). \end{array}$$

The set of equivalence classes $C(M; Y)$ is topologized as the obvious quotient of

$$\coprod_{k \geq 0} \mathcal{F}^k(M) \times_{\mathfrak{S}_k} Y^k.$$

It is filtered by subspaces $C^k(M; Y)$, $k \geq 0$, consisting of the classes represented by pairs (Q, f) with $\#Q \leq k$. Thus $C^k(M; Y)$ is a quotient of $\mathcal{F}^k(M) \times_{\mathfrak{S}_k} Y^k$.

As a trivial, but important, example, for $Y = S^0$ we have a natural identification:

$$C^k(M; S^0) = \coprod_{0 \leq i \leq k} C^i(M),$$

and $C^k(M; S^0) = C^{k-1}(M; S^0) \sqcup C^k(M)$.

For a general space Y, the complement of the closed subspace $C^{k-1}(M; Y)$ in $C^k(M; Y)$ is the space $\mathcal{F}^k(M) \times_{\mathfrak{S}_k} (Y - \{*\})^k$, which fibres over $C^k(M)$. To exploit this description of the complement we need to know that $C^{k-1}(M; Y)$ is included in $C^k(M; Y)$ as a neighbourhood deformation retract.

Lemma 14.8 *There is a deformation retraction $r_t : U \to C^k(M; Y)$ of an open neighbourhood U of $C^{k-1}(M; Y)$ in $C^k(M; Y)$ onto the closed subspace $C^{k-1}(M; Y)$.*

We are assuming that Y is a pointed compact ENR; all that we really need for the proof is that Y be well-pointed. Let $s_t : V \to Y$ be a deformation retraction of an open neighbourhood of the basepoint $*$ to the basepoint: $s_0(x) = x$, $s_1(x) = *$, and $s_t(*) = *$; and let $\rho : Y \to [0, 1]$ be a continuous function which has zero-set $\{*\}$ and takes the value 1 outside V. We can now define a continuous function $\psi : C^k(M; Y) \to [0, 1]$ by setting

$$\psi(Q, f) := \min\{\rho(f(x)) \mid x \in Q\}.$$

The zero-set of ψ is exactly $C^{k-1}(M; Y)$. Let U be the open subset $\psi^{-1}[0, 1)$. We retract U onto $C^{k-1}(M; Y)$ by pushing the internal structure of a particle x with $\rho(f(x))$ minimal towards the basepoint. To carry this out, choose a continuous function $h : [0, 1] \times [0, 1) \to [0, 1]$ such that $h(1, u) = 0$, $h(t, u) = t$ for $t \leq u$. Now define $r_t(Q, f)$ to be (Q, g), where

$$g(x) = \begin{cases} s_{h(t, \psi(Q, f))}(f(x)) & \text{if } f(x) \in V, \\ f(x) & \text{if } f(x) \notin V. \end{cases}$$

It follows from the lemma that the inclusion $C^{k-1}(M; Y) \subseteq C^k(M; Y)$ is a closed cofibration. The quotient $C^k(M; Y)/C^{k-1}(M; Y)$ is obtained by collapsing the basepoints in the pointed fibre bundle

$$\mathcal{F}^k(M) \times_{\mathfrak{S}_k} \textstyle\bigwedge^k Y \to C^k(M). \tag{14.9}$$

We note as a consequence:

Lemma 14.10 *If Y is connected, then the inclusion $C^k(M; Y) \to C(M; Y)$ is a k-equivalence (that is, induces an isomorphism on homotopy groups π_i for $i < k$ and an epimorphism for $i = k$). More generally, if $\pi_i(Y) = 0$ for $i < c$, then the inclusion is a $(k + 1)c - 1$-equivalence.*

For if $X \to B$ is a pointed homotopy fibre bundle with fibres having vanishing homotopy groups $\pi_i(X_b) = 0$ in dimensions $i < c$, then the pointed space X/B is $(c - 1)$-connected. (Indeed, an inductive argument over cells of the base using the relative exact sequence shows easily that the stable homotopy groups in dimensions $< c$ are zero. A similar induction using the van Kampen theorem establishes simple connectivity.) We apply this result to the inclusions $C^{j-1}(M; Y) \to C^j(M; Y)$ for $j > k$.

Combinatorial models of loop spaces

It is straightforward to introduce the structure space Y into the definition of the map ϕ using the Pontrjagin–Thom construction to define

$$\phi : C(E; Y) \to \mathrm{map}^*(E^+, E^+ \wedge Y)$$

for a Euclidean space E and, in general,

$$
\begin{array}{ccc}
C(M - \partial M; Y) & \to & {}_c\Gamma(M - \partial M; (\tau M)^+_M \wedge_M (M \times Y)) \\
\phi: \quad \simeq \downarrow & & \downarrow \simeq \qquad\qquad (14.11) \\
C(M; Y) & \to & \Gamma(M, \partial M; (\tau M)^+_M \wedge_M (M \times Y))
\end{array}
$$

We shall use the same symbol ϕ for both horizontal maps in the diagram. The basic theorem, which provides a combinatorial, configuration space, model of the loop space $\Omega\Sigma Y$ and more generally $\Omega^n\Sigma^n Y$, was formulated and proved over a period of twenty years by numerous authors: from James (1955) [77] to McDuff (1975) [107].

Theorem 14.12 *If Y is connected, then the map ϕ above, (14.11), is a pointed homotopy equivalence.*

The idea of the proof, as conceived by McDuff, is to begin from the basic example (Example 14.5): $C(D(E), \partial D(E)) \simeq E^+$, and use a gluing argument.

Example 14.13. The special case $M = D(\mathbb{R})$ reduces to the original theorem of [77]. By Example 14.2, the configuration space model $C(\mathbb{R}; Y)$ is homotopy equivalent to the reduced product (James) construction

$$J(Y) = \coprod_{k \geq 0} Y^k / \sim,$$

described in Part I, Section 20. As a set this is the free monoid on the pointed set Y (with the basepoint $*$ as identity); as a space it is filtered by subspaces $J^k(Y)$, the quotient of Y^k, corresponding to $C^k(\mathbb{R}; Y)$. The tangent bundle τM is trivial and

$$\Gamma(M, \partial M; (\tau M)_M^+ \wedge_M (M \times Y)) = \mathrm{map}^*(D(\mathbb{R})/S(\mathbb{R}), \mathbb{R}^+ \wedge Y) = \Omega\Sigma Y.$$

Example 14.14. Taking $M = D(\mathbb{R}^n)$, with $\tau M = M \times \mathbb{R}^n$ again trivial we obtain the model

$$C(\mathbb{R}^n; Y) \simeq \mathrm{map}^*(D(\mathbb{R}^n)/S(\mathbb{R}^n), (\mathbb{R}^n)^+ \wedge Y)$$

for the n-fold loop space $\Omega^n\Sigma^n Y$ due to May [103].

Example 14.15. The case of the closed manifold $M = S^1$ was considered by Bödigheimer and Madsen [15] following work of Carlsson and Cohen [21]. Again the manifold is parallelizable: $\tau S^1 = S^1 \times \mathbb{R}$, and we obtain a homotopy equivalence

$$\phi : C(S^1; Y) \to \mathrm{map}(S^1, \Sigma Y) = \mathcal{L}(\Sigma Y)$$

to the space of free loops on the suspension of Y.

The fibrewise theory

As we have already observed, there is very little that needs to be done to extend these constructions and the fundamental theorem to fibrewise spaces. The combinatorial models will be *fibrewise configuration spaces*. We work over a compact ENR B. Let ξ, playing the rôle of the vector space E, be a real vector bundle of dimension $n \geq 1$ over B, and let $M \to B$ be a (locally trivial) compact fibrewise manifold (possibly with non-empty boundary). The internal structure will be described by a pointed homotopy fibre bundle Y over B, with fibre of the homotopy type of a finite complex; we recall that Y is homotopy well-pointed. Its pull-back to M is written as p^*Y.

Performing the various constructions fibrewise we obtain pointed homotopy fibre bundles:

$$\mathcal{C}_B^k(M - \partial M; Y) \quad \text{and} \quad {}_c\Gamma_B(M - \partial M; (\tau_B M)_M^+ \wedge_M p^*Y)$$

with respective fibres at $b \in B$

$$C^k(M_b - \partial M_b; Y_b) \quad \text{and} \quad {}_c\Gamma(M_b - \partial M_b; (\tau M_b)_{M_b}^+ \wedge_{M_b} Y_b)$$

and fibrewise homotopy equivalent to

$$\mathcal{C}_B^k(M; Y) \quad \text{and} \quad \Gamma_B(M, \partial M; (\tau_B M)_M^+ \wedge_M p^*Y),$$

respectively. And the fibrewise quotient of successive terms in the filtration can be expressed as:

$$\mathcal{C}_B^k(M; Y)/_B \mathcal{C}_B^{k-1}(M; Y) = (\mathcal{F}_B^k(M) \times_{(B \times \mathfrak{S}_k)} \bigwedge_B^k Y)/_B \mathcal{F}_B^k(M). \quad (14.16)$$

(Here the balanced product is the orbit space of the action of the fibrewise group $B \times \mathfrak{S}_k$ over B.)

Using the fibrewise Pontrjagin–Thom construction we obtain compatible fibrewise pointed maps, for $k \geq 0$:

$$\begin{array}{ccc}
\mathcal{C}_B^k(M - \partial M; Y) & \xrightarrow{\phi^{(k)}} & {}_c\Gamma_B(M - \partial M; (\tau_B M)_M^+ \wedge_M p^*Y) \\
\simeq \downarrow & & \downarrow \simeq \\
\mathcal{C}_B^k(M; Y) & \xrightarrow[\phi^{(k)}]{} & \Gamma_B(M, \partial M; (\tau_B M)_M^+ \wedge_M p^*Y)
\end{array} \quad (14.17)$$

If $M \to B$ is a disc-bundle $D(\xi)$, then $\Gamma_B(M, \partial M; (\tau_B M)_M^+ \wedge_M p^*Y)$ reduces to the fibrewise loop space $\mathrm{map}_B^*(\xi_B^+, \xi_B^+ \wedge_B Y)$, and we have a map

$$\phi^{(k)} : \mathcal{C}_B^k(\xi; Y) \to \mathrm{map}_B^*(\xi_B^+, \xi_B^+ \wedge_B Y). \quad (14.18)$$

From Dold's theorem we can read off the fibrewise generalization of Theorem 14.12.

Theorem 14.19 *Suppose that $Y \to B$ has connected fibres. Then*

$$\phi : \mathcal{C}_B(M; Y) \to \Gamma_B(M, \partial M; (\tau_B M)_M^+ \wedge_M p^*Y)$$

is a fibrewise pointed homotopy equivalence.

Obstruction theory, Proposition 2.15, leads to a more precise finite version.

Corollary 14.20 *Suppose that B is a finite complex of dimension $\leq m$, that A is a subcomplex of B, and that $X \to B$ is a pointed homotopy fibre bundle* *with fibres of the homotopy type of finite complexes of dimension $\leq l$. Then, if the fibres of $Y \to B$ have vanishing homotopy groups in dimensions $< c$, where $c \geq 1$, the induced map*

$$\phi^{(k)} : \pi^{-i}_{(B,A)}[X; \mathcal{C}^k_B(M; Y)] \to \pi^{-i}_{(B,A)}[X; \Gamma_B(M, \partial M; (\tau_B M)^+_M \wedge_M p^* Y)]$$

is bijective for $i < (k+1)c - 1 - (m+l)$, surjective for $i \leq (k+1)c - 1 - (m+l)$.

Example 14.21. Let us consider the fibrewise form of the reduced product (James) construction. Suppose that λ is a Euclidean line bundle over B, and take $M = D(\lambda)$. The theorem gives a combinatorial model for the twisted fibrewise loop space $\mathrm{map}^*_B(\lambda^+_B, \lambda^+_B \wedge_B Y)$. Let P denote the principal $\mathbb{Z}/2$-bundle $S(\lambda)$ with the involution -1. From Example 14.2 we see that the ordered configuration space $\mathcal{F}^k_B(\lambda)$ is fibrewise homeomorphic to

$$(P \times_{\mathbb{Z}/2} (\mathfrak{S}_k \times (0, \infty)^{k-1})) \times_B \lambda,$$

where the involution in $\mathbb{Z}/2$ acts on $\mathfrak{S}_k$ as left multiplication by the order-reversing permutation of $\{1, \ldots, k\}$ and on the $(k-1)$-fold product by reversing the order of the factors. This space is fibrewise homotopy equivalent to $P \times_{\mathbb{Z}/2} \mathfrak{S}_k$, and $\mathcal{C}^k_B(\lambda; Y)$ is fibrewise homotopy equivalent to a fibrewise pointed space, which we shall denote by $J^k_B(\lambda; Y)$, with fibre $P_b \times_{\mathbb{Z}/2} J^k(Y_b)$ at $b \in B$. Here the involution in $\mathbb{Z}/2$ acts on $J^k(Y_b)$ by reversing the order of the factors.

Hence we obtain a concrete combinatorial approximation

$$\phi^{(k)} : J^k_B(\lambda; Y) \to \mathrm{map}^*_B(\lambda^+_B, \lambda^+_B \wedge_B Y)$$

(when Y has connected fibres) to the fibrewise loop space.

When λ is the trivial bundle $B \times \mathbb{R}$ we omit it from the notation and write simply $J^k_B(Y)$. A detailed study of this case can be found in Part I, Section 20.

Example 14.22. Next consider a 2-dimensional Euclidean vector bundle η over B, writing $P \to B$ for the associated principal $O(2)$-bundle and M for the circle-bundle $S(\eta)$. The fibrewise tangent bundle $\tau_B M$ is the pull-back of the determinant line bundle λ of η over B. (For the pull-back of η splits as a direct sum of $\tau_B M$ and a trivial line bundle.) So we have maps

$$\phi^{(k)} : \mathcal{C}^k_B(S(\eta); Y) \to \mathrm{map}_B(S(\eta), \lambda^+_B \wedge_B Y).$$

According to the discussion in Example 14.4 the ordered configuration space bundle $\mathcal{F}^k_B(S(\eta))$ is fibrewise homotopy equivalent to

$$P \times_{O(2)} (S^1 \times_{\mathbb{Z}/k} \mathfrak{S}_k),$$

while the bundle $\mathcal{C}^k_B(S(\eta))$ is fibrewise homotopy equivalent to the circle-bundle $P \times_{O(2)} (S^1/(\mathbb{Z}/k))$.

A stable splitting of a kth power

The technique that we describe next is essentially that used by Cohen in [23]; see also [25] for the case $k = 2$.

Consider, to begin with, a pointed space Y (of the homotopy type of a finite complex, say). Let N denote the space Y with the basepoint $*$ forgotten, and form a pointed space N_+ by adjoining a new basepoint. Then we have maps of pointed spaces

$$p : N_+ \to S^0 \quad \text{and} \quad q : N_+ \to Y,$$

the first induced by the projection of N to a point and the second extending the identity map on $Y = N \subset N_+$, and their sum gives a stable equivalence $p + q : N_+ \to S^0 \vee Y$. Indeed, the homotopy equivalence is realized after a single suspension:

$$\mathbb{R}^+ \wedge N_+ \to \mathbb{R}^+ \wedge (S^0 \vee Y). \tag{14.23}$$

More precisely, there is a commutative diagram of cofibrations and stable maps:

$$
\begin{array}{ccccc}
S^0 & \xrightarrow{\ j\ } & N_+ & \xrightarrow{\ q\ } & Y \\
{\scriptstyle =}\downarrow & & {\scriptstyle p+q}\downarrow & & {\scriptstyle =}\downarrow \\
S^0 & \longrightarrow & S^0 \vee Y & \longrightarrow & Y
\end{array}
$$

in which j is given by the inclusion of the basepoint $*$ in $Y = N$.

The k-fold (smash) power of the map $p + q$, (14.23), gives an $\mathfrak{S}_k$-equivariant homotopy equivalence

$$E^+ \wedge (N \times \ldots \times N)_+ \to E^+ \wedge \left(\bigvee_{0 \le i \le k} Z_i \right),$$

where E is the k-dimensional permutation representation of $\mathfrak{S}_k$ and the ith summand Z_i is a wedge of $\binom{k}{i}$ copies of the i-fold product $Y \wedge \cdots \wedge Y$ indexed by the i-element subsets of $\{1, \ldots, k\}$. (In particular, $Z_0 = S^0$.) This gives us immediately a fibrewise stable splitting theorem.

Proposition 14.24 *Let $P \to B$ be a principal $\mathfrak{S}_k$-bundle over a compact ENR B, and let Y be a pointed space (of the homotopy type of a finite complex). Then, writing N for Y regarded as a space without basepoint, we have a fibrewise stable decomposition:*

$$P \times_{\mathfrak{S}_k} (N^k)_{+B} \xrightarrow{\ \simeq\ } \bigvee_{0 \le i \le k} P \times_{\mathfrak{S}_k} Z_i.$$

By collapsing fibrewise basepoints we deduce a stable splitting theorem for the total space.

Corollary 14.25 *There is a stable splitting of spaces:*

$$(P \times_{\mathfrak{S}_k} N^k)_+ \xrightarrow{\ \simeq\ } \bigvee_{0 \le i \le k} (P \times_{\mathfrak{S}_k} Z_i)/B.$$

Remark 14.26. This result, with $Y = \mathbb{T}$ the circle group, is used in [30] to give a proof of Miller's theorem, Proposition 13.9, by reduction to a maximal torus. The product $T := N^k$ is a torus of rank k, filtered by the subspaces

$$T^{(i)} := \{(z_1, \ldots, z_k) \in T \mid \#\{j \mid z_j \neq 1\} \leq i\},$$

with the quotient $T^{(i)}/T^{(i-1)}$ a wedge of $\binom{n}{i}$ i-spheres. We regard T as the standard maximal torus of $U(k)$ with normalizer $N(T)$; the Weyl group $W(T) = N(T)/T$ is just the permutation group $\mathfrak{S}_k$. Applying Corollary 14.25 to the principal $W(T)$-bundle $P = U(k)/T \to U(k)/N(T) = B$ over the space of maximal tori in $U(k)$, we obtain a stable splitting of the space $U(k) \times_{N(T)} T$. This space is related to $U(k)$ by the standard Weyl map, which is one-to-one on regular elements and has cohomological degree equal to 1. For the derivation of the splitting of $U(k)$ the reader is referred to [30].

The stable splitting of configuration spaces

The notes of Milnor [112], written soon after James's construction of the reduced product space, initiated the study of stable splittings of configuration spaces. The main result, however, is due to Snaith [126]. Subsequent refinements may be found in [22], [23] and [14].

We begin with the case in which the base $B = *$ is trivial.

Theorem 14.27 *Let M be a compact manifold and Y be a pointed compact ENR. Then there is a stable equivalence (of spectra):*

$$\bigvee_{k \geq 1} (\mathcal{F}^k(M) \times_{\mathfrak{S}_k} \wedge^k Y)/C^k(M) \xrightarrow{\simeq} C(M; Y).$$

Remark 14.28. The reason for the splitting, and the essence of the proof, can be seen by looking at the case in which $Y = N_+$ for some compact ENR N. Then we have a decomposition of the space $C^k(M; Y)$ as a disjoint union:

$$C^k(M; N_+) = \coprod_{0 \leq i \leq k} \mathcal{F}^i(M) \times_{\mathfrak{S}_i} N^i.$$

The theorem is established by showing that each step of the filtration is split.

Lemma 14.29 *The cofibration sequence:*

$$C^{k-1}(M; Y) \to C^k(M; Y) \to C^k(M; Y)/C^{k-1}(M; Y)$$

admits a stable splitting (defined after some finite suspension depending on, and increasing with, k).

The splitting is obtained from Corollary 14.25 with $\mathcal{F}^k(M) \to C^k(M)$ as the principal $\mathfrak{S}_k$-bundle $P \to B$. (The argument does not actually need the base to be compact, but one does need to know that the vector bundle over $C^k(M)$ associated to the permutation representation E of $\mathfrak{S}_k$ is of finite type.) In the diagram below, s is the splitting map provided by Corollary 14.25, r is given by projecting the k-fold cartesian to the smash product, and q is the surjective map appearing in the description of $C^k(M; Y)$.

$$
\begin{array}{ccc}
(\mathcal{F}^k(M) \times_{\mathfrak{S}_k} \wedge^k Y)/C^k(M) & \xrightarrow{\;s\;} & (\mathcal{F}^k(M) \times_{\mathfrak{S}_k} Y^k)/C^k(M) \\
{\scriptstyle =}\downarrow & \swarrow_r & \downarrow q \\
(\mathcal{F}^k(M) \times_{\mathfrak{S}_k} \wedge^k Y)/C^k(M) & \longleftarrow & C^k(M; Y)
\end{array}
$$

The composition $q \circ s$ provides the required splitting:

$$
C^k(M; Y) \to C^k(M; Y)/C^{k-1}(M; Y)
$$

and establishes the stable decomposition:

$$
C^k(M; Y) \xrightarrow{\;\cong\;} C^{k-1}(M; Y) \vee \left(C^k(M; Y)/C^{k-1}(M; Y) \right).
$$

When the space Y is connected we may combine Theorem 14.27 with Theorem 14.12 to obtain a stable splitting theorem for loop spaces.

Example 14.30. In the special case $M = D(\mathbb{R})$ we get Milnor's stable equivalence:

$$
\bigvee_{k \geq 1} \wedge^k Y \xrightarrow{\;\cong\;} \Omega \Sigma Y.
$$

Example 14.31. More generally, with $M = D(\mathbb{R}^n)$, we have Snaith's splitting of the n-fold loop space

$$
\bigvee_{k \geq 1} (\mathcal{F}^k(\mathbb{R}^n) \times_{\mathfrak{S}_k} \wedge^k Y)/C^k(\mathbb{R}^n) \xrightarrow{\;\cong\;} \Omega^n \Sigma^n Y,
$$

and taking a limit as $n \to \infty$

$$
\bigvee_{k \geq 1} (E\mathfrak{S}_k \times_{\mathfrak{S}_k} \wedge^k Y)/B\mathfrak{S}_k \xrightarrow{\;\cong\;} \Omega^\infty \Sigma^\infty Y.
$$

Example 14.32. The case of the closed manifold $M = S^1$ leads to the splitting theorem of Carlsson and Cohen [21] for the space of free loops on the suspension of a connected space Y:

$$
\bigvee_{k \geq 1} (S^1 \times_{\mathbb{Z}/k} \wedge^k Y)/(S^1 \times_{\mathbb{Z}/k} *) \xrightarrow{\;\cong\;} \mathcal{L}(\Sigma Y).
$$

Fibrewise stable splittings

The generalization to the fibrewise theory is purely formal and is complicated only by the notation.

Proposition 14.33 *Let $M \to B$ be a compact fibrewise manifold over a compact ENR B, and let $Y \to B$ be a pointed homotopy fibre bundle with fibres of the homotopy type of a finite complex. Then there are natural fibrewise stable splittings:*

$$\bigvee_{1 \leq i \leq k} C_B^i(M; Y)/_B C_B^{i-1}(M; Y) \xrightarrow{\simeq} C_B^k(M; Y).$$

Let us apply this in conjunction with the basic theorem, Theorem 14.19, when the fibres of Y are connected, to our standard examples.

We will begin with the fibrewise reduced product (James) construction: $M = B \times D(\mathbb{R})$. The homotopy fibre bundles $J_B^k(Y)$, defined in Example 14.21, have fibrewise quotients

$$J_B^k(Y)/_B J_B^{(k-1)}(Y) = \bigwedge_B^k Y.$$

Proposition 14.34 *There is a natural stable splitting:*

$$J_B^k(Y) \simeq_B J_B^{(k-1)}(Y) \vee_B \bigwedge_B^k Y,$$

and hence

$$J_B^k(Y) \simeq \bigvee_{i=0}^k {}_B \bigwedge_B^i Y.$$

Especially interesting is the case of a sphere-bundle: $Y = \zeta_B^+$, for some real vector bundle ζ. Then we have a stable decomposition of $\Omega_B \Sigma_B \zeta_B^+$ as a wedge of sphere-bundles $(k\zeta)_B^+$:

$$\bigvee_{k \geq 1} {}_B (k\zeta)_B^+ \xrightarrow{\simeq} \Omega_B \Sigma_B(\zeta_B^+). \tag{14.35}$$

The fibrewise homology of this fibrewise Hopf space is investigated in the next section (Proposition 15.28).

For the case that M is a disc-bundle: $M = D(\xi)$, let us again specialize and take $Y = \zeta_B^+$.

Proposition 14.36 *Let ξ and ζ be real vector bundles over a compact ENR B. Then there is a fibrewise stable splitting:*

$$\bigvee_{B \atop k \geq 1} C_B^k(\xi)_B^{\eta_k \otimes \zeta} \xrightarrow{\simeq} \mathrm{map}_B^*(\xi_B^+, \xi_B^+ \wedge_B \zeta_B^+),$$

where η_k is the k-dimensional real vector bundle over the fibrewise configuration space $C_B^k(\xi)$ associated to the permutation representation of $\mathfrak{S}_k$.

Finally, we consider the example $M = D(\eta)$ (Example 14.22), where η is a 2-dimensional Euclidean bundle with associated principal $O(2)$-bundle $P \to B$. Then one has a fibrewise stable splitting:

$$\bigvee_{B \atop k \geq 1} (P \times_{(B \times \mathbb{Z}/k)} \textstyle\bigwedge_B^k Y) \xrightarrow{\simeq} \mathrm{map}_B(S(\eta), \lambda_B^+ \wedge_B Y), \tag{14.37}$$

and the associated stable splitting of spaces obtained by factoring out the fibrewise basepoints B. (Here again the balanced product is formed as the orbit space of an action of the fibrewise group $B \times \mathbb{Z}/k$.) Passing to the classifying space $\mathbf{BT}$ as a limit of complex projective spaces and taking η to be the complex Hopf line bundle and Y to be a trivial bundle $B \times Z$, we obtain, following [15], the Carlsson–Cohen stable splitting:

$$\bigvee_{k \geq 1} (\mathbf{E}(\mathbb{Z}/k) \times_{\mathbb{Z}/k} \textstyle\bigwedge^k Y)/\mathbf{B}(\mathbb{Z}/k) \xrightarrow{\simeq} (\mathbf{ET} \times_{\mathbf{T}} \mathcal{L}(\Sigma Z))/\mathbf{BT},$$

where $\mathbb{T}$ acts on the free loop space $\mathcal{L}(\Sigma Z)$ by rotating loops.

The fibrewise EHP-sequence

To conclude this section we discuss the fibrewise generalization of the classical EHP-sequence following the approach taken by Milgram [110]. Let X and Y be pointed homotopy fibre bundles over a finite complex B of dimension $\leq m$, and let ζ be a real vector bundle over B. Suppose that the fibres of $X \to B$ have the homotopy type of pointed finite complexes of dimension $\leq l$. We assume, as in Corollary 14.20, that the fibres of Y are connected and, more precisely, have vanishing homotopy groups in dimensions $< c$, where $c \geq 1$.

The suspension map

$$\pi_B^0[X; Y] \to \pi_B^0[\zeta_B^+ \wedge_B X; \zeta_B^+ \wedge_B Y] = \pi_B^0[X; \mathrm{map}_B^*(\zeta_B^+, \zeta_B^+ \wedge_B Y)]$$

is traditionally denoted by E in this context, as the initial letter of the German *Einhängung*. The EHP-sequence is concerned with the obstruction to desuspension.

Now by the main theorem, Theorem 14.19, we may replace the mapping space $\mathrm{map}_B^*(\zeta_B^+, \zeta_B^+ \wedge_B Y)$ by the configuration space model $C_B(\zeta; Y)$. In a range of dimensions we can even use the second step of the filtration, since

$$\phi^{(2)} : \pi_B^{-i}[X; C_B^2(\zeta; Y)] \to \pi_B^{-i}[X; \operatorname{map}_B^*(\zeta_B^+, \zeta_B^+ \wedge_B Y)]$$

is, by Corollary 14.20, bijective for $i < 3c - 1 - (m + l)$ (and surjective for $i = 3c - 1 - (m + l)$). There is a natural fibrewise homotopy equivalence: $Y \to C_B^1(\zeta; Y)$, given by labelling the zero vector in a fibre by an element of Y, and the suspension corresponds via ϕ to the inclusion:

$$g : Y \to C_B^2(\zeta; Y).$$

We thus reduce the desuspension problem, in low dimensions, to consideration of the homotopy exact sequence:

$$\ldots \to \pi_B^{-i}[X; F_B(g)] \to \pi_B^{-i}[X; Y] \to \pi_B^{-i}[X; C_B^2(\zeta; Y)] \to \ldots$$

involving the homotopy-fibre $F_B(g)$ of g (Proposition 2.7).

On the other hand, we have an easy description of the homotopy-cofibre $C_B(g)$ as the fibrewise quotient $C_B^2(\zeta; Y)/_B C_B^1(\zeta; Y)$, by Lemma 2.2 and the fibrewise version of Lemma 14.8. So, using (14.9), we can replace $C_B(g)$ by the fibrewise pointed space over B obtained from the balanced product $S(\zeta) \times_{B \times \mathbb{Z}/2} (Y \wedge_B Y)$, where the $\mathbb{Z}/2$ action is antipodal on the sphere-bundle and interchanges the two factors of $Y \wedge_B Y$, by collapsing the subspace $P(\zeta) = S(\zeta) \times_{B \times \mathbb{Z}/2} B$. Let us write this fibrewise pointed space as (the quadratic construction) $\mathbf{Q}_\zeta(Y)$.

Remark 14.38. In the important special case in which $Y = \eta_B^+$, for some real vector bundle η, is a sphere-bundle, we can identify $\mathbf{Q}_\zeta(Y)$ with the fibrewise Thom space $P(\zeta)_B^{\eta \oplus (H \otimes \eta)}$, where H is the Hopf line bundle over the real projective bundle $P(\zeta)$.

As we noted in our discussion of the Serre exact sequence, there is a natural fibrewise pointed map (2.17)

$$f : F_B(g) \to \Omega_B(C_B(g))$$

from the homotopy-fibre to the loop-space on the homotopy-cofibre. According to Proposition 2.18, this is an equivalence in a range:

$$f_* : \pi_B^{-i-1}[X; F_B(g)] \to \pi_B^{-i}[X; C_B(g)]$$

is bijective if $i \le 3(c - 1) - (m + l)$.

Putting the pieces together, we obtain the fibrewise EHP-sequence:

Proposition 14.39 *Under the hypotheses described in the text there is an exact EHP-sequence:*

$$\pi_B^{-i}[X; Y] \xrightarrow{E_\zeta} \pi_B^{-i}[\zeta_B^+ \wedge_B X; \zeta_B^+ \wedge_B Y] \xrightarrow{H_\zeta} \omega_B^{-i}\{X; \mathbf{Q}_\zeta(Y)\}$$

$$\xrightarrow{P_\zeta} \pi_B^{-i+1}[X; Y] \to \cdots$$

$$\cdots \xrightarrow{E_\zeta} \pi_B^0[\zeta_B^+ \wedge_B X; \zeta_B^+ \wedge_B Y] \xrightarrow{H_\zeta} \omega_B^0\{X; \mathbf{Q}_\zeta(Y)\}$$

for $i \le 3(c - 1) - (m + l)$.

Here the Freudenthal suspension theorem (Corollary 3.19) allows us to write the term involving $\mathbf{Q}_\zeta(Y)$ as a stable homotopy group.

The geometric Hopf invariant

In the treatment just described the maps H_ζ and P_ζ lack geometric content. The *Hopf invariant* H_ζ, which gives the obstruction to desuspension, can be described more explicitly in terms of $\mathbb{Z}/2$-equivariant homotopy theory and a squaring construction. (See, for example, Appendix A in [25]. The account which follows is based on an unpublished manuscript by A.L. Cook, M.C. Crabb and W.A. Sutherland to which reference is made in [24].)

Given a class $x \in \pi^0_B[\zeta^+_B \wedge_B;\ \zeta^+_B \wedge_B Y]$, we shall construct quite explicitly the (stable) Hopf invariant $H_\zeta(x) \in \omega^0_B\{X;\ \mathbf{Q}_\zeta(Y)\}$. To begin with, suppose that B is a point, so that X and Y are just pointed spaces, and write E, instead of ζ, for a finite-dimensional real vector space (with an inner product).

Consider a map $x : E^+ \wedge X \to E^+ \wedge Y$. Its square

$$x \wedge x : (E^+ \wedge E^+) \wedge (X \wedge X) \to (E^+ \wedge E^+) \wedge (Y \wedge Y)$$

is $\mathbb{Z}/2$-equivariant with respect to the involution which switches the factors. Here, as elsewhere, we change the order of factors in a smash product in the obvious way. Now it is an easy, but fundamental, observation that the $\mathbb{Z}/2$-module $E \oplus E$, with this involution is isomorphic to the module $E \oplus (L \otimes E)$, where L, as in Section 11, is the module $\mathbb{R}$ with the involution -1:

$$E \oplus E \to E \oplus (L \otimes E)\ :\ (u, v) \mapsto (u + v, u - v). \qquad (14.40)$$

(We think of $L \otimes E$ as E with the involution $v \mapsto -v$.) Composing $x \wedge x$ with the diagonal inclusion of X in $X \wedge X$ we thus obtain a $\mathbb{Z}/2$-map

$$p = (x \wedge x) \circ (1 \wedge \Delta_X) : (L \otimes E)^+ \wedge E^+ \wedge X \to (L \otimes E)^+ \wedge E^+ \wedge (Y \wedge Y).$$

The smash product of the identity on $(L \otimes E)^+$ with x, followed by the diagonal $Y \to Y \wedge Y$, gives a second equivariant map

$$q = (1 \wedge \Delta_Y) \circ (1 \wedge x) : (L \otimes E)^+ \wedge E^+ \wedge X \to (L \otimes E)^+ \wedge E^+ \wedge (Y \wedge Y).$$

Notice that, if x is the suspension of a map $X \to Y$, then $p = q$. In any case, p and q agree on the subspace $0^+ \wedge E^+ \wedge X$ fixed by the action of $\mathbb{Z}/2$. For the fixed-subspaces in the correspondence (14.40) are the diagonal $\Delta(E) \subseteq E \oplus E$ and the subspace $E \oplus 0 \subseteq E \oplus (L \otimes E)$. The Hopf invariant is defined by a difference construction as an obstruction to the existence of an equivariant homotopy between p and q which is constant on the fixed-subspace.

The difference construction is quite explicit. Let us extract the basic structure. We have pointed spaces $V = E^+ \wedge X$ and $W = (L \otimes E)^+ \wedge E^+ \wedge (Y \wedge Y)$, a finite-dimensional vector space $F = L \otimes E$, and two maps $p,\ q : F^+ \wedge V \to W$, which agree on the subspace $0^+ \wedge V (= V)$. Their *difference class* $\delta(p, q)$ is a map $\Sigma(S(F)_+) \wedge V \to W$. Pictorially, F^+ is the join $S^0 * S(F)$. We map $\Sigma(S(F)_+)$ to two copies of the join stuck together along S^0. To be analytical, let $\phi : [0, 1) \to [0, \infty)$ be a homeomorphism (so $\phi(0) = 0$). The map

$$[0, 1] \times S(F) \times V \to W \; : \; (t, u, v) \mapsto \begin{cases} p(\phi(2t - 1)u, v) & \text{for } \frac{1}{2} \le t \le 1, \\ q(\phi(1 - 2t)u, v) & \text{for } 0 \le t \le \frac{1}{2}, \end{cases}$$

determines $\delta(p, q)$ on the quotient.

In this way we obtain a canonical equivariant homotopy class

$$h_E(x) : \Sigma(S(L \otimes E)_+) \wedge E^+ \wedge X \to (L \otimes E)^+ \wedge E^+ \wedge (Y \wedge Y). \quad (14.41)$$

This is one manifestation of the Hopf invariant. To realize it in a more familiar, non-equivariant, form, we pass to the orbit space $P(E) = S(L \otimes E)/(\mathbb{Z}/2)$, the real projective space of E. Let us abbreviate $P(E)$ to P and as before write H (to be distinguished from the Hopf invariant) for the Hopf line bundle. Let $Z \to P$ denote the fibrewise pointed space associated to $Y \wedge Y$. The map $h_E(x)$ determines a fibrewise map

$$\Sigma_P(P \times (E^+ \wedge X)) \to (H \otimes E)_P^+ \wedge_P (P \times E^+) \wedge_P Z$$

and so a stable class in

$$\omega_P^{-1}\{P \times X; (H \otimes E)_P^+ \wedge_P Z\}.$$

But the tangent bundle τP is the quotient of $H \otimes E$ by a trivial line bundle: $(B \times \mathbb{R}) \oplus \tau P = H \otimes E$. So, by Proposition 12.43, this group is identified with

$$\omega^0\{X; Z/P\}.$$

Since the quotient Z/P is $\mathbf{Q}_E(Y)$, we have produced finally the *stable Hopf invariant*

$$H_E(x) \in \omega^0\{X; \mathbf{Q}_E(Y)\}. \quad (14.42)$$

Remark 14.43. One can perform the implicit duality construction quite explicitly at the equivariant level. The sphere $S(L \otimes E)$ is embedded in $L \otimes E$ with trivial normal bundle $S(L \otimes E) \times \mathbb{R}$. By an equivariant Pontrjagin–Thom construction we get a duality map

$$i : (L \otimes E)^+ \wedge S(L \otimes E)_+ \to \Sigma(S(L \otimes E)_+) \wedge S(L \otimes E)_+,$$

which is well-defined as an equivariant homotopy class. Composing this with $h_E(x)$ smashed with the identity on $S(L \otimes E)_+ \wedge X$, we obtain a canonical (unstable) class

$$(L \otimes E)^+ \wedge X \to (L \otimes E)^+ \wedge S(L \otimes E)_+ \wedge (Y \wedge Y),$$

and hence a stable equivariant class in

$$\mathbf{z}/2\,\omega^0\{X; S(L \otimes E)_+ \wedge (Y \wedge Y)\}.$$

A standard result of equivariant stable homotopy theory identifies this equivariant group with the non-equivariant group in (14.42). (Indeed, that equivariant result is very closely related to the fibrewise result (Proposition 12.43) that we have used here. See, for example, [99].)

Only minor modifications are needed to construct the fibrewise Hopf invariant from a fibrewise map $x : \zeta_B^+ \wedge_B X \to \zeta_B^+ \wedge_B Y$. One can write down an equivariant fibrewise map

$$h_\zeta(x) : \Sigma_B(S(L \otimes \zeta)_{+B}) \wedge_B \zeta_B^+ \wedge_B X \to (L \otimes \zeta)_B^+ \wedge_B \zeta_B^+ \wedge_B (Y \wedge_B Y)$$

and construct from it a stable class $H_\zeta(x) \in \omega_B^0\{X; \mathbf{Q}_\zeta(Y)\}$. The result Proposition 12.43 will be replaced by the generalization Proposition 12.41. (Alternatively, one can use the equivariant fibrewise theory.)

By inspection of the definition, one can see that the geometric Hopf invariant H_ζ is compatible with suspensions, that is, there is a commutative square:

$$
\begin{array}{ccc}
\pi_B^0[\zeta_B^+ \wedge_B X; \zeta_B^+ \wedge_B Y] & \xrightarrow{\;H_\zeta\;} & \omega_B^0\{X; \mathbf{Q}_\zeta(Y)\} \\
\downarrow & & \downarrow \\
\pi_B^0[(\eta \oplus \zeta)_B^+ \wedge_B X; (\eta \oplus \zeta)_B^+ \wedge_B Y] & \xrightarrow{\;H_{\eta \oplus \zeta}\;} & \omega_B^0\{X; \mathbf{Q}_{\eta \oplus \zeta}(Y)\}
\end{array}
$$

for any real vector bundle η over B. The left- and right-hand vertical maps are given by the fibrewise suspension and the inclusion $\mathbf{Q}_\zeta(Y) \subseteq \mathbf{Q}_{\eta \oplus \zeta}(Y)$ respectively.

To show that the geometrically defined Hopf invariant coincides with that defined using the configuration space model, it suffices, by exploiting the stable splitting theorem and naturality, to check the following special case.

Lemma 14.44 *Let $N \to B$ be a homotopy fibre bundle with fibres of the homotopy type of finite complexes. Write*

$$X = (S(L \otimes \zeta) \times_{B \times \mathbb{Z}/2} (N \times_B N))_{+B}, \quad Y = N_{+B},$$

so that $X = \mathbf{Q}_\zeta(Y)$, and let $x : X \to \mathrm{map}_B^*(\zeta_B^+, \zeta_B^+ \wedge_B Y)$ be given by the basic construction $\phi^{(2)}$, (14.17). Then the geometric Hopf invariant

$$H_\zeta(x) \in \omega_B^0\{X; \mathbf{Q}_\zeta(Y)\}$$

is the identity map.

This computation can be further simplified by a change of base. Lifting from B to the projective bundle $P(\zeta)$ and using the compatibility of the Hopf invariant with the fibrewise suspension, one reduces to the case in which ζ is a line bundle.

Chapter 4. Homology Theory

15 Fibrewise homology

In this section we present fibrewise homology as a 'bi-variant' theory, for which we shall use cohomological indexing, in a way which emphasizes the parallel with the description of fibrewise stable homotopy theory given in Section 3. The base space B is, unless noted otherwise, a compact ENR, and X and Y are fibrewise pointed spaces over B.

The definition

Fibrewise homology (and cohomology) groups with coefficients in a commutative ring, with identity, R (probably $\mathbb{Z}$, $\mathbb{Q}$ or a finite field) will be defined by generalizing the classical definition of homology as a representable theory using Eilenberg–MacLane spaces, which we write as $K(R, n)$. We write simply 'H' for homology, leaving the coefficient ring to be inferred from the context.

Define the R-module of fibrewise *homology maps* over B from X to Y by

$$H^0_B\{X; Y\} := \varinjlim_{n \geq 0} \pi^0_B[\Sigma^n_B X; (B \times K(R, n)) \wedge_B Y],$$

where the maps in the direct system are given by taking the fibrewise suspension and composing with the standard maps $\Sigma K(R, n) \to K(R, n+1)$ (adjoint to the homotopy equivalences $K(R, n) \to \Omega K(R, n+1)$). Then, just as for stable homotopy theory, we can define the $\mathbb{Z}$-graded theory $H^i_B\{X; Y\}$, $i \in \mathbb{Z}$, by

$$H^i_B\{X; Y\} = H^0_B\{(B \times \mathbb{R}^N)^+_B \wedge X; (B \times \mathbb{R}^{N+i})^+_B \wedge Y\},$$

for any integer $N \geq 0$ with $N + i \geq 0$.

It is immediate that $H^i_B\{X; B \times S^0\}$ is the usual (reduced) cohomology module $\tilde{H}^i(X/B)$ of the pointed space X/B. More generally, using this notation without the subscript B for the classical bi-variant modules, we have:

Proposition 15.1 *For a fibrewise pointed space X over a compact ENR B and pointed space F, there is a natural equivalence:*

$$H^*_B\{X; B \times F\} \to H^*\{X/B; F\}.$$

We can, thus, think of $H_B^*\{X; B \times S^0\}$ as the *fibrewise cohomology* of X over B. Now $\tilde{H}_j(F) = H^{-j}\{S^0; F\}$ ($= H^0\{S^j; F\}$ for $j \geq 0$) is the homology of the pointed space F. Passing to the fibrewise theory, we should, therefore, consider $H_B^*\{B \times S^0; Y\}$ to be the *fibrewise homology* of Y. We do not introduce a special notation for these modules, nor a change of sign in the indexing, so that fibrewise homology in this sense is mostly in negative dimensions.

If $A \subseteq B$ is a closed sub-ENR, then we can define relative homology modules

$$H_{(B,A)}^i\{X; Y\}$$

by considering fibrewise maps which are zero over A; and if B is not necessarily compact, we can define fibrewise homology with compact support

$$_cH_B^i\{X; Y\}$$

in terms of fibrewise maps which are zero outside a compact subset of the base.

For a map $\alpha : B' \to B$ of compact ENRs we have a *pull-back* homomorphism:

$$\alpha^* : H_B^*\{X; Y\} \to H_{B'}^*\{\alpha^*X; \alpha^*Y\}.$$

The formal properties of fibrewise homology theory translate directly from the properties of fibrewise stable homotopy theory. We state the results appropriate to a compact base B; the modifications necessary for pairs (B, A) or a non-compact base are routine. Given fibrewise pointed spaces X, Y and Z over B, there is a *composition* of homology maps over B:

$$\circ : H_B^0\{Y; Z\} \otimes_R H_B^0\{X; Y\} \to H_B^0\{X; Z\}, \qquad (15.2)$$

which extends to a composition product on the graded modules:

$$\circ : H_B^i\{Y; Z\} \otimes_R H_B^j\{X; Y\} \to H_B^{i+j}\{X; Z\}.$$

For fibrewise pointed spaces $X' \to B$ and $Y' \to B$ we have the *smash product*:

$$\wedge : H_B^i\{X; Y\} \otimes H_B^{i'}\{X'; Y'\} \to H_B^{i+i'}\{X \wedge_B X'; Y \wedge_B Y'\}$$

with the usual properties. (Traditionally, different manifestations of these products go under the names of 'cup', 'cap' and 'slant' product.)

In particular, we see that $H_B^*\{X; Y\}$ is a graded module over the cohomology ring

$$H_B^*\{B \times S^0; B \times S^0\} = H^*(B)$$

of the base, and that the products are $H^*(B)$-bilinear.

From the composition we obtain a category whose objects are the fibrewise pointed spaces over B with morphisms from X to Y the (module) of homology maps $H_B^0\{X; Y\}$. We refer to this category as the *homology category over B*.

The suspension isomorphism

Built into the definition we have chosen are suspension isomorphisms

$$H_B^*\{X; Y\} \to H_B^*\{\Sigma_B X; \Sigma_B Y\} \to \cdots \to H_B^*\{\Sigma_B^n X; \Sigma_B^n Y\} \to \cdots$$

To obtain the general suspension isomorphism stated below we shall revise our definition to one more in keeping with the definition (Definition 3.3) of a fibrewise stable map.

Proposition 15.3 *For a finite-dimensional real vector bundle ζ over B, there is a suspension isomorphism*

$$1\wedge \quad : H_B^i\{X; Y\} \to H_B^i\{\zeta_B^+ \wedge_B X; \zeta_B^+ \wedge_B Y\},$$

given by the smash product with the identity $1 \in H_B^0\{\zeta_B^+; \zeta_B^+\}$.

Let ξ be an n-dimensional real vector bundle over B. Then one can define an associated *fibrewise Eilenberg–MacLane space* $K_B(R, \xi)$, which is a pointed fibre bundle with fibre $K(R, n)$. We shall give an explicit construction (essentially following Segal [121]) shortly. And there are fibrewise homotopy equivalences:

$$K_B(R, \xi) \to \operatorname{map}_B^*(\zeta_B^+, K_B(R, \zeta \oplus \xi)),$$

with adjoint

$$s : \zeta_B^+ \wedge_B K_B(R, \xi) \to K_B(R, \zeta \oplus \xi), \tag{15.4}$$

for any vector bundle ζ (generalizing the familiar homotopy equivalence $K(R, n) \to \Omega^k K(R, k + n)$).

Now we can think of a homology map $X \to Y$ over B, in just the same way as we defined a stable map, in terms of representatives (f, ξ), where f is a fibrewise pointed map

$$\xi_B^+ \wedge_B X \to K_B(R, \xi) \wedge_B Y.$$

Still following the account in Section 3, we introduce the equivalence relation generated by (i) homotopy, (ii) stability and (iii) vector bundle isomorphism. The stability condition says that the pair (f, ξ) is equivalent to the pair $((s \wedge 1) \circ (1 \wedge f), \zeta \oplus \xi)$:

$$\zeta_B^+ \wedge_B \xi_B^+ \wedge_B X \xrightarrow{1\wedge f} \zeta_B^+ \wedge_B K_B(R, \xi) \wedge_B Y \xrightarrow{s\wedge 1} K_B(R, \zeta \oplus \xi) \wedge_B Y,$$

with s as in (15.4). Because every finite-dimensional bundle over the compact base B is a direct summand of a trivial bundle, the equivalence classes give the elements of $H_B^0\{X; Y\}$ defined above using only suspension by trivial bundles.

The Hurewicz transformation

The construction of the Hurewicz map from stable homotopy to homology proceeds exactly as in the classical theory. The identity $1 \in R$ gives a fibrewise pointed map $B \times S^0 \to B \times R = K_B(R, B \times 0)$, where R has the discrete topology, and hence a map

$$\xi_B^+ \to K_B(R, \xi)$$

for any vector bundle ξ. A fibrewise map $\xi_B^+ \wedge_B X \to \xi_B^+ \wedge_B Y$, representing a stable map $X \to Y$, thus gives by composition a map $\xi_B^+ \wedge_B X \to K_B(R, \xi)$, representing a homology map. The transformation

$$\omega_B^*\{X; Y\} \to H_B^*\{X; Y\}$$

so defined is a natural transformation from the stable homotopy category over B to the homology category over B; it is compatible with all the structure, such as products and pull-backs, that we have considered.

The cofibre exact sequences

The arguments described in Section 3 for stable homotopy establish the usual cofibre exact sequences on the left and right in homology.

Proposition 15.5 *Associated to the two cofibre sequences $X' \to X \to X''$ and $Y' \to Y \to Y''$ over B there are long exact sequences:*

$$\cdots \to H_B^i\{X''; Y\} \to H_B^i\{X; Y\} \to H_B^i\{X'; Y\}$$
$$\xrightarrow{\delta} H_B^{i+1}\{X''; Y\} \to \cdots$$

and

$$\cdots \to H_B^i\{X; Y'\} \to H_B^i\{X; Y\} \to H_B^i\{X; Y''\}$$
$$\xrightarrow{\delta} H_B^{i+1}\{X; Y'\} \to \cdots$$

of $H^*(B)$-module homomorphisms.

The relative and Mayer–Vietoris sequences

Homotopy local triviality is needed for the results involving the excision property, Lemma 2.10.

Proposition 15.6 *Let X and Y be pointed homotopy fibre bundles over a compact ENR, and let $A' \subseteq A$ be closed sub-ENRs of B. Then there is a long exact sequence of graded $H^*(B)$-modules*

$$\cdots \to H_{(B,A)}^i\{X; Y\} \to H_{(B,A')}^i\{X; Y\} \to H_{(A,A')}^i\{X_A; Y_A\}$$
$$\xrightarrow{\delta} H_{(B,A)}^{i+1}\{X; Y\} \to \cdots .$$

Proposition 15.7 *Suppose that B is a union of compact sub-ENRs B_1 and B_2 whose intersection A is a sub-ENR. Then, for two pointed homotopy fibre bundles X and Y over B, one has a long exact homology Mayer–Vietoris sequence*

$$\cdots \to H_B^i\{X; Y\} \to H_{B_1}^i\{X_{B_1}; Y_{B_1}\} \oplus H_{B_2}^i\{X_{B_2}; Y_{B_2}\} \to H_A^i\{X_A; Y_A\}$$
$$\xrightarrow{\delta} H_B^{i+1}\{X; Y\} \to \cdots$$

of $H^*(B)$-modules.

Rational stable homotopy theory

In the classical theory the Hurewicz map to $\mathbb{Q}$-homology determines stable homotopy modulo torsion. By induction over the cells of the base, using the relative exact sequence and the five-lemma, we obtain a homological description of the rational fibrewise stable theory.

Proposition 15.8 *Let X and Y be pointed homotopy fibre bundles, with fibres of the homotopy type of finite complexes, over a compact ENR B. Then the Hurewicz transformation to homology, H, with rational coefficients gives an isomorphism*

$$\omega_B^*\{X; Y\} \otimes \mathbb{Q} \to H_B^*\{X; Y\}$$

of finite-dimensional $\mathbb{Q}$-vector spaces.

The Dold–Thom theorem

For the remainder of this section the fibrewise spaces considered will be (pointed) homotopy fibre bundles with fibres of the homotopy type of finite complexes.

In [50] Dold and Thom gave a description of the integral homology groups of a pointed space as the homotopy groups of the infinite symmetric product. The extension of this result to the fibrewise theory is essentially routine. The two papers [121] and [122] of Segal are recommended as supplementary reading.

We start with the case $R = \mathbb{Z}$, and define the *fibrewise infinite symmetric product* $A_B(Y) \to B$ with fibre at $b \in B$ the free Abelian monoid on the fibre Y_b with the basepoint as zero element. It is filtered by subspaces

$$B = A_B^0(Y) \subseteq Y = A_B^1(Y) \subseteq \cdots \subseteq A_B^k(Y) \subseteq \cdots \subseteq A_B(Y),$$

where the fibre $A^k(Y_b)$ of $A^k(Y)$ consists of the sums of at most k elements of Y_b. Each $A_B^k(Y)$, topologized as a quotient of $Y \times_B \cdots \times_B Y$ (k factors), is a pointed homotopy fibre bundle.

There is a suspension map: $\xi_B^+ \wedge_B A_B^k(Y) \to A_B^k(\xi_B^+ \wedge_B Y)$ described in the fibres at a point by: $[x, y_1 + \ldots + y_k] \mapsto [x, y_1] + \ldots + [x, y_k]$, with adjoint

$$A_B^k(Y) \to \mathrm{map}_B^*(\xi_B^+, A_B^k(\xi_B^+ \wedge_B Y)), \tag{15.9}$$

and, similarly, a map

$$A_B^k(\xi_B^+) \wedge_B Y \to A_B^k(\xi_B^+ \wedge_B Y). \tag{15.10}$$

Now the original Dold–Thom theorem asserts that, for a connected pointed space Y, the integral homology groups $\tilde{H}_n(Y)$, for $n \geq 1$, can be computed as the homotopy groups $\pi_n(A(Y))$. We see, in particular, that $A(S^n)$, for $n \geq 1$, is an Eilenberg–MacLane space $K(\mathbb{Z}, n)$. It follows that $A_B(\xi_B^+)$, for a vector bundle ξ of dimension $n \geq 1$, is an Eilenberg–MacLane bundle $K_B(\mathbb{Z}, \xi)$ as used above to define fibrewise homology, and we have a map:

$$\pi_B^0[\xi_B^+ \wedge_B X; A_B(\xi_B^+) \wedge_B Y] \to H_B^0\{X; Y\},$$

which is bijective if n is sufficiently large. (We recall that B is a retract of a finite complex and that the fibres of X have the homotopy type of finite complexes.) Using the transformation (15.9), we obtain for n large a map from $H_B^0\{X; Y\}$ to $\pi_B^0[\xi_B^+ \wedge_B X; A_B(\xi_B^+ \wedge_B Y)]$, which by the classical Dold–Thom theorem and obstruction theory (Proposition 2.15) will be a bijection when n is sufficiently large. The maps are compatible with suspension as in Proposition 15.3 and (15.10), and we have established:

Theorem 15.11 (Dold–Thom theorem). *There is, for any real vector bundle ξ, a natural map*

$$\pi_B^0[\xi_B^+ \wedge_B X; A_B(\xi_B^+ \wedge_B Y)] \to H_B^0\{X; Y\},$$

which is an isomorphism for dim ξ *sufficiently large (greater than some bound depending on the dimension of B and of the fibres of X and on the connectivity of the fibres of Y).*

For a general ring of coefficients R the definition of $A_B(Y)$ has to be modified to form the free R-module on each fibre.

The Thom isomorphism

We now turn to the calculation of fibrewise homology modules. The Thom isomorphism theorem can be interpreted as a classification of pointed sphere-bundles in the fibrewise homology category. Working first with $\mathbb{Z}$-coefficients, let ξ be an oriented n-dimensional real vector bundle over B. Then the Thom class

$$u \in \tilde{H}^n(B^\xi) = H^n\{B^\xi; S^0\} = H_B^0\{\xi_B^+; B \times (\mathbb{R}^n)^+\}$$

is an isomorphism in the homology category from ξ_B^+ to the trivial pointed sphere-bundle $B \times S^n$. Its inverse is given by the Thom class of the virtual bundle $-\xi$. Two such isomorphisms differ by multiplication by a unit in the ring $H^0(B)$.

More generally:

Proposition 15.12 *Let ξ and η be real vector bundles over B. Then ξ_B^+ and η_B^+ are isomorphic in the $\mathbb{Z}$-homology category over B if and only if $\dim \xi = \dim \eta$ and $w_1(\xi) = w_1(\eta)$. In that case, isomorphisms are in 1–1 correspondence with isomorphisms of the orientation bundles: $\sigma(\xi) \to \sigma(\eta)$.*

In characteristic 2, any two pointed sphere-bundles ξ_B^+ and η_B^+ of the same dimension are isomorphic in the homology category over B, and the automorphism group of a sphere-bundle is trivial.

H-free fibrewise spaces

Definition 15.13 We shall say that a pointed homotopy fibre bundle $X \to B$ (with fibre of the homotopy type of a finite complex) is *H-free* over B if it is isomorphic in the fibrewise homology category to a trivial bundle $B \times F$ with fibre F a wedge of spheres.

This condition can be reformulated in classical language by using the Leray–Hirsch lemma.

Lemma 15.14 *The pointed homotopy fibre bundle $X \to B$ is H-free over B if and only if there exist classes*

$$e_i \in \tilde{H}^{n_i}(X/B), \quad 1 \le i \le m$$

which restrict to a basis of the cohomology (understood to be free) of each fibre X_b, $b \in B$. Such classes $e_1, \dots, e_m$ form a basis of $\tilde{H}^*(X/B)$ over the graded ring $\mathcal{R} := H^*(B)$.

For a class $e \in \tilde{H}^n(X/B)$ can be read as a homology map $X \to B \times S^n$. Classes $e_1, \dots, e_m$ as in the statement above describe a homology map

$$X \to B \times (S^{n_1} \vee \dots \vee S^{n_m})$$

over B. A homology map which restricts to an isomorphism on fibres is an isomorphism, by induction over cells of the base using the relative exact sequence or a Mayer–Vietoris argument.

If $X \to B$ and $X' \to B$ are both H-free over B, we obtain immediately a Künneth theorem for fibrewise cohomology:

$$\tilde{H}^*((X \wedge_B X')/B) = \tilde{H}^*(X/B) \otimes_{\mathcal{R}} \tilde{H}^*(X'/B). \tag{15.15}$$

Homology maps admit a similar description in terms of fibrewise cohomology or homology:

Lemma 15.16 *If $X \to B$ and $Y \to B$ are H-free pointed homotopy fibre bundles, then*

$$H_B^*\{X; Y\} = \mathrm{Hom}_{\mathcal{R}}^*(\tilde{H}^*(Y/B), \tilde{H}^*(X/B))$$
$$= \mathrm{Hom}_{\mathcal{R}}^*(H_B^*\{B \times S^0; X\}, H_B^*\{B \times S^0; Y\}).$$

In particular, there is duality between homology and cohomology over B:

$$H_B^*\{B \times S^0; Y\} = \mathrm{Hom}_{\mathcal{R}}^*(\tilde{H}^*(Y/B), \mathcal{R}).$$

The Künneth theorem generalizes to describe homology maps between smash products:

Lemma 15.17 *Suppose that the four pointed homotopy fibre bundles X, X', Y and Y' over B are H-free. Then there is a Künneth isomorphism*

$$H_B^*\{X \wedge_B X'; Y \wedge_B Y'\} = H_B^*\{X; Y\} \otimes_{\mathcal{R}} H_B^*\{X'; Y'\}.$$

Example 15.18. As an example, take $X = P(\xi)_{+B}$, where $P(\xi)$ is the projective bundle of an n-dimensional complex vector bundle ξ over B. Then $H_B^*\{X; B \times S^0\}$ is just the integral cohomology ring $H^*(P(\xi))$, which as an algebra over $\mathcal{R}$ is $\mathcal{R}[x]/(x^n - c_1 x^{n-1} + \ldots + (-1)^n c_n)$, where x is the Euler class of the Hopf line bundle and $c_i \in \mathcal{R}$ is the ith Chern class of ξ. We may take x^i, $0 \le i < n$ as a basis. The fibrewise homology $H_B^*\{B \times S^0; X\}$ has a dual basis:

$$y_i \in H^{-2i}\{B \times S^0; X\} \qquad (0 \le i < n)$$

with

$$\langle y_j, x^i \rangle = \begin{cases} 1 & \text{if } i = j \\ 0 & \text{if } i \ne j. \end{cases}$$

The Serre spectral sequence

Serre's spectral sequence relating the cohomology of X/B to the cohomology of the base B and the cohomology of the fibres of the pointed homotopy fibre bundle $X \to B$ over a finite complex is a synthesis of the information contained in the various relative exact sequences (Proposition 15.6) involving the skeleta of B. Its generalization to fibrewise homology maps between pointed homotopy fibre bundles X and Y is direct. The homology maps between fibres of X and Y, with the action of the fundamental groupoid of B determined by Proposition 1.11, form a local coefficient system of R-modules on B:

$$\mathcal{H}_B^i\{X; Y\} = (H^i\{X_b; Y_b\})_{b \in B}.\tag{15.19}$$

Proposition 15.20 *Let* $X \to B$ *and* $Y \to B$ *be pointed homotopy fibre bundles, with fibres of the homotopy type of finite complexes, over a compact ENR B. Then there is a Serre spectral sequence with*

$$E_2^{i,j} = H^i(B; \mathcal{H}_B^j\{X; Y\})$$

converging to $H_B^*\{X; Y\}$.

(In the description of the E_2-term the first 'H' is to be read in the usual sense as cohomology with coefficients in the local system; elsewhere 'H' is R-cohomology.)

The Eilenberg–Moore spectral sequences

In major work on the foundations of fibrewise homology theory [125] Smith showed how the Eilenberg–Moore spectral sequence could be generalized to a Künneth spectral sequence for homology over a base. We assume in this subsection that the coefficient ring R is a field. The aim is to compute vector spaces such as $H_B^*\{X \times_B X'; B \times S^0\}$ or $H_B^*\{X; Y\}$ starting from the fibrewise cohomology (or homology) of X and X' or Y respectively. It is clear that some restriction will be necessary. Take, for example, X to be the fibrewise one-point compactification of the Hopf line bundle over the circle $B = S^1$, with $R = \mathbb{Q}$. Then $\tilde{H}^*(X/B) = 0$, but $H_B^0\{X; X\} = \mathbb{Q}$. The next lemma describes the fibrewise version of an acyclic space.

Lemma 15.21 *The following conditions on the pointed homotopy fibre bundle X are equivalent.*
(i) $1 = 0 \in H_B^0\{X; X\}$;
(ii)   *X is isomorphic in the fibrewise homology category over B to the fibrewise space* $B \to B$;
(iii) $\tilde{H}^*(X_b) = 0$ *for all* $b \in B$.

Clearly condition (i) is preserved by pull-backs $\alpha : B' \to B$ and, in particular, by restriction to a fibre. Thus (i) implies (iii). From (iii) it follows that $H^*\{X_b; X_b\} = 0$, by Lemma 15.16 say. Then we deduce (i) from the Serre spectral sequence, or equivalently by an induction over cells of the base.

To exclude cases of the type considered in the example above, where the fibrewise cohomology $\tilde{H}^*(X/B)$ vanishes although X is not fibrewise acyclic, we impose the following condition on the fundamental group of B.

Lemma 15.22 *Suppose that the fundamental group of each component of B is trivial if the field R has characteristic 0, or is a finite p-group if R has* *characteristic* $p > 0$. *Then a pointed homotopy fibre bundle* $X \to B$ *with* $\tilde{H}^*(X/B) = 0$ *is isomorphic in the homology category over* B *to the fibrewise space* $B \to B$.

We may assume that B is connected and a finite complex. Suppose that the cohomology of the fibre of X is non-zero in dimension d and is zero in lower dimensions. Write $X^{(n)}$ for the restriction of X to the n-skeleton $B^{(n)}$ of B. Then $\tilde{H}^i(X^{(1)}/B^{(1)})$ is 0 for $i < d$ and non-zero for $i = d$ from the monodromy exact sequence (the relative exact sequence of the pair $(B^{(1)}, B^{(0)} = *)$). For if a finite p-group acts on a non-zero finite dimensional $\mathbb{F}_p$-vector space the fixed subspace is non-zero. But now $\tilde{H}^i(X^{(n+1)}/B^{(n+1)})$, for $n \geq 1$, maps isomorphically to $\tilde{H}^i(X^{(n)}/B^{(n)})$ for $i \leq d$. So the cohomology module $\tilde{H}^d(X/B)$ is non-trivial. This completes the proof.

To any pointed homotopy fibre bundle $X \to B$ we associate a pointed trivial bundle

$$\mathbf{F}_B X := B \times (X/B),$$

with the projection onto the first factor B, equipped with a fibrewise pointed map

$$i : X \to \mathbf{F}_B X \, : \, x \mapsto (p(x), [x]). \tag{15.23}$$

It is an elementary exercise, using the Künneth theorem over the field R, to verify:

Lemma 15.24 *The fibrewise pointed space* $\mathbf{F}_B X$ *is* H-free *and the map* $i : X \to \mathbf{F}_B X$ *induces an epimorphism* $\tilde{H}^*((\mathbf{F}_B X)/B) \to \tilde{H}^*(X/B)$ *in cohomology.*

The cofibre of $i : X \to \mathbf{F}_B$ is another pointed homotopy fibre bundle, with fibres of the stable homotopy type of finite complexes, to which we can again apply the construction $\mathbf{F}_B$. In this way we produce an explicit free resolution of X, and the Eilenberg–Moore spectral sequences are manufactured by the standard machinery. (See, for example, [68].)

Proposition 15.25 *Let* X, X' *and* Y *be pointed homotopy fibre bundles, with fibres of the homotopy type of finite complexes, over the compact ENR* B *satisfying the condition of Lemma 15.22 on the fundamental group depending on the characteristic of the coefficient field* R. *Then there exist natural convergent Eilenberg–Moore spectral sequences:*

$$E_2^{i,j} = \mathrm{Tor}_{\mathcal{R}}^{i,j}(\tilde{H}^*(X/B), \tilde{H}^*(X'/B)) \Rightarrow \tilde{H}^*((X \wedge_B X')/B),$$

$$E_2^{i,j} = \mathrm{Ext}_{\mathcal{R}}^{i,j}(\tilde{H}^*(Y/B), \tilde{H}^*(X/B)) \Rightarrow H_B^*\{X; Y\}.$$

We recall that $\mathcal{R}$ is the graded R-algebra $H^*(B)$. The spectral sequences have been stated in the form involving the fibrewise cohomology groups: $H_B^*\{X; B \times S^0\} = \tilde{H}^*(X/B)$. There are related spectral sequences involving the less familiar fibrewise homology groups $H_*^B\{B \times S^0; X\}$.

The Pontrjagin ring of a fibrewise Hopf space

Before we continue, let us remove the restriction that the coefficient ring R be a field. Suppose that M is a fibrewise Hopf space over B with associative product $m : M \times_B M \to M$ and identity $e : B \to M$. Then the Pontrjagin multiplication

$$m_* : H_B^*\{B \times S^0; M_{+B}\} \otimes H_B^*\{B \times S^0; M_{+B}\} \to H_B^*\{B \times S^0; M_{+B}\} \quad (15.26)$$

gives the fibrewise homology module $H_B^*\{B \times S^0; M_{+B}\}$ the structure of an algebra over $H^*(B)$, with identity given by e_*.

Example 15.27. Consider the bundle of unitary groups $M = U(\xi) \to B$ associated to an n-dimensional Hermitian vector bundle ξ over B. The first term $R^1(\xi)$ of the filtration (13.5) is the suspension $\Sigma_B(P(\xi)_{+B}) \subseteq U(\xi)$ of the complex projective bundle of ξ and its fibrewise homology, with $\mathbb{Z}$-coefficients, generates the homology of $U(\xi)$. Indeed, if we write

$$x_i \in H_B^{-(2i-1)}\{B \times S^0; \Sigma_B(P(\xi)_{+B})\}$$

for the suspension of the class y_{i-1} introduced in Lemma 15.17, then the Pontrjagin ring $H_B^*\{B \times S^0; U(\xi)_{+B}\}$ is an exterior algebra over $H^*(B)$ on the n generators x_i, $1 \le i \le n$. The homology $H_B^*\{B \times S^0; R^k(\xi)_{+B}\}$ of the kth term of the filtration is included in the homology of $U(\xi)_{+B}$ as the submodule spanned by the exterior powers $\Lambda^r[x_1, \dots, x_n]$, $0 \le r \le k$, and the homology of the quotient $R^k(\xi)/_B R^{k-1}(\xi)$ is thus identified with $\Lambda^k[x_1, \dots, x_n]$. (The fibrewise homology of the fibrewise loop space $\Omega_B U(\xi)$ can be described in a similar way. Details can be found in [28], where the result is used to determine the structure of the cohomology ring $H^*(\Omega_B U(\xi))$.)

The fibrewise homology of $\Omega_B \Sigma_B \zeta_B^+$

We consider as an interesting example an oriented real vector bundle ζ of even dimension $2n$, $n \ge 1$, over a finite complex B and describe the fibrewise homology and cohomology, with integer coefficients, of the bundle $\Omega_B \Sigma_B \zeta_B^+ \to B$.

Proposition 15.28 *As Pontrjagin algebra over $\mathcal{R} := H^*(B)$, the fibrewise homology of $\Omega_B \Sigma_B \zeta_B^+$ is a polynomial algebra*

$$H_B^*\{B \times S^0; (\Omega_B \Sigma_B \zeta_B^+)_{+B}\} = \mathcal{R}[z]$$

on one generator z in dimension $-2n$, and admits a natural Hopf algebra structure with comultiplication

$$\Delta z = z \otimes 1 + e(z \otimes z) + 1 \otimes z,$$

where $e \in H^{2n}(B)$ is the Euler class of ζ.

To define the generator z we use the standard inclusion of ζ_B^+ in $\Omega_B \Sigma_B \zeta_B^+$. The fibrewise homology group of ζ_B^+ is identified by the Thom isomorphism:

$$H_B^*\{B \times S^0; \zeta_B^+\} = H^*\{B \times S^0; B \times (\mathbb{R}^{2n})^+\} = H^{*+2n}(B)$$

and is dual over $\mathcal{R}$ to the free $\mathcal{R}$-module $\tilde{H}^*(B^\varsigma)$ generated by the Thom class $u \in \tilde{H}^{2n}(B^\varsigma)$. This identifies

$$H_B^*\{B \times S^0; (\zeta_B^+)_{+B}\} = H_B^*\{B \times S^0; B \times S^0\} \oplus H_B^*\{B \times S^0; \zeta_B^+\}$$

with $\mathcal{R}1 \oplus \mathcal{R}z$, where z is the generator corresponding under duality to the Thom class. The comultiplication is determined, by duality, from the identity: $u \cdot u = e \cdot u$ in cohomology.

Remark 15.29. In the filtration by fibre bundles of the combinatorial model $J_B(\zeta_B^+)$ of $\Omega_B \Sigma_B \zeta_B^+$

$$J_B^0(\zeta_B^+) = B \subseteq J_B^1(\zeta_B^+) = \zeta_B^+ \subseteq \cdots \subseteq J_B^k(\zeta_B^+) \subseteq \cdots$$

the fibrewise homology of $J_B^k(\zeta_B^+)$ is the $\mathcal{R}$-submodule consisting of the polynomials of degree at most k.

The cohomology of $\Omega_B \Sigma_B \zeta_B^+$ can be read off by duality. Let us write a_i, $i \geq 0$, for the basis over $\mathcal{R}$ dual to the z^i. We shall use a formal power series to encode the duality pairing:

$$\langle z^k, a_i a_j \rangle = \langle \Delta(z^k), a_i \otimes a_j \rangle = \langle (z \otimes 1 + e(z \otimes z) + 1 \otimes z)^k, a_i \otimes a_j \rangle.$$

Proposition 15.30 *The ring $H^*(\Omega_B \Sigma_B \zeta_B^+)$ is free as an $\mathcal{R}$-module, with basis a_i, $i \geq 0$, and multiplication described by the identity $a_i a_j = \sum A_{i,j;k} a_k$ where*

$$\sum A_{i,j;k} X^i Y^j = (X + Y + eXY)^k$$

in the formal power series ring $\mathcal{R}[[X, Y]]$, that is,

$$A_{i,j;k} = \frac{k!}{(k-i)!\,(i+j-k)!\,(k-j)!}\, e^{i+j-k}$$

if $i, j \leq k \leq i + j$, and $A_{i,j;k} = 0$ otherwise. In particular, $a_0 = 1$.

An example: the space of free loops on $\mathbb{C}P^n$

As a concrete application of this computation, and to emphasize once again the way in which fibrewise methods may be used to tackle classical problems, we shall compute the integral cohomology ring of the space of free loops $\mathcal{L}\mathbb{C}P^n$ on complex projective space $\mathbb{C}P^n$. We begin with a general observation on mapping spaces.

Remark 15.31. Let F be a pointed space. Then the space $\mathrm{map}(F, B)$ of (unpointed) maps from F to B fibres over B by evaluation at the basepoint of F. There is a natural fibrewise homeomorphism

$$\mathrm{map}(F, B) \quad \cong \quad \mathrm{map}^*_B(B \times F, B \times B)$$
$$\searrow \quad \swarrow$$
$$B$$

where $B \times B \to B$ is the fibrewise pointed space with projection $(a, b) \mapsto a$ and basepoint (b, b) in the fibre at b. As a special case, taking F to be the circle $\mathbb{T}$ we have a fibrewise homeomorphism:

$$\mathcal{L}B \to \Omega_B(B \times B)$$

over B, where the projection from the free loop space $\mathcal{L}B \to B$ is evaluation at $1 \in \mathbb{T}$. We recall, from Remark 11.22, that if B is a closed manifold $B \times B \to B$ is a pointed fibre bundle.

Let us consider first the space $\mathcal{L}S(\mathbb{C}^{n+1})$ of free loops on the sphere $S(\mathbb{C}^{n+1}) = S^{2n+1}$. Write ζ for the n-dimensional complex vector bundle over $B := S(\mathbb{C}^{n+1})$ with fibre at v the orthogonal complement of $\mathbb{C}v$. Thus ζ is the pull-back of the tangent bundle to the complex projective space under the projection $S(\mathbb{C}^{n+1}) \to P(\mathbb{C}^{n+1}) = \mathbb{C}P^n$. We can then identify the trivial sphere-bundle $B \times S(\mathbb{C}^{n+1}) \to B$ with $S(\mathbb{R}v \oplus i\mathbb{R}v \oplus \zeta) \to B$. Using stereographic projection we can express this as the fibrewise one-point compactification $(i\mathbb{R} \oplus \zeta)^+_B$, so that the diagonal section corresponds to the section at infinity.

Now $\mathcal{L}S(\mathbb{C}^{n+1})$ fibres, by evaluation at the basepoint $1 \in S^1$, over $S(\mathbb{C}^{n+1})$, and we can identify this bundle, by the discussion above, with the fibrewise loop-space $\Omega_B(\Sigma_B \zeta^+_B) \to B$.

We can use similar ideas to describe the space of free loops $\mathcal{L}\mathbb{C}P^n$. It fibres, by evaluation at $1 \in S^1$, over $\mathbb{C}P^n$. Let us now write B for this base $\mathbb{C}P^n$ and ζ for its complex tangent bundle. The fibre over a point $L \in \mathbb{C}P^n$, that is, a 1-dimensional subspace L of $V := \mathbb{C}^{n+1}$, is the loop-space $\Omega P(V)$, where the complex projective space $P(V)$ of V has basepoint $P(L)$. Now $P(V)$ is naturally identified, by tensoring with the dual L^*, with $P(L^* \otimes V) = P(\mathbb{C} \oplus \zeta_L)$. The basepoint is now given by $P(\mathbb{C})$.

Next, we lift loops in complex projective space to paths in the covering sphere. Thus a loop in $P(\mathbb{C} \oplus \zeta_L)$ can be identified with a path $\omega : [0, 1] \to S(\mathbb{C} \oplus \zeta_L)$ with $\omega(0) = (1, 0)$ and $\omega(1) \in S(\mathbb{C})$. By mapping ω to its endpoint $\omega(1) \in S^1$, we obtain a fibrewise fibration over B

$$\mathcal{L}\mathbb{C}P^n \quad \longrightarrow \quad B \times S^1$$
$$\searrow \quad \swarrow \qquad\qquad (15.32)$$
$$B$$

with fibrewise fibre $\Omega_B \Sigma_B \zeta^+_B$ (as in the discussion above). We need to know the monodromy map

$$M : \Omega_B \Sigma_B \zeta_B^+ \to \Omega_B \Sigma_B \zeta_B^+.$$

The pull-back from S^1 to $[0,1]$ can be trivialized by composing paths with a standard path from the basepoint in the suspension factor. We thus see that M is loop-multiplication by the degree 1 loop in ΩS^1.

We use the fibrewise fibration (15.32) over $B \times S^1$ to obtain a cohomology exact sequence:

$$\cdots \xrightarrow{M^* - 1} H^{i-1}(\Omega_B \Sigma_B \zeta_B^+) \longrightarrow H^i(\mathcal{L}CP^n)$$

$$\longrightarrow H^i(\Omega_B \Sigma_B \zeta_B^+) \xrightarrow{M^* - 1} H^i(\Omega_B \Sigma_B \zeta_B^+) \longrightarrow \cdots;$$

this is the fibrewise version of the monodromy exact sequence for a bundle over the circle. Since the cohomology of $\Omega_B \Sigma_B \zeta_B^+$, calculated by Proposition 15.30, is concentrated in even dimensions, this completely describes the cohomology of the free loop-space.

We need to calculate M^*. As seen above, the monodromy M is given by loop-multiplication by the degree 1 loop in ΩS^1. This loop determines a homology class in $H^0\{S^0; (\Omega S^1)_+\}$ coming from $(1,1) \in \mathbb{Z} \oplus \mathbb{Z} = H^0\{S^0; (S^0)_+\}$ by the inclusion of $S^0 = \{\pm 1\}$ in $\Omega S^1 = \Omega \Sigma(S^0)$ which maps 1 to the trivial loop and -1 to the standard loop of degree 1.

Now consider the map

$$H^*_B\{B \times S^0; (0_B^+)_{+B}\} = \mathcal{R} \oplus \mathcal{R} \to H^*_B\{B \times S^0; (\zeta_B^+)_{+B}\} = \mathcal{R} \oplus \mathcal{R}z$$

induced by the inclusion $0 \hookrightarrow \zeta$ of the zero vector bundle into ζ. Since the inclusion of the zero-section corresponds to multiplication by the Euler class, we see that the element $(1,1)$ is mapped to $(1, ez)$.

In homology the monodromy is thus given by Pontrjagin multiplication by $1 + ez$. Taking the dual, we find that $M^* - 1$ maps a_i to ea_{i-1} for $i \geq 1$, and $a_0 = 1$ to zero.

Now the cohomology of the projective space $\mathcal{R} = \mathbb{Z}[[x]]/(x^{n+1})$ is generated by the Euler class x of the Hopf line bundle. The Euler class of the tangent bundle is $e = (n+1)x^n$ and $e^2 = 0$. We can read off the structure of $H^*(\mathcal{L}CP^n)$ as follows. Write $\beta_i \in H^{2in+1}(\mathcal{L}CP^n)$, $i \geq 0$, for the image of a_i in the monodromy exact sequence above and $\alpha_i \in H^{2in+2}(\mathcal{L}CP^n)$ for the unique class mapping to xa_i.

Proposition 15.33 *The cohomology ring $H^*(\mathcal{L}CP^n)$ is generated as an algebra over $H^*(CP^n) = \mathbb{Z}[[x]]/(x^{n+1})$ by the classes $\alpha_i \in H^{2in+2}(\mathcal{L}CP^n)$, $i \geq 1$, and $\beta_i \in H^{2in+1}(\mathcal{L}CP^n)$, $i \geq 0$, defined above, subject to the relations:*

$$\alpha_i \alpha_j = \frac{(i+j)!}{i!\,j!}\, x\alpha_{i+j},$$

$$\alpha_i \beta_j = \frac{(i+j)!}{i!\,j!}\, x\beta_{i+j}, \qquad \begin{array}{c} x^n \alpha_i = 0, \\ (n+1)x^n \beta_i = 0, \end{array}$$

$$\beta_i \beta_j = 0.$$

The fibrewise homology of $\Omega_B \Sigma_B Y$ in general

The discussion above for sphere-bundles of even dimension generalizes to the case of a pointed homotopy fibre bundle $Y \to B$ (with fibres of the homotopy type of a finite complex) which is H-free over B.

Proposition 15.34 *In the setting described above, the Pontrjagin algebra of $\Omega_B \Sigma_B Y$ is the tensor algebra*

$$H_B^* \{ B \times S^0; \, \Omega_B \Sigma_B Y \} = \bigoplus_{k \geq 0} \otimes_\mathcal{R}^k \mathcal{M}$$

of the free module $\mathcal{M} := H_B^* \{ B \times S^0; Y \}$ over $\mathcal{R}$. It admits a natural Hopf algebra structure with comultiplication determined by

$$\Delta m = m \otimes 1 + \delta(m) + 1 \otimes m, \quad (m \in \mathcal{M}),$$

where $\delta : \mathcal{M} \to \mathcal{M} \otimes_\mathcal{R} \mathcal{M}$ is induced by the diagonal map $Y \to Y \wedge_B Y$.

The splitting of the fibrewise homology as a module follows at once from the stable homotopy splitting, Proposition 14.34. To check the algebra structure it is necessary to look at the precise definition of the splitting. This lifts the computation, in the notation $N = Y$ for Y regarded as an unpointed fibrewise space that we used in (14.23), to the fibrewise pointed space

$$J_B(N_{+B}) = \coprod_{k=0}^{\infty} N^k$$

with the multiplication defined by concatenation.

As a special case one obtains the analogue of Remark 15.29 for an oriented vector bundle ζ of odd dimension. One has to replace the usual formal power series ring $\mathcal{R}[[X, Y]]$ by the super-commutative version in which $XY = -YX$. Computations are simplified somewhat by the fact that $2e = 0$; we leave the derivation of a formula for the coefficients $A_{i,j;k}$ as an exercise.

Duality and fixed-point theory

The discussion of duality and fixed-point theory in Section 10 translates effortlessly from stable homotopy to homology. From duality structure maps

$$\mathrm{i} \in \omega_B^0 \{ B \times S^0; \, X^* \wedge_B X \} \quad \text{and} \quad \mathrm{e} \in \omega_B^0 \{ X^* \wedge_B X; \, B \times S^0 \}$$

in stable homotopy the Hurewicz transformation supplies homology maps

$$\mathrm{i} \in H_B^0 \{ B \times S^0; \, X^* \wedge_B X \} \quad \text{and} \quad \mathrm{e} \in H_B^0 \{ X^* \wedge_B X; \, B \times S^0 \} \quad (15.35)$$

with the same formal properties, Definition 10.8.

The fibrewise homology $H^*_B\{B \times S^0; Y\}$ of a pointed homotopy fibre bundle Y is equal to the fibrewise cohomology $H^*_B\{Y^*; B \times S^0\}$, that is $\tilde{H}^*(Y^*/B)$, of the stable dual Y^*. More generally, we can write

$$H^*_B\{X; Y\} = H^*_B\{X \wedge_B Y^*; B \times S^0\} = \tilde{H}^*((X \wedge Y^*)/B),$$

and so rephrase statements about fibrewise homology maps in terms of standard cohomology modules. The Serre spectral sequence, Proposition 15.20, reduces in this way to its classical (cohomological) form.

Remark 15.36. When performing homology calculations it suffices to work in the fibrewise homology category. We could call a fibrewise stable space X^* an H-*dual* of X if there are given elements **i** and **e** as in (15.35) with the requisite properties, as in Definition 10.8, in homology. This leads naturally to a definition of a *fibrewise Poincaré duality space* which is H-dual to a suspension of itself.

When the fibrewise space X is H-free, so also is the dual X^*, and the cohomology modules are algebraically dual:

$$\tilde{H}^*_B(X^*/B) = \mathrm{Hom}^*_{\mathcal{R}}(\tilde{H}^*(X/B), \mathcal{R}).$$

So the definitions (Proposition 10.17 and Definition 10.18) of the Lefschetz trace and transfer translate into purely algebraic form, and we obtain the following result of Dold [48], generalizing the classical interpretation, or even definition, of the Lefschetz index as the trace in homology.

Proposition 15.37 *Let the pointed homotopy fibre bundle* $X \to B$, *with fibres of the homotopy type of finite complexes, be H-free. Let $f \in H^*_B\{X; X\}$ be a homology self-map. Then the Lefschetz trace $T_B(f, X) \in H^*(B) = \mathcal{R}$ is the algebraic (super) trace of the $\mathcal{R}$-linear map*

$$f^* : \mathcal{M} := \tilde{H}^*(X/B) \to \mathcal{M},$$

and the Lefschetz transfer $\tilde{T}_B(f, X)$ in $H^*_B\{B \times S^0; X\} = \mathrm{Hom}^*_{\mathcal{R}}(\mathcal{M}, \mathcal{R})$ is given by

$$x \in \mathcal{M} \quad \mapsto \quad \mathrm{tr}\{y \mapsto f^*(x \cdot y) : \mathcal{M} \to \mathcal{M}\}.$$

We have already noted the algebraic form of the transfer in Remark 10.19. Of course, the homology trace $T_B(f, X) \in H^0(B)$ for a fibrewise self-map $f : X \to X$ is determined by its restriction to a fibre, if B is connected.

Example 15.38. As an example, take $X = M_{+B}$, where $M = P(\xi)$ is the projective bundle of an n-dimensional complex vector bundle ξ over B, $n > 1$. Then, as in Lemma 15.17, the integral cohomology ring $H^*(M)$ is

$$\mathcal{R}[x]/(x^n - c_1 x^{n-1} + \ldots + (-1)^n c_n).$$

The homology Lefschetz trace $\tilde{T}_B(1, X)$ is given by:

$$x^k \mapsto s_k \; : H^*(M) \to H^*(B),$$

where s_k is the characteristic class defined by $\sum_{i=1}^n y_i^k$ if ξ is split as a sum of line bundles $\lambda_1 \oplus \ldots \oplus \lambda_n$ with Chern classes $c_1(\lambda_i) = y_i$. (See, for example, [48].)

Borel cohomology

Borel's definition of equivariant cohomology, for a compact Lie group G, is a classical fibrewise construction. Let E and F be compact pointed G-ENRs. Then, for any principal G-bundle $P \to B$ over a compact ENR B we can form the fibrewise pointed spaces (indeed pointed fibre bundles) $X := P \times_G E$ and $Y := P \times_G F$ over B and consider the fibrewise cohomology modules $H_B^i\{X; Y\}$. The equivariant *Borel cohomology* is the inverse limit

$$H_G^i\{E; F\} = \varprojlim_{P \to B} H_B^i\{P \times_G E; P \times_G F\} \qquad (15.39)$$

over the category of such principal G-bundles P. (Some clash of notation between the equivariant and fibrewise theories is, alas, almost inevitable.) To be more careful, let $B_0 \subseteq B_1 \subseteq \ldots \subseteq \mathbf{B}G$ be the standard filtration of the Milnor classifying space of G and let $P_n \to B_n$ be the restriction of the universal bundle $\mathbf{E}G \to \mathbf{B}G$. It suffices to consider the subcategory consisting of these bundles and the inclusion maps, so that

$$H_G^i\{E; F\} = \varprojlim_{n \geq 0} H_B^i\{P_n \times_G E; P_n \times_G F\}.$$

Moreover, for fixed i this inverse limit stabilizes for large n.

The suspension isomorphism which one requires for an equivariant theory follows from the fibrewise suspension isomorphism (Proposition 15.3). For consider a finite-dimensional real G-module V. The natural suspension isomorphism for the associated vector bundle $\zeta := P \times_G V$ over B produces the equivariant suspension isomorphism

$$1\wedge \; : H_G^i\{E; F\} \xrightarrow{\cong} H_G^i\{V^+ \wedge E; V^+ \wedge F\}.$$

Example 15.40. As an example let us compute the $U(n)$-equivariant integral homology and cohomology of the space $U(n)$ with the adjoint action of the group. The coefficient ring $H_{U(n)}^*(*)$ is, by definition, the cohomology $H^*(\mathbf{B}U(n))$ of the classifying space, that is, the formal power series ring $\mathbb{Z}[[c_1, \ldots, c_n]]$ on the Chern classes of the universal bundle. (We have written the graded ring as a formal power series ring because we defined Borel cohomology as an inverse limit, but the graded ring sees only the homogeneous components.) The results described in Example 15.27 determine the Borel homology of the group $U(n)$ as a Pontrjagin ring:

$$H^*_{U(n)}\{S^0; U(n)_+\} = \mathbb{Z}[[c_1, \ldots, c_n]] \otimes \Lambda^*[x_1, \ldots, x_n],$$

where c_i has (cohomological) degree $2i$ and x_i has degree $-(2i-1)$. The homology in each degree is finitely generated as a $\mathbb{Z}$-module. We can now compute the Borel cohomology as the dual, by Lemma 15.16:

$$H^*_{U(n)}\{U(n)_+; S^0\} = \mathbb{Z}[[c_1, \ldots, c_n]] \otimes \Lambda^*[e_1, \ldots, e_n]$$

for certain classes e_i of degree $2i-1$. (To be exact, we may fix $\langle x_i, e_i \rangle = 1$ and $\langle u, e_i \rangle = 0$ for any other monomial u in the generators $x_1, \ldots, x_n$.) In non-equivariant terms this is the cohomology of the space

$$\mathbf{E}U(n) \times_{U(n)} U(n),$$

which may be thought of as the space $\mathcal{L}(\mathbf{B}U(n))$ of free loops on the classifying space of $U(n)$ or as the classifying space $\mathbf{B}(\mathcal{L}U(n))$ of the loop group. (See, for example, [4] or [38].)

Other theories

The other classical cohomology theories, notably periodic complex K-theory and the connective k-theory, and complex cobordism MU, can be extended to fibrewise theories by following along the lines of the treatment we have given for ordinary cohomology.

For example, we can define fibrewise periodic K-theory as

$$K^0_B\{X; Y\} := \varinjlim_{n \geq 0} \pi^0_B[\Sigma^{2n}_B X; (B \times (\mathbb{Z} \times \mathbf{B}U)) \wedge_B Y],$$

where the maps in the direct system are given by the Bott periodicity $\mathbb{Z} \times \mathbf{B}U \to \Omega^2(\mathbb{Z} \times \mathbf{B}U)$.

When X and Y are fibre bundles with compact ENR fibres, these groups are Kasparov KK-groups [97] over the C^*-algebra $\mathcal{C}(B)$ of continuous functions $B \to \mathbb{C}$. Algebraically, one replaces the fibrewise pointed space $p: X \to B$ over B by the homomorphism $p^*: \mathcal{C}(B) \to \mathcal{C}(X)$ of commutative C^*-algebras (with identity). This paves the way for the non-commutative theory studied by Kasparov in which one considers central homomorphisms $\mathcal{C}(B) \to R$ to a general C^*-algebra (with identity) R.

In [122] Segal gives an illuminating treatment of connective k-theory using C^*-algebra methods.

Connective complex k-theory

We shall explain how to define the fibrewise k-group $k_B^0\{X; Y\}$ when X and Y are pointed fibrewise compact ENRs over a compact ENR B.

Suppose first of all that the base $B = *$ is a point, so that X and Y are pointed compact ENRs. Roughly speaking, one should think of k-theory maps $X \to Y$, that is elements of $k^0\{X; Y\}$, in terms of C^*-algebra homomorphisms

$$C(E^+ \wedge Y) \to \text{End}_{C(E^+ \wedge X)} P,$$

where P is a finitely generated projective Hilbert $C(E^+ \wedge X)$-module, for some finite-dimensional real vector space E.

To be more precise, let H be a finite-dimensional complex Hilbert space, and define $F^H(Y)$ to be the space of C^*-algebra homomorphisms

$$\rho : C(Y, *) \to \text{End}(H),$$

where $C(Y, *)$, as in Remark 5.40, is the ideal of complex-valued functions vanishing at the basepoint. The space $F^H(Y)$ is topologized in the usual way as a mapping-space with the compact-open topology. We take the zero homomorphism as basepoint. Given such a representation ρ of the C^*-algebra (without identity) on H, let us write H_y ($y \in Y$) for the subspace of H on which $\rho(f)$ acts as multiplication by $f(y)$, for each $f \in C(Y, *)$. By the spectral theory of commutative C^*-algebras, H_y is zero for all but finitely many points $y \in Y$ and H is the orthogonal direct sum

$$H = \bigoplus_{y \in Y} H_y.$$

Associated to ρ is the essentially finite sum

$$\dim \rho := \sum_{y \in Y} (\dim H_y) y,$$

which gives an element of the infinite symmetric product $A(Y)$. Thus we obtain a continuous map

$$\dim : F^H(Y) \to A^{\dim H}(Y) \subseteq A(Y).$$

Segal shows in [122] that a pointed map $X \to F^H(Y)$ determines a class in $k^0\{X; Y\}$ (lifting a naturally defined class in the periodic Kasparov KK-group $K^0\{X; Y\}$) and proves that by adequate suspension one obtains every k-theory class in this way. For a real vector space E and a complex Hilbert space H, the natural map

$$\pi^0[E^+ \wedge X; F^H(E^+ \wedge Y)] \longrightarrow k^0\{E^+ \wedge X; E^+ \wedge Y\} \xleftarrow{\cong} k^0\{X; Y\}$$

is a bijection for $\dim E$ and $\dim H$ sufficiently large.

We turn now to the fibrewise theory. Suppose that Y is a pointed fibre bundle with each fibre a compact ENR. (This simplifies the exposition, and is no great restriction, because it follows from Lemma 5.18 that every fibrewise compact ENR is locally a fibrewise retract of such a bundle.) Let μ be a finite-dimensional complex Hilbert bundle over B. Then, by using the construction described above on each fibre we can construct a fibre bundle $F_B^\mu(Y)$, with fibre $F^{\mu_b}(Y_b)$ at $b \in B$. Furthermore, we have a fibrewise pointed map

$$\dim : F_B^\mu(Y) \to A_B(Y) \tag{15.41}$$

given by the dimension on fibres.

Proposition 15.42 (Segal's definition of k-theory). *Let X and Y be fibrewise pointed compact ENRs over a compact ENR B. Then for a real vector bundle ξ and complex Hilbert bundle μ over B there is a natural transformation*

$$\pi_B^0[\xi_B^+ \wedge_B X; F_B^\mu(\xi_B^+ \wedge_B Y)] \to k_B^0\{X; Y\},$$

and this transformation is an equivalence if ξ and μ are of sufficiently high dimension.

Granted the existence of the natural transformation, the fact that it is an equivalence in high dimensions follows from Dold's theorem and Segal's original result.

As in our definition in Section 3 of fibrewise stable homotopy groups it would suffice to look only at trivial bundles ξ. We note again that this fibrewise formulation lends itself to immediate equivariant generalization.

The dimension transformation (15.41) corresponds to the so-called *Thom map*:

$$k_B^*\{X; Y\} \to H_B^*\{X; Y\}$$

from connective k-theory to integral homology.

Cobordism

Our last example is unitary cobordism. One can simply set, in the original notation of tom Dieck [41],

$$U_B^0\{X; Y\} := \varinjlim_{n \geq 0} \pi_B^0[\Sigma_B^{2n} X; (B \times MU(n)) \times_B Y],$$

where $MU(n)$ is the Thom space of the universal n-dimensional complex vector bundle, ζ_n, over $BU(n)$. In the fibrewise theory it is more natural, however, to consider not just trivial bundles $B \times \mathbb{C}^n$, but all complex (Hermitian) vector bundles ξ over B. The bundle of groups $U(\xi)$ has a fibrewise classifying space $\mathbf{B}_B U(\xi)$, and there is a corresponding fibrewise Thom space

$\mathbf{M}_B U(\xi)$ of a universal bundle ζ_ξ. Then one can define $U_B^0\{X; Y\}$ as a set of equivalence classes of fibrewise maps

$$\xi_B^+ \wedge_B X \to \mathbf{M}_B U(\xi) \wedge_B Y.$$

Remark 15.43. In fact, the fibrewise classifying space is trivial: there is an equivalence

$$\mathbf{B}_B U(\xi) \cong B \times \mathbf{B}U(n),$$

where $n = \dim \xi$, under which the universal bundle ζ_ξ corresponds to $\xi \otimes \zeta_n$. To be exact, this is a fibrewise homotopy equivalence, but not (unless ξ is trivial) a fibrewise pointed homotopy equivalence: the basepoint section of $\mathbf{B}_B U(\xi)$ corresponds to the section of the trivial bundle given by the classifying map $B \to \mathbf{B}U(n)$ of ξ. We leave the verification of these facts as a final exercise for the reader.

References

1. J.F. Adams, *On the groups J(X)-III*, Topology **3** (1965), 193–222.
2. M.F. Atiyah, *Thom complexes*, Proc. London Math. Soc. **11** (1961), 291–310.
3. M.F. Atiyah, *Power operations in K-theory*, Quart. J. Math. Oxford **17** (1966), 165–193.
4. M.F. Atiyah and R. Bott, *The Yang-Mills equations over Riemann surfaces*, Phil. Trans. Royal Soc. London A **308** (1982), 523–615.
5. M.F. Atiyah and I.M. Singer, *The index of elliptic operators IV*, Ann. of Math. **93** (1971), 119–138.
6. T. Bartsch, *A global index for bifurcation of fixed points*, J. Reine Angew. Math. **391** (1988), 181–197.
7. H.J. Baues, Algebraic Homotopy, Cambridge University Press, 1989.
8. J.C. Becker, *On the existence of A_k-structures on stable vector bundles*, Topology **9** (1970), 367–384.
9. J.C. Becker and D.H. Gottlieb, *The transfer map and fiber bundles*, Topology **14** (1975), 1–12.
10. J.C. Becker and D.H. Gottlieb, *Transfer maps for fibrations and duality*, Comp. Math. **33** (1976), 107–133.
11. A.J. Berrick, *The Samelson ex-product*, Quart. J. Math. Oxford **27** (1976), 173–180.
12. I. Berstein and T. Ganea, *On the homotopy-commutativity of suspensions*, Illinois J. Math. **6** (1962), 336–340.
13. I. Berstein and P.J. Hilton, *Category and generalized Hopf invariants*, Illinois J. Math. **12** (1968), 421–432.
14. C.-F. Bödigheimer, *Stable splittings of mapping spaces*, in Algebraic Topology, Proc. Seattle 1985, Lecture Notes in Math. **1286** (1987), 174–187.
15. C.-F. Bödigheimer and I. Madsen, *Homotopy quotients of mapping spaces and their stable splitting*, Quart. J. Math. Oxford **39** (1988), 401–409.
16. P.I. Booth and R. Brown, *On the application of fibred mapping spaces to exponential laws for bundles, ex-spaces and other categories of maps*, Gen. Topology and its Appl. **8** (1978), 165–179.
17. P.I. Booth and R. Brown, *Spaces of partial maps, fibred mapping spaces and the compact-open topology*, Gen. Topology and its Appl. **8** (1978), 181–195.
18. K. Borsuk, Theory of Retracts, PWN Polish Scientific Publishers, Warsaw, 1966.
19. G.E. Bredon, Introduction to Compact Transformation Groups, Academic Press, 1972.
20. E.H. Brown, A proof of the Adams conjecture, manuscript, 1973.
21. G. Carlsson and R. Cohen, *The cyclic groups and the free loop space*, Comment. Math. Helv. **62** (1987), 423–449.
22. F.R. Cohen, J.P. May and L.R. Taylor, *Splitting of certain spaces CX*, Math. Proc. Camb. Phil. Soc. **84** (1978), 465–496.

23. R.L. Cohen, *Stable proofs of stable splittings*, Math. Proc. Camb. Phil. Soc. **88** (1980), 149–151.

24. A. Cook and M.C. Crabb, *Fibrewise Hopf structures on sphere-bundles*, Proc. London Math. Soc. **48** (1993), 365–384.

25. M.C. Crabb, $\mathbb{Z}/2$-Homotopy Theory, Cambridge University Press, 1980.

26. M.C. Crabb, *On the stable splitting of $U(n)$ and $\Omega U(n)$*, in Algebraic Topology Barcelona 1986 (ed. J. Aguadé and R. Kane) Lecture Notes in Math. **1298** (1987), 35–53.

27. M.C. Crabb, *The Fuller index and T-equivariant stable homotopy theory*, Astérisque **191** (1990), 71–86.

28. M.C. Crabb, Fibrewise homology, in preparation.

29. M.C. Crabb and D.L. Gonçalves, *On the space of matrices of given rank*, Proc. Edinburgh Math. Soc. **32** (1989), 99–105.

30. M.C. Crabb, J.R. Hubbuck and J.A.W. McCall, *Stable summands of $U(n)$*, Proc. Royal Soc. Edinburgh **127A** (1997), 963–973.

31. M.C. Crabb and I.M. James, *Fibrewise configuration spaces*, Bol. Soc. Mat. Mexicana **37** (1992), 83–97.

32. M.C. Crabb and K. Knapp, *James quasi-periodicity for the codegree of vector bundles over complex projective spaces*, J. London Math. Soc. **35** (1987), 353–366.

33. M.C. Crabb and K. Knapp, *James numbers*, Math. Ann. **282** (1988), 395–422.

34. M.C. Crabb and K. Knapp, Applications of non-connective $\mathrm{Im}(J)$-theory, in Handbook of Algebraic Topology, 463–503, Elsevier Science, 1995.

35. M.C. Crabb and S.A. Mitchell, *The loops on $U(n)/O(n)$ and $U(2n)/Sp(n)$*, Math. Proc. Camb. Phil. Soc. **104** (1988), 783–809.

36. M.C. Crabb and A.J.B. Potter, The Fuller index, in preparation.

37. M.C. Crabb and W.A. Sutherland, *The space of sections of a sphere-bundle I*, Proc. Edinburgh Math. Soc. **29** (1986), 383–403.

38. M.C. Crabb and W.A. Sutherland, The homotopy type of gauge groups, in preparation.

39. J.-P. Dax, *Étude homotopique des espaces de plongements*, Ann. Scient. Éc. Norm. Sup. **5** (1972), 303–377.

40. J. Derwent, *A note on numerable covers*, Proc. Amer. Math. Soc. **19** (1968), 1130–1132.

41. T. tom Dieck, *Bordism of G-manifolds and integrality theorems*, Topology **9** (1970), 345–358.

42. T. tom Dieck, *Partitions of unity in homotopy theory*, Comp. Math. **23** (1971), 153–167.

43. T. tom Dieck, Transformation Groups, De Gruyter, 1987.

44. T. tom Dieck, K.H. Kamps and D. Puppe, Homotopietheorie, Lecture Notes in Math. **57**, Springer, 1970.

45. A. Dold, *Partitions of unity in the theory of fibrations*, Ann. of Math. **78** (1963), 223–255.

46. A. Dold, Lectures on Algebraic Topology, Springer, 1972.

47. A. Dold, *The fixed point index of fibre preserving maps*, Inv. Math. **25** (1974), 281–297.

48. A. Dold, *The fixed point transfer of fibre preserving maps*, Math. Zeit. **148** (1976), 215–244.

49. A. Dold and D. Puppe, Duality, trace and transfer, in Proc. of the Internat. Conf. on Geom. Topology, 81–102, Warszawa, 1980.

50. A. Dold and R. Thom, *Quasifaserungen und unendliche symmetrische Produkte*, Ann. of Math. **67** (1958), 239–281.

51. E. Dror Farjoun, Cellular Spaces, Null Spaces and Homotopy Localization, Lecture Notes in Math. **1622**, Springer, 1996.
52. P.F. Duvall and L.S. Husch, *Embedding finite covering spaces into bundles*, Topology Proc. **4** (1979), 361–370.
53. W.G. Dwyer and J. Spalinski, Homotopy theories and model categories, in Handbook of Algebraic Topology, 73–126, Elsevier Science, 1995.
54. W. Dwyer, M. Weiss and B. Williams, A parametrized index theorem for the algebraic K-theory Euler class, preprint, 1995.
55. E. Dyer and S. Eilenberg, *Globalizing fibrations by schedules*, Fund. Math. **130** (1988), 125–136.
56. B. Eckmann and P.J. Hilton, *Operators and cooperators in homotopy theory*, Math. Annalen **141** (1960), 1–21.
57. B. Eckmann and P.J. Hilton, *Homotopy groups of maps and exact sequences*, Comment. Math. Helv. **34** (1960), 271–304.
58. M.G. Eggar, *The piecing comparison theorem*, Indag. Math. **35** (1973), 320–330.
59. M.G. Eggar, *Ex-homotopy theory*, Comp. Math. **27** (1973), 185–195.
60. M.G. Eggar, *On structure-preserving maps between spaces with cross-sections*, J. London Math. Soc. **7** (1973), 303–311.
61. E. Fadell, *Homotopy groups of configuration spaces and the string problem of Dirac*, Duke Math. J. **29** (1962), 231–242.
62. E. Fadell and L. Neuwirth, *Configuration spaces*, Math. Scand. **10** (1962), 111–118.
63. P. Fantham, I.M. James and M. Mather, *On the reduced product construction*, Canadian Math. Bull. **39** (1996), 385–389.
64. T. Ganea, *Lusternik-Schnirelmann category and cocategory*, Proc. London Math. Soc. **10** (1960), 623–639.
65. T. Ganea, *On the homotopy suspension*, Comment. Math. Helv. **43** (1968), 225–234.
66. T. Ganea, *Cogroups and suspensions*, Invent. Math. **9** (1970), 185–197.
67. A. Granas, *The Leray-Schauder index and the fixed point theory for arbitrary ANRs*, Bull. Soc. Math. France **100** (1972), 209–228.
68. J.P.C. Greenlees, *Generalized Eilenberg-Moore spectral sequences for elementary abelian groups and tori*, Math. Proc. Camb. Phil. Soc. **112** (1992), 77–89.
69. A. Grothendieck and J.A. Dieudonné, Eléments de Géométrie Algébrique, I, Springer, 1971.
70. A. Haefliger, *Plongements différentiables dans le domaine stable*, Comment. Math. Helv. **37** (1963), 155–176.
71. D. Haibao and E. Rees, *Parametrized Morse theory and non-focal embeddings*, Bull. London Math. Soc. **25** (1993), 598–602.
72. V.L. Hansen, *Embedding finite covering spaces into trivial bundles*, Math. Ann. **236** (1978), 239–243.
73. K.A. Hardie and K.H. Kamps, *Track homotopy over a fixed space*, Glasnik Mat. **24** (1989), 161–179.
74. A. Heller, *Relative homotopy theory*, J. London Math. Soc. **44** (1991), 537–552.
75. M.J. Hopkins, M. Mahowald and H. Sadofsky, *Constructions of elements in Picard groups*, Contemp. Math. **158** (1994), 89–126.
76. J.R. Hubbuck, *On homotopy commutative H-spaces*, Topology **8** (1969), 119–126.
77. I.M. James, *Reduced product spaces*, Ann. of Math. **62** (1955), 170–197.
78. I.M. James, *Bundles with special structure*, Ann. of Math. **89** (1969), 359–390.
79. I.M. James, *Ex-homotopy theory I*, Illinois J. Math. **15** (1971), 324–337.

80. I.M. James, *On sphere-bundles with certain properties*, Quart. J. Math. Oxford **22** (1971), 353–370.

81. I.M. James, *Alternative homotopy theories*, Enseign. Math. (2) **23** (1977), 221–247.

82. I.M. James, *On fibre spaces and nilpotency*, Math. Proc. Camb. Phil. Soc. **84** (1978), 57–60.

83. I.M. James, *On category, in the sense of Lusternik–Schnirelmann*, Topology **17** (1978), 331–348.

84. I.M. James, *On fibre spaces and nilpotency II*, Math. Proc. Camb. Phil. Soc. **86** (1979), 215–217.

85. I.M. James, General Topology and Homotopy Theory, Springer, 1985.

86. I.M. James, Fibrewise Topology, Cambridge University Press, 1990.

87. I.M. James, *Fibrewise coHopf spaces*, Glasnik Mat. **27** (1992), 183–190.

88. I.M. James, Fibrewise reduced product spaces, in Adams Memorial Symposium on Algebraic Topology I, 179–185, Cambridge University Press, 1992.

89. I.M. James, *On fibrewise homotopy theory*, CRM Proceedings and Lecture Notes **6** (1994), 61–74.

90. I.M. James, *Numerical invariants of fibrewise homotopy type*, Contemp. Math. **188** (1995), 121–135.

91. I.M. James, *Fibrewise compactly-generated spaces*, Publ. Research Inst. for Math. Sci. Kyoto Univ. **31** (1995), 45–61.

92. I.M. James, *Fibrewise complexes*, Progress in Math. **136** (1996), 193–199.

93. I.M. James and J.R. Morris, *Fibrewise category*, Proc. Roy. Soc. Edinburgh **119A** (1991), 177–190.

94. I.M. James and J.H.C. Whitehead, *The homotopy theory of sphere-bundles over spheres. I*, Proc. London Math. Soc. (3) **4** (1954), 196–218.

95. I.M. James and Ping Zhang, *Fibrewise transformation groups and fibre bundles*, North-east Chinese Math. J. **8** (1992), 263–274.

96. D.S. Kahn and S.B. Priddy, *Applications of the transfer to stable homotopy theory*, Bull. Amer. Math. Soc. **78** (1972), 981–987.

97. G.G. Kasparov, *Equivariant KK-theory and the Novikov conjecture*, Invent. Math. **91** (1988), 147–201.

98. D. Lever, Relative topology, in Categorical Topology, 337–374, Heldermann (Berlin), 1984.

99. L.G. Lewis, J.P. May and M. Steinberger, Equivariant Stable Homotopy Theory, Lecture Notes in Math. **1213**, Springer, 1986.

100. W.S. Massey, *The homotopy type of certain configuration spaces*, Bol. Soc. Math. Mexicana **37** (1992), 355–365.

101. M.R. Mather, Paracompactness and partitions of unity, Mimeographed notes, 1964.

102. M.R. Mather, *Pull-backs in homotopy theory*, Can. J. of Math. **28** (1976), 225–263.

103. J.P. May, The geometry of iterated loop spaces, Lecture Notes in Math. **271**, Springer, 1972.

104. J.P. May, Classifying Spaces and Fibrations, Memoirs of the Amer. Math. Soc., **155**, 1975.

105. J.P. May, *Fibrewise localization and completion*, Trans. Amer. Math. Soc. **268** (1980), 127–146.

106. J.F. McClendon, *Higher order twisted cohomology operations*, Invent. Math. **7** (1969), 183–214.

107. D. McDuff, *Configuration spaces of positive and negative particles*, Topology **14** (1975), 91–107.

108. H. Meiwes, *On fibrations and nilpotency — some remarks upon two articles by I.M. James*, Manuscripta Math. **39** (1982), 263–270.
109. A. Meyerhoff and T. Petrie, *Quasi equivalence of G modules*, Topology **15** (1976) 69–75.
110. R.J. Milgram, Unstable homotopy from the stable point of view, Lecture Notes in Math. **368**, Springer, 1974.
111. H. Miller, *Stable splittings of Stiefel manifolds*, Topology **24** (1985), 411–419.
112. J. Milnor, *On the construction FK (Lecture notes, Princeton University, 1956)*, in Algebraic Topology, A Student's Guide, by J.F. Adams, London Math. Soc. Lecture Note Series **4** (1972), 119–136.
113. J. Milnor, *Construction of universal bundles II*, Ann. of Math. **63** (1956), 430–436.
114. J. Milnor, *Microbundles I*, Topology **3, Suppl.** (1964), 53–80.
115. K.C. Min and S.J. Lee, *Fibrewise convergence and exponential laws*, Tsukuba J. Math **16** (1992), 53–62.
116. G. Mislin, Wall's finiteness obstruction, in Handbook of Algebraic Topology, 1261–1291, Elsevier Science, 1995.
117. G.D. Mostow, *Cohomology of topological groups and solvmanifolds*, Ann. of Math. **73** (1961), 20–48.
118. J.L. Noakes, *Self-maps of sphere bundles I*, J. Pure Appl. Algebra **10** (1977), 95–99.
119. D. Puppe, *Homotopiemengen und ihre induzierten Abbildungen I*, Math. Zeit. **69** (1958), 299–344.
120. H. Scheerer, *On H-spaces over a base*, Coll. Mat. Barcelona **36** (1985), 219–228.
121. G. Segal, *The multiplicative group of classical cohomology*, Quart. J. Math. Oxford **26** (1975), 289–293.
122. G. Segal, *K-homology theory and algebraic K-theory*, in K-theory and operator algebras, Lecture Notes in Math. **575** (1977), 113–127.
123. W. Singhof, *On the Lusternik–Schnirelmann category of Lie groups*, Math. Z. **145** (1975), 111–116.
124. W. Singhof, *On the Lusternik–Schnirelmann category of Lie groups: II*, Math. Z. **151** (1976), 143–148.
125. L. Smith, Lectures on the Eilenberg–Moore spectral sequence, Lecture Notes in Math. **134**, Springer, 1970.
126. V.P. Snaith, *A stable decomposition of $\Omega^n S^n X$*, J. London Math. Soc. **7** (1974), 577–583.
127. E. Spanier, Algebraic Topology, McGraw-Hill, 1966.
128. N.E. Steenrod, The Topology of Fibre Bundles, Princeton University Press, 1951.
129. D. Sullivan, *La classe d'Euler réelle d'un fibré vectoriel à groupe structural $SL_n(\mathbb{Z})$ est nulle*, C.R. Acad. Sc. Paris A **281** (1975), 17–18.
130. A.M. Sunderland, Fibrewise coHopf spaces, D. Phil. thesis, Oxford, 1993.
131. R. Thom, *L'homologie des espaces fonctionnels*, Colloque de topologie algébrique tenue à Louvain 1956, 26–39.
132. H. Ulrich, Fixed point theory of parametrized equivariant maps, Lecture Notes in Math. **1343**, Springer, 1978.
133. J.E. West, *Mapping Hilbert cube manifolds to ANRs: A solution of a conjecture of Borsuk*, Ann. of Math. **106** (1977), 1–18.
134. G.W. Whitehead, The homotopy suspension, Colloque de topologie algébrique tenue à Louvain 1956, 89–95.
135. J.H.C. Whitehead, *Combinatorial homotopy*, Bull. Amer. Math. Soc. **55** (1949), 213–245.

Index